持續改善III——頂級企業實戰秘笈

曾惠珍　黃國強　廖寶城　林煒彬

陳勤業　羅秉強　林絢琛　著

Business 048

持續改善III
——頂級企業實戰秘笈

作者：曾惠珍、黃國強、廖寶城、林煒彬
　　　陳勤業、羅秉強、林絢琛

總編輯：阮佩儀

編輯：藍天圖書編輯組

設計：4res

出版：紅出版（藍天圖書）

地址：香港灣仔道 133 號卓凌中心 11 樓

出版計劃查詢電話：(852) 2540 7517

電郵：editor@red-publish.com

網址：http://www.red-publish.com

印刷查詢電郵：gary@hklabel.com

香港總經銷：聯合新零售（香港）有限公司

台灣總經銷：貿騰發賣股份有限公司

　　　　　　新北市中和區立德街 136 號 6 樓

　　　　　　(886) 2-8227-5988

　　　　　　http://www.namode.com

出版日期：2021 年 11 月

圖書分類：企業／管理

國際書號：978-988-8743-45-2

定價：港幣 100 元正／新台幣 400 圓正

持續改善

選堂

目錄

Foreword

It is a great honor to introduce this book of research articles. First, it is an opportunity to congratulate the authors for their excellent research that demonstrates their capabilities as accomplished researchers. Second, this book will be very helpful to describe to Westerners the diversity and quality of the research conducted in China.

The studies assembled in this book have been conducted by members of One World Research Group. All have completed a post-doctoral research program. The research articles assembled in this book will sure be an inspiration and a guide for young doctoral students.

Indeed, these articles make a great contribution to the advancement of many disciplines and furthermore they make a significant contribution by linking many disciplines that are too often isolated in our universities. These professors came from different disciplines. One way to make progress in research is to link two domains and to link two international groups.

Why do people enter in a postdoctoral formal research structure? It is to push further the research done in the doctoral program so that it can lead to international publications. I have been honored to guide many Chinese postdoctoral participants.

Nowadays, research can no more be locked solely in libraries and should be available under divers formats among which a book is certainly one of the most useful to broadcast in a significant manner a group of researchers. These articles are brought together and are conveniently arranged by the editor after proper verification. The editor role is ingrate as it uses a lot of energy to synchronize efforts of many authors into the elaboration of a synergy.

Today, business is international due to globalization and opening of markets. In order to succeed, this strategy must be supported by exchange of ideas and knowledge between countries. A book like this one fulfills very well this task as it is intended to a large audience with various interests; as such this book is a witness of China's engagement toward research and an ambassador of its scientific community.

The articles will interest readers in many Western and Chinese universities, research centers, and business communities. I am sure that it will encounter a large popularity in these environments.

Dr. Prosper Bernard
Honorary President of the International Postdoctoral Association.
Professor and Former Vice Vector at the University of Québec in Montréal.
Chairman of the University Consortium of the Americas.
Co-founder of the One World Research Group.
President of the International MBA Association.
May 2021

Foreword

I will start out this foreword by putting these outstanding papers into a grander perspective – at the beginning of the 21st century and on the both coasts of the Pacific Ocean. In this particular time and location, the torrents of globalization, the flux of innovation, the growing aspiration of the East, the increasing self-doubt of the West, make the topics and conclusions of these papers ever so poignant and compelling to read.

The collection of papers, "Continuous Improvement towards world class Construction Industry secret", "A Study on Innovation and e-Service Quality for Developing e-Retailing Mass Entrepreneurship", "Enhancement of Corporate Governance in Small to Medium Size Non-Governmental Organizations (NGO) in Hong Kong from the Board's Perspective" and "Sustainable Development Water Resource in Hong Kong Special Administrative Region, China" by Drs. Philip Chan, Eddie Liu, Marianna Tsang, Keith Wong, Wilson Lam, Patrick Wong and Hely Law. all intertwine with the common threads of innovation, technology, globalization and the rise of China as the major economic power. The research papers not only describe the results of their post-doctoral research projects, but also mark the unique moment in history. For the last three centuries, the innovations in the world are mostly one-sided – they happened in the West, started with the industrial revolution in Europe, to the latest computing technology in Silicon Valley. Well, no more! The waves of innovation is moving east gradually, by people that understand the importance of global connection, appreciate the advances in technology and capable of conducting solid, timely research – like the four authors, who, with modest advises from me, have accomplished a set of remarkable researches after studying at several top universities in California such as Stanford University, University of California at Berkeley and at the Institute for Innovation and Economic Development that I founded and direct at the California State University Monterey Bay.

This is a remarkable set of research papers by a group of dedicated researchers. Their research projects have not only demonstrated their ingenuity and hard work, they provide clear evidence that through decades of connection between the East and the West, the innovation is taking roots in China. A Must Read! I'll say. Well done! Drs. Philip Chan, Eddie Liu, Marianna Tsang and Keith Wong, Wilson Lam, Patrick Wong and Hely Law!

Eric Y. Tao. Ph.D.
Professor and Director
Institute for Innovation and Economic Development at
California State University Monterey Bay
Honorary Advisor, International Postdoctoral Association
June 2021
Carmel, California, USA

Foreword

The business sector in Hong Kong and the Mainland of China has been dramatically transformed over the past two decades. Guangdong-Hong Kong-Macao Greater Bay Area, Belt and Road Initiative, globalisation, deregulation, rising customer expectation, volatile business environment and intensified competition all present new challenges and opportunities for businesses, particularly in the era of the COVID-19 pandemic. The issue of sustainable growth has become imminent for almost all business entities.

This book captures the central themes of performance management and organization development of business organizations in seven postdoctoral papers from California State University (Monterey Bay), University of Quebec at Montreal, University of Hong Kong and Chinese Academy of Governance, together with more than ten cases of experience sharing. The authors give a comprehensive account of the process of continuous improvement spiral in multi-national corporations. With the research focusing on participative learning and individual enlightenment, the concepts of learning organization and personal mastery are widely discussed. They form the backbone of the excellent research findings in the postdoctoral study.

I would like to express my appreciation to the seven authors of this volume for their research effort and dedication to the process of continuous improvement in management. This study embodies collaboration between members of an organization to achieve enduring success, and represents an extension of the "Hong Kong Spirit" in the business setting.

I also find the contents of this publication a complement to the initiative of lifelong learning for people in general. The self-improvement spiral conveys the important message of continuous learning in life. It is a process which fosters self-understanding, self-management and self-actualisation.

It is a wonderful experience reading this book, and I gladly recommend it to both business people and non-business people. You will find something in it for you.

Jeff Sze

Mr Jeff SZE
Political Assistant to Secretary for Education
Education Bureau,
The Government of the HKSAR
May 2021

Preface

It has always been the vision of business leaders to build a learning organization. Who wouldn't want to work for a company that has created a culture of learning from and even embracing mistakes as opportunities to gain knowledge? Or pushing innovation through bold experimentation? Or fostering and investing in development for its employees? The common questions are: how to make this happen and what are the secrets of success?

There is an old saying: "To do a good job, one must first sharpen one's tool." The continuous triumph of a company comes from its employees as well as its customers. For the company to keep customers satisfied, it is vital to have competent and committed staff, and a self-improvement culture to achieve enduring greatness.

A learning organization makes up of employees skilled at creating, acquiring, and transferring knowledge. These people could help their firms cultivate tolerance, foster open discussion, and think holistically and systemically. Such learning organizations would be able to adapt to the unpredictable more quickly than their competitors could.

This book, which is the recent work of seven Postdoctoral Fellows, Visiting Scholars, Doctorates from California State University Monterey Bay (CSUMB), University of Quebec at Montreal (UQAM), Azteca University, Hong Kong University and National Academy of Governance (NAG), may provide the answers. Through action research efforts with typical multinational companies in the emerging Chinese market, the authors explore the in-depth details of turning traditional companies into market-oriented and dynamic learning organizations.

Readers should be inspired through reading these nine postdoctoral and doctoral papers. I gladly recommend this book to people in the academic fields and the business sectors.

W. K. Lo

Ir Dr the Honorable LO WAI KWOK, SBS, MH, JP
Honorary Advisor of International Postdoctoral Association (www.IPostdocA.org)
Member, CPPCC National Committee
Member of Legislative Council (Functional Constituency – Engineering)
Chairman, Business and Professionals Alliance for Hong Kong
June, 2021

Preface

Dramatic changes have taken place globally and locally during the last few years. The US-China trade war, the conflict between unilateralism and multilateralism. The local social turmoil. The severity of COVID-19. The major overhaul of the political scene in Hong Kong. The recent major revamp of the HKSAR administration system to ensure stability and prosperity in the long term, and, most important of all, the success of "One Country Two Systems"

Despite all these new changes which he have never experienced before, Hong Kong remains as one of the four international financial centres in the world, according to the 29th Global Financial Index, apart from ever well known status as the freest city in the world.

This book has a good collection of postdoctoral and doctoral papers from top universities as well as research studies of the operation of global enterprises. The findings and comments from these works will equip the reader to appreciate the ever rapid-changing world business and economic scenes.

I would be please to recommend this book to anyone whether he is in the business sector, or he simply wishes to extend his life-long learning interest.

Raymond Ho

Ir Dr RAYMOND HO CHUNG-TAI, SBS, MBE, SBStJ, JP
Honorary Advisor of International Postdoctoral Association
Member of Legislative Council, HKSAR (1996-2012)
Deputy to the National People's Congress, PRC (10th and 11th terms)
June 2021

序

現代企業的繁榮建立在創新和效率之上。技術、銷售和管理技術是對成功來說都至為關鍵，然而，「人」是成功的主要驅動力。只有那些能為員工提供持續幫助、優化並改善其理念、技巧和生產力的公司，才能在 21 世紀日益激烈的全球化競爭中佔據優勢。

1913 年雨果·孟斯特伯格出版了其開創性著作《心理學與工業效率》，其將心理學原理應用於培訓和行為研究，自那時起西方工業便做出調整來應對變革的挑戰，並在過去一個世紀中實現了持續增長和不斷創新。而中國在其進行現代化進程時，在建立本土工業的嘗試中卻舉步維艱。造成中國的落後和失敗有諸多原因，未能認識到個體的獨特性和創造性就是其中的原因之一。其未能將心理學作為人的研究的一部分，尤其是在創新的過程中沒有幫助個人解放創新的潛力和協同能力，這是另外一個原因。我很高興，本書的作者，不僅發現了行為研究在改善中國的工業中所扮演的重要角色，還將其應用到電梯工業，尤其是將其應用到改善中國境內世界最大的電梯公司的產品和服務上。

七位博士後，訪問學者和博士包括美國蒙特利灣加州州立大學（CSUMB），加拿大魁北克大學（UQAM），香港大學（HKU），中國國家行政學院和阿茲特克大學等，作者從事教學、管理、醫療、建築、製造、一帶一路、互聯網、仲裁，調解和智慧財產權工作，並為實現全要素生產率增長的行為研究和培訓系統提供指導。另外，他敏銳地認識到並鼓勵應用儒家智慧在人類和學習方面的作用，並努力將其發揚光大。在過去的二十年，根據聯合國教科文組織有關人類未來一系列的宣言精神，全世界很多著名的學者致力於此。儒家思想具有整體性，它和中國古代思想關於人類是不斷地自我反省和自我調整的過程的理念相互呼應。人類是自由和奮鬥的。生活的目的在於愛和幸福，其是通過自我實現和為他人服務是實現的。面對著 21 世紀初期這麼多的自然災害和人造災害，人類必須放棄個人主義和貪婪，接受我們的生存必須依靠普遍意識並實踐孔子的智慧這一事實。

中國的企業面臨著來自國內外的史無前例的困難和挑戰，在此之際，本書的出版是非常及時的。本書是在中國的土地上為了提高企業總生產力所做的深入細緻的實驗，其對科學管理的建議將幫助中國的經理們作出改變並獲得成功。我謹向各個領域的領導者推薦此書。

初志農教授

初志農教授
中央人民政府駐香港特別行政區聯絡辦公室
教育科技部前部長
2021 年 6 月

序

　　寒風拂面，季節轉換，又到了需要添加衣裳的季節。從 2008 年 9 月 15 日震驚世界的「雷曼兄弟破產」到「希臘金融危機」事件以來，全球陷入「全世界同步衰退」的世界經濟，雖然我國在政府一系列強有力的宏觀經濟調控政策及各方努力下仍異軍突起，保持增長，但仍然背負著巨大任務前行的形式更加嚴峻！

　　然而，從企業層面上試著詳細看一下經濟狀況，隨著國家經濟發展結構的轉型，恐怕大家的做法都是以無論怎樣方式都要獲得利潤，增加銷售額，參與全球化競爭，提升國際競爭力為目標，提升「產品技術創新」，努力致力於提升「中國製造」的產品品質並「降低成本」；或者企業家們從戰略的層面上對「持續改善革新」進行了重新的認識，在改善活動上付出了前所未有的努力並創造出來成果。

　　在經濟全球化的今天，品質問題更是企業活動的生命線，出現品質問題與其說只是個別企業的問題，更反映一個國家的大大品質實際狀況，因而其資訊迅速傳遍世界。

　　本書作者為幾位頂尖博士後，從事品質管理實戰的運用與研究十餘年。對品質管理理論及實踐經驗進行了詳細的總結和歸納，廣大品質管理工作者和愛好者可以將此書作為操作案頭手冊，亦可作為企業培訓教材，在工作中隨時可以從本書中找到對應的管理知識，工具，方法和案例解說，學以致用，以幫助企業提升全員的品質意識和管理技能從而提升產品品質增強企業的競爭力。此乃一本質量人的必讀書！

俱孟軍
新華社亞太總分社原社長
2021 年 6 月

Prof. Dr. TSANG Wai Chun Marianna
Postdoc
曾惠珍教授、博士後

Commercial Experience 工商界經驗

曾惠珍教授現為 TWC Management Limited 迪德施管理顧問有限公司創辦人及董事長。她同時任職上市公司獨立非執行董事超過廿年，及審核委員會主席超過十年。

曾教授先後在各大商業機構及專業機構就職，擁有特許公司秘書，企業管治，相關法律，會計，財務、金融、稅務及其他企業管理顧問工作經驗超過 35 年，先後參與世界各地公司上市工作及上市公司秘書服務，包括香港、美國、德國及英國上市的秘書、財務或顧問服務。曾教授業務亦涵蓋國際私募基金管理服務及協助客戶申請香港證監會金融業務的牌照。

曾教授除是一位成功企業家，獲獎無數外，並獲得多項學業及專業資格與學術成果。

曾教授同時熱心義務參與教育及社區服務超過 30 年，公益範圍涵蓋廣闊，包括各類形的慈善團體或非牟利機構，政府資助的或私人慈善團體，橫跨香港本地及海外慈善機構。她所服務的慈善機構性質涵蓋大學、中學、小學之管理層、學生、校友各類的公益服務、政府機構、禁 / 戒毒、扶貧、弱勢社群、傷殘、環保、綜合社會服務、工商會、調解、法規、專業學術團體等。

Academic Qualifications 學歷

1) Hong Kong Baptist University (Formerly Hong Kong Baptist College), Diploma in Secretarial Management with "Distinction" and placed on "President's Honors Roll"
   香港浸會大學（前香港浸會學院）秘書管理系文憑，以優異成績名列「校長榮譽」榜

2) Heriot-Watt University, UK, Master of Business Administration (MBA)
   英國 Heriot-Watt 大學工商管理碩士

3) Honorary Doctorate in Corporate Governance, Sabi University
   法國 SABI 大學榮譽博士（企業管治）

4) Post Doctoral Fellowship, Advanced Research Program jointly by California State University, Monterey Bay, Stanford University and U.C. Berkeley University
博士後，美國加州州立大學蒙特利灣分校，美國史丹福大學及美國加利福尼亞大學柏克萊分校合辦之高深研究項目

Academic Achievements 學術成果

1) The First Hong Kong Baptist University Distinguished Alumni Award 2014
香港浸會大學第一屆傑出校友獎 2014

2) HK Baptist University MBA Awards Management Committee Chairman/ Member 2000-2014
香港浸會大學工商管理碩士課程獎勵計劃獎項遴選委員會主席／成員 2000-2014

3) Member, Board of Review (HK Inland Revenue Ordinance) The Government of HKSAR 2010-2015
香港稅務上訴委員會 2010-2015

4) Hong Kong Productivity Council, 3th Seed Program, Panel Judge Chairman HK 2019
香港生產力促進局第三屆「種籽聯盟」體驗計劃評審主席 2019

5) Special Researcher of the Committee of Experts of National Academy of Governance Government Economics Research Center
中國國家行政學院政府經濟研究中心專家委員會特聘研究員

6) Adjudicator, post doctoral research, the National Academy of Governance Government Economics Research Center 2019
中國國家行政學院政府經濟研究中心博士後論文答辯專家評委 2019

7) Session Chairman, The 18th ANQ Congress 2020
2020 年第 18 屆 ANQ Congress，分組主席

8) HK Baptist University Century Club Sponsorship Scheme Panel Chairman/Member
香港浸會大學尚志會贊助計劃審核委員會主席／成員

9) HK Baptist University Knowledge Transfer Award Panel Judge
香港浸會大學知識轉移獎項評委

10) Adjunct Professor in Corporate Governance, Sabi University
法國 SABI 大學兼任教授（企業管治）

11) Arbitrator, Quanzhou, China, Arbitration Committee
中國泉州市泉州委員會仲裁員
As Chief Arbitrator 首席仲裁員 in a recent case.

12) Guest Professor (Corporate Governance), The Bo Ya Organization of Peking University – Yuanpei Business School
北大博雅元培商學院客座教授（企業管治）

13) Vice president of Business Innovation Research Institute of China Business Economics Association
中國商業經濟學會商業創新研究院副院長

Professional Qualifications 專業資格

1) Fellow, The Taxation Institute of Hong Kong (FTIHK)
香港稅務學會

2) Associate, The Institute of Chartered Secretaries & Administrators UK (ACIS)
英國特許秘書及行政人員公會

3) Associate, The Hong Kong Institute of Company Secretaries (ACS) (name changed to The Hong Kong Institute of Chartered Secretaries, Aug 2005)
香港特許秘書公會

4) Fellow Member, Society of Registered Financial Planners (FHKRFP)
香港註冊財務策劃師協會

5) Certified Tax Advisor, The Taxation Institute of Hong Kong
香港註冊稅務師

6) HKICMe Certified Mediator (Finance), Fellow, Hong Kong Institute of Community Mediation
香港社區調解學會

7) Member, Chartered Institute of Arbitrators UK (MCIArb)
英國特許仲裁師公會會員

8) Senior Associate CIP (Certified Insurance Professional), Australian and New Zealand Institute of Insurance and Finance (ANZIIF)
澳大利亞及新西蘭 保險與金融學會

9) Member, The Association of Business Executives (MABE)
美國工商管理專業人員協會

10) Chartered member, Chartered Institute of Personnel and Development (CIPD)
英國特許人事發展協會

11) Member, Australian Human Resources Institute (AHRI)
澳洲人力資源協會

12) Member, Institution of Occupational Safety and Health UK (IOSH)
英國職業安全及健康學會

13) Fellow, Institute of Financial Accountants UK (FFA)
英國註冊財務會計師公會

14) Fellow, Institute of Public Accountants Australia (FIPA)
澳洲公共會計師公會

15) Fellow member, Association of International Accountants UK (FAIA)
英國國際會計師公會

16) Fellow member, China Association of Chief Financial Officers (AAIA)
中國總會計師協會

17) Member, The Hong Kong Association of Financial Advisors Limited (HKAFA)
香港財務顧問協會

18) Tier 3 member, Professional Paralegal Register UK (PPR)
英國註冊專業法務員

19) Chartered Governance Professional, The Institute of Chartered Secretaries and Administrators and The Hong Kong Institute of Chartered Secretaries (name changed to The Chartered Governance Institute, Nov 2019)
特許企業管治人員協會

20) Fellow Member, The International PostDoctor Association
國際博士後協會

21) Chief Legal Officer (Senior), All-China federation of industry & commerce talent exchange and service center
首席法務師（高級），全國工商聯人才交流服務中心

Achievements / Awards 成就／獎項

1) Various scholarships awarded by Hong Kong Baptist College (now Hong Kong Baptist University)
香港浸會學院（現香港浸會大學）各項獎學金

2) Placed on "President's Honors Roll" by Hong Kong Baptist College (now Hong Kong Baptist University)
   名列香港浸會學院（現香港浸會大學）「校長榮譽榜」

3) Placed on the list of "Who's Who in the HKSAR 2004"
   名列「香港特區名人錄 2004」

4) Placed on the list of "Who's Who in Asia 2004"
   名列「亞洲名人錄 2004」

5) "2009Asia – Pacific the most Creative Chinese Entrepreneurial Leaders · Female Entrepreneur" awarded by APCE
   「2009 亞太最具創造力華商領袖 · 女企業家獎」

6) "Remarkable Contribution of Asia – Pacific Chinese Entrepreneurial Leaders Forum 2009" awarded by APCE
   「2009 亞太華商領袖論壇卓越貢獻大獎」

7) "2010Asia – Pacific the most Responsibility Chinese Entrepreneurial Leaders · Female Entrepreneur" awarded by APCE
   「2010 亞太最具社會責任感華商領袖 · 女企業家獎」

8) 2010 Interview by ATV program "Code of Success"
   亞洲電視節目《成功秘笈》專訪人物

9) Various awards by IBC, Cambridge England including the publication of 2011 Outstanding Intellectuals of the 21st Century

10) "Hong Kong's Most Valuable Companies 2011" awarded by Mediazone Ltd.
    「香港最具價值公司獎 2011」

11) Winner of the Prestigeous Best Showcase Winner Awards 2011 awarded by Mediazone Ltd.

12) Over 20 various awards by various China organizations, including the (translation: China Distinguished Commercial leader 2010), etc.
    國內獎項超過 20 個不同的獎項，包括中華傑出商業領袖

13) Success Institute of Higher Education and Professional Training, Certified Mediator – Qualifying Program, placed on "Distinction" list 2012
    認證調解員並名列於優異榜 2012

14) Recognition of Corporate Partnership awarded by Tung Wah Group of Hospitals
    企業伙伴合作 2013（東華三院）

15) 2015 Interview by ATV program
    亞洲電視《世說論語》專訪人物

16) "Most Valuable Services Award in Hong Kong 2015" awarded by Mediazone Ltd.
    香港最優秀服務獎

17) "Hong Kong's Most Valuable Companies 2016" awarded by Mediazone Ltd.
    香港最具價值公司獎 2016

18) "Hong Kong's Most Valuable Companies 2017" awarded by Mediazone Ltd.
    香港最具價值公司獎 2017

19) "Certificate of Appreciation, 1st SEED Programme 2017" awarded by Hong Kong Productivity Council
    香港生產力促進局「種籽聯盟」計劃嘉許狀

20) 「2018 全球華人楷模年度人物」由全球華人春節聯歡晚會組織委員會頒發

21) Happy Company of 2013-2021 awarded by Promoting Happiness Index Foundation, Hong Kong Productivity Council
    香港生產力促進局開心公司 2013-2021

22) "Hong Kong's Most Outstanding Business Awards" awarded by Corphub
    香港最優秀企業大獎 2019

23) "Greater Bay Area Outstanding Leader Award 2018" awarded by Metrodaily & Metro Prosperity
    大灣區傑出領袖（由都市日報、都市盛世頒發）

24) 2020, Senior Fellow Social Enterprise Research Academy, SERA, HK
    社會企業研究院資深院士

25) 2020, Hong Kong Baptist University Foundation Honorary Vice-President
    香港浸會大學基金榮譽副主席

26) Forum on new business models in China, Innovative persona of business models in China
    中國商業新模式高峰論壇，中國商業模式創新人物

27) 2021, The Third Guangdong-Hong Kong-Macao Greater Bay Area Outstanding Entrepreneurship Inheritage Award
    第三屆粵港澳大灣區優秀企業家傳承大獎

28) 2021, Minister of Hong Kong Inheritage Foundation
    香港傳承基金總理

Community Service Achievements 社區義務公職／公益服務

Hong Kong Baptist University (Current) 香港浸會大學（現任）：

Knowledge Transfer BEST Committee Member

知識轉移處創業支援與培訓計劃諮詢委員會委員

Hong Kong Baptist University Foundation Honorary Vice-President

大學基金榮譽副主席

CN Yang Hall, Hall Fellow

楊振寧堂顧問

Business Management Society, Advisor

工商管理學會顧問

Hong Kong Baptist University (Past) 香港浸會大學（曾任）：

Century Club, President
尚志會會長

Hong Kong Baptist University Foundation Alumni Committee member
大學基金校友委員會委員

School of Business, Honorary Associate
工商管理學院榮譽院使

Court Member
諮議會成員

Alumni Association, President
校友會會長

Advisory Board on Graduate Employment, Board Member
畢業生就業輔導顧問委員會顧問成員

Others (Current) 其他（現任）

Bradury Chun Lei Primary School, Management Committee Member
循理會白普理基金循理會小學校董（同時曾任校長遴選委員會）

Pui Hong Self-Help Association, Hon. Advisor
香港培康聯會榮譽顧問

Perfect Fellowship, Advisor
全備團契顧問

Y's Men International, Y's Men Club of Bauhinia, Honorary Auditor
紫荊國際聯青社義務核數師

Free Methodist Mei Lam Primary School, Management Committee Member
循理會美林小學校董（曾任署理校監及現任校長遴選委員會）

International Education Association, Director
國際教育協會董事

Scout Association of Hong Kong North Kwai Chung District, Hon. President
香港童軍總會北葵涌區區務委員會名譽會長

Regulatory Committee member, The Hong Kong Independent Non-Executive Director Association
香港獨立非執行董事協會監管委員會會員

Member of RadHealath
輻射生機學會會員

Chairman of doctorate division, Shenzhen-Hong Kong Institute of Professional Management Association Limited
深港專業管理學會博士後—PhD 博士組主席

Chairman of The Belt and Road Initiative council, Shenzhen-Hong Kong Institute of Professional Management Association Limited
深港專業管理學會 一帶一路議會—主席

Zhu's Scalp Acupuncture Hong Kong Association, Honorary Finance Advisor
朱明清頭皮針生命醫學香港學會榮譽財務顧問

Dashun Foundation Think Tank member
大舜基金智囊團成員

Others (Past) 其他（曾任）

Hong Kong Government, Juror in High Court of Hong Kong
香港政府高等法院陪審員

The Public Relations Association of Hong Kong, Board member
香港公關學會董事

Y's Men Club of Bauhinia, President
紫荊國際聯青社社長

The International Association of Lion Club, Youth Outreach & Drug Awareness Committee, Committee Member
國際獅子總會青年拓展及禁毒警覺委員會委員

Y's Men International, South East Asia Region, Hong Kong District, Deputy District Governor
國際聯青社亞洲區域香港地區副總監

Y's Men International, Regional Service Director
國際聯青社地區服務總監

Fu Min Association, Director
福民董事

Hong Kong Small and Medium Enterprises Association, Director
香港中小型企業聯合會會董

Hong Kong Institute of Community Mediation, Member of Council of Governor
香港社區調解學會管治委員會理事

City University of Hong Kong, Society of Investment and Finance, Honorary Student Advisor
香港城市大學學生會投資及財務學會學生顧問

The Institute of Clerks of Works and Construction Inspectorate (Hong Kong), Honorary Auditor
工程監督及建設監理學會（香港）義務核數師

Hong Kong Productivity Council, 1st & 4th Seed Program Corporate sponsor/ mentor
香港生產力促進局第一屆及第四屆「種籽聯盟」體驗計劃贊助企業／師友導師

Enhancement of Corporate Governance in Small to Medium Size Non-Governmental Organizations (NGO) in Hong Kong from the Board's Perspective

Tsang Wai Chun, Marianna[1,2,3,*] and Philip K.I. Chan[1,2,3,4]

[1] Stanford University, Stanford, California, USA 94305
[2] University of California, Berkeley, Berkeley, CA, USA 94720-1234
[3] California State University, Monterey Bay, 100 Campus Center, Seaside, CA, USA 93955-8001
[4] Stanford, Berkeley, Monterey Postdoc Advisor
*marianna.tsang@gmail.com

Abstract

HK Regulators have been advocating sound corporate governance in HK and in the past decade, HK has seen a flux of statutory and non-statutory regulations in enhancing the level of corporate governance in commercial world.

While HK has quite a substantial number of NGOs (non-government, non-profit making and/or charitable organizations), corporate governance seems not to be advocated as their commercial counterparts. Except those big NGOs of an international nature, or governed under various ordinances, or under Government's subvention etc, quite often, the small to medium size NGOs ("SM NGOs") are not quite properly governed.

This paper attempts to address this issue in SM NGOs, then explore ways for proper corporate governance or enhancement.

This paper firstly sets out the principle/theoretical framework for corporate governance, under various authoritative bodies.

Then the legislative framework is set out under the Hong Kong Companies Ordinance ("HKCO") for an NGO to be incorporated and maintenance compliance. Particular discussion will be on the reporting exemption and business review exemption granted under the HKCO.

Other related legal considerations shall also be set out, such as that from the Inland Revenue Department for charities, the Narcotics Division of Security Bureau on money laundering and terrorist financing etc.

Simple survey questionnaire shall be sent out to SM NGOs for empirical results. Several case studies shall be made to illustrate the importance of corporate governance.

The researcher shall make conclusion and finally list out simple and easy to follow recommendations for the Board of SM NGOs to adopt proper corporate governance in view of the SM NGOs limited resources. References shall be made from guidelines from HKICPA, Hong Kong Social Welfare Department, Hong Kong Institute of Chartered Secretaries Guidance Notes, Hong Kong Council of Social Service, National Council of Non-Profits USA etc.

The researcher also recommends that the "exemption reporting" and exemption from providing "business review" under the HKCO not to be adopted but full reporting to be adopted where SM NGO holds significant public funds/interest.

Key words: NGO corporate governance, public interest/ stakeholders, ethical conduct, risk management and internal control, accountability.

1. Introduction

Hong Kong is one of the most important business centers in the world. Hong Kong remains ranking second globally in the International Institute for Management Development's (IMD) World Competitiveness Yearbook (WCY) 2019 and a well known financial hub in Asia despite the recent social unrest. Both governmental and non-governmental regulators have been advocating sound corporate governance in HK to go in line with the global trend. In the past decade, Hong Kong has seen a flux of both statutory and non-statutory regulations especially from the HK Stock Exchange and the Securities & Futures Commission with the aim of enhancing the level of corporate governance in commercial world. Commercial corporations have been implementing measures to improve corporate governance, and regulators have been taking a closer look on monitoring corporate governance among the commercial corporations, be it via legislation or via regulatory bodies.

2. The Issue

However, the researcher (having over 30 years working experience both in corporate governance field as well as voluntary services in HK NGOs), observe that while Hong Kong has quite a substantial number of NGOs (non-government, non-profit making and / or charitable organization), corporate governance seems not to have a place of emphasis in these NGOs as other commercial limited companies (limited by shares) perhaps because the NGOs are mostly voluntary organizations. Except those big NGOs of an international nature or those governed under various ordinances, quite often, the small to medium size NGOs ("SM NGOs") are not quite properly governed or need improvements or enhancement in corporate governance etc,. Similar to their big NGO counterparts, these SM NGOs are also usually voluntary organizations with public interests or objectives as a charity or other well being missions. Hong Kong is indeed well known to have significant number of NGOs, be it big or small, providing public voluntary charity services to supplement the Government's support. NGOs have been making contribution with a great impact to the community. Good corporate governance is especially important in these SM NGO which does not have direct Government or other regulatory body's supervision, and which are public companies utilizing public funds: .

This paper attempts to address the issue of this missing gap phenomenon in SM NGOs as both regulators and non-regulators have not particularly attended to corporate governance on these NGOs, with the aim of raising the awareness of corporate governance therein and thereby raising the value of these SM NGOs. This research shall define SM NGO as those with annual revenue of below or around HK$25 million as small to medium size. This research then explore ways to set up proper corporate governance system or enhancement thereof in these NGO using the simple and/ or small scale approach to follow bearing in mind SM NGOs have limited or restricted resources and professional expertise.

3.　Research Methodology

This paper will first of all run through literature review, to find out the concept of corporate governance and its evolution, and then set out the various principle, regulatory and legal framework. After that, get empirical data from survey on SM NGOs Case studies will be make to illustrate the important of corporate governance. Then draw conclusion on the situation of corporate governance of the Board or governing body in SM NGOs in Hong Kong and make recommendations on the guidelines of implementing proper but simple to follow corporate governance system or enhancement therein.

4.　Principle and Regulatory Framework

4.1 Background information/Evolution of corporate governance

Important work in governance was conducted in late 1980's and early 1990's by inter alia, the National Commission on Fraudulent Financial Reporting ("Treadway Commission") in the USA, and the Committee on the Financial Aspects of Corporate Governance ("Cadbury Committee") in the UK. In October 2001, Enron broke. This was a US energy company employing 20,000 people. It used creative accounting, fraud and corruption to claim it had US$101 billion in revenue. There was use of special purpose vehicle for off balance sheet liabilities. The concern of corporate scandal led to the 2002 US Sarbanes-Oxley Act. Under the post Enron circumstances, corporate governance was further brought to the global level when the Organization for Economic Co-operation and Development ("OECD") Principles of corporate Governance 1999, was updated and expanded in 2004.

4.2 Principle & regulatory framework

An authoritative source for the definition of corporate governance can be found in the OECD Principles of Corporate Governance, which were revised again as the G20/OECD Principles of Corporate Governance (2015). The Principles offer the following definition.

OECD's Preamble on the Principles of Corporate Governance states that: *"Corporate Governance involves a set of relationships between an organization's management, its board, its shareholders and other stakeholders. Corporate governance also provides the structure through which the objectives of the organization are set, and the means of attaining those objectives and monitoring performance are determined."*

The report of the **Cadbury Committee 1992 ("Cadbury Report")** identified three commonly accepted fundamental principles of good corporate governance. They are – Openness, Integrity and Accountability. They were used to apply to the corporate governance of listed companies and with a particular emphasis on financial reporting. However, they have been accepted to have a wider application as being relevant not only to financial reporting but more broadly to a company's key processes, practices and communications.

As to the key qualities on personal level that are required of governing board members and senior management of public sector bodies in order to fully discharge their responsibilities, notably contained in its first report of the Nolan Committee, published in May 1995. The **Nolan Committee** identified key personal qualities, well known as the **Seven Principles of Public Life**, which are *Selflessness, Integrity, Objectivity, Accountability, Openness, Honesty and Leadership*. It is thought that emphasis should be placed on the personal qualities since they are the foundation upon which good governance is built.

In 2001, the **International Federation of Accountants ("IFAC")** issued a report entitled "Governance in the public Sector: A Governing Body Perspective **("IFAC Study")**. The IFAC Study provides a comprehensive international benchmark for public sector corporate governance which states that: *"public sector entities have to satisfy a complex range of political, economic and social objectives, which subject them to a different set of external constraints. They are also subject to forms of accountability to various stakeholders, which are different to those that a company in the private sector has to its shareholders, customers etc.".* (stakeholders) *"each with a legitimate interest in public sector entities, but not necessarily with any ownership rights".*

A regulatory body in Hong Kong, Hong Kong Institute of Certified Public Accountants has issued a publication **HKICPA Corporate Governance for Public Bodies – A Basic Framework (2004).**

HKICPA applies the various Fundamental Principles and underlying Personal Qualities in the following main areas of the governance of public sector organizations extracted as below.

- *Standards of Behaviour*
- *Organizational Structures and Processes*

- *Risk Management and Control*
- *Accountability, Reporting and Disclosure*

HKICPA guideline then derives a set of recommendations on public sector corporate governance. The scope of the recommendations is adapted from the IFAC study and quoted as below.

Table 1 – Regulatory framework of corporate governance

Extracted from HKICPA Corporate Governance for the public Bodies – A Basic Framework (2004)

Standards of Behaviour		

• Ethical conduct
   ➤ Personal Qualities
   ➤ Leading by example
   ➤ Integrity, honesty and objectivity
   ➤ Openness and accountability
   ➤ Selflessness and dealing with conflicts of interest
• Codes of conduct		

Organizational Structures and Processes	**Risk Management and Control**	**Accountability, Reporting and Disclosure**

• Regulatory accountability
• Accountability for public monies
• Communication with stakeholders
• Roles and responsibilities
   ➤ Governing board
   ➤ Chairman
   ➤ Non-executive members
   ➤ Executive management
   ➤ Human resources and
     remuneration policies
   ➤ Staff training
	• Internal control
	• Risk management
	• Internal audit
	• Audit committee
	• External audit
	• Budgeting and financial
	  management
		• Internal reporting
		• External reporting
		• Use of appropriate accounting
		  policies and standards
		• Performance measures
		• External reporting
		• Use of appropriate accounting
		  policies and standards
		• Performance measures

5. Legal framework

In Hong Kong, the available legal structures for an NGO could be:

- a trust
- a society established under the Societies Ordinance
- a company under the Hong Kong law namely the Hong Kong Companies Ordinance (Cap 622) ("HKCO") (usually incorporated as a company limited by guarantee), or
- a statutory body formed under specific ordinance of the Hong Kong Government.

The use of a trust structure is normally requires the trustees to have very high personal responsibilities at law. The ruling council should be registered under the Registered Trustees

Incorporation Ordinance. Registering as a society under the Societies Ordinance has the main drawback of unlimited liability. The most appropriate structure and commonly used by HK NGOs, be it big or small, to minimise liabilities of directors would be a company limited by guarantee incorporated under the HKCO. This paper shall therefore look into the legislative framework under the HKCO for an SM NGO to be incorporated and maintenance compliance as a limited company by guarantee, which is explained as follows.

The constitutional document required under the HKCO is the Articles of Association ("AA"). The AA constitutes a contract between the company, its members and among themselves. The AA sets out the objectives of the organization, rules, protocols applicable to the company, including relating to membership, member meetings, directors appointment and responsibilities, directors meetings, authorisations to deal with day-to-day business of the company, committees, company secretary etc.

In Hong Kong, for an NGO to be established as a company limited by guarantee, apart from the write up of the AA, the basic legal setup and maintenance requirements for a company limited by guarantee include:

5.1 Under the Hong Kong Companies Ordinance (the following sections are sections under the HKCO)

- appointing at least two directors (Section 453);

- the directors must be individuals (Section 456), who are at least 18 years old (Section 459); and not undischarged bankrupts (Section 480)

- directors must adhere to their fiduciary duties. The Hong Kong Companies Registry has issued "A Guide on Director's Duties" in March 2014. The fiduciary duties of directors are now codified in the company law. Under Section 465 of the HKCO, a director is legally required to conduct in a manner comparable to that of a reasonably diligent person with (i) the general knowledge, skill and experience that may be reasonably expected of a person carrying out the functions carried out by the director.

- Appoint a resident company secretary (individual or body corporate). The company secretary is a responsible officer as with the director and can be liable for the company not complying with the compliance obligations of the company (Sections 3(2) and 474-475).

- Have at least one member.

- Must have a registered office in Hong Kong

- Directors must minute directors and members' meetings and the minutes kept for 10 years. The minutes, signed by the chairperson, serve as evidence of the matters therein unless proven to the contrary. The written records without a meeting of directors should be provided to the company and are effective as with minutes of physical meetings.

- Subject to additional requirements under the AA, where a director is directly or indirectly interested in a transaction, arrangement or contract, or a proposed transaction, arrangement or contract, that is significant in relation to the company's business, and the director's interest is material, the director must declare the nature and extent of the director's interest to the other directors in accordance with certain requisite procedures (Sections 536-542). This requirement also extends to shadow or alternate directors (Section 540)

- Maintain proper accounting records: Directors must prepare for each financial year, financial statements that comply with the requirements of the HKCO (Section 379). Directors need to arrange a practicing auditor to audit on these statements (section 405). The directors should send 21 days before the annual general meeting the financial statements for the financial year, the directors' report for the financial year and the auditor's report on those financial statements. (Sections 357(2), 429-430, 571 and 576), and lay before the annual general meeting for adoption by members of the company.

- A company must, in respect of each financial year of the company, hold a general meeting as its annual general meeting (AGM) within nine months after the end of its accounting reference period, but in any event not more than 18 months from incorporation or 15 months from the last meeting (Sections 369 and 431).

- A company could dispense with the requirement for holding of annual general meetings altogether, by passing a written resolution or a resolution at a general meeting by all members. The Company then is required to file a copy of the written resolution to the HK Companies Registry within 15days of the passing of the written resolution. (Section 613). In this respect, the HKCO specifies special requirements to protect members of the company.

- Directors are required to keep various registers, including register of members, register of directors etc.

- Information as to trusts are not to be entered in the members register, but companies are now required to keep a significant controllers register identifying beneficial owners holding more than 25% interest in the company, which is subject to narrow exceptions. This applies to all companies within the chain of title from the company to its beneficial owners (Sections 634 and 653H).

- must submit an annual return to the HK Companies Registry ("CR") together with certified true copies of the relevant financial statements, directors' report and auditor's report within 42 days after the annual return date, which is nine months after the end of the company's accounting reference period.

- There are statutory provisions for various filings to the CR, including notifying CR of any change in the company's particulars, such as change of directors, secretary, registered office, and other particulars of the guarantee company within a certain time period.

Under the Model Articles under HKCO of a Company Limited by Guarantee (https://www. cr.gov.hk/en/companies_ordinance/docs/AA_Sample_D.pdf) – the day-to-day 'business and affairs of the company are managed by the directors, who may exercise all the powers of the company'. This means that the board of directors is also in charge of the day-to-day operational governance of the NGO.

However, under Model Article 4, it is stated that the directors may, 'if they think fit, delegate any powers that are conferred on them under these articles to any person or committee' on any terms and conditions they deem fit and to alter or revoke the delegation. However, the important point is that delegation does not absorb the Board of its responsibility to be accountable for the actions of its delegates.

There are many statutory provisions of the HKCO where non-compliance could amount to an offence that carries with it fines from Level 1 of HK$2,000 to Level 6 of HK$100,000 (in addition to any daily default fines, and at times, imprisonment). The fine levels are not stated in the HKCO, they are set out in the Criminal Procedure Ordinance (Cap 221), Schedule 8.

5.2 Under the Hong Kong Inland Revenue Ordinance (Cap 112) ("IRO")

• NGOs need to submit annual profits tax return together with the audited financial statements to the Inland Revenue Department. They shall be exempted from paying tax on surplus if they obtained tax exemption status as a charity under Section 88 of IRO.

5.3 Under the Hong Kong Business Registration Ordinance (Cap 310) ("BRO")

• Unless NGO obtained tax exemption status under Section 88 of the IRO, NGO needs to apply business registration licence under the BRO, and pay business licence fee annually.

5.4 Financial Statement & Director Report – Exemption Reporting under the law requirement of the Hong Kong Company Ordinance

• Section 380(4)(6) of the HKCO provides that the financial statement of a company must be prepared in the applicable accounting standard. A statement in financial statement confirms that they have been prepared in accordance with applicable accounting standards and if not, provide the particulars and reason. However a "small guarantee company" or a guarantee company that is the holding company of a "group of small guarantee companies" is automatically qualified for simplified reporting if its total annual revenue or aggregate total annual revenue (as the case may be) does not exceed HK$25 million. The accounting standards that will be appliable to a company falling within the reporting exemption is Small and Medium-sized Entity-Financial Reporting Standard ("SME-FRS") and Small and Medium-sized Entity-Financial Reporting Framework ("SME-FRF") issued or specified by the HKICPA which is the body prescribed in the Companies (Accounting Standards (Prescribed

Body)) Regulation for issuing or specifying the applicable accounting standards under section 380(8)(a).

Exemptions reporting or simplified reporting from full reporting means:-

- No requirement to disclose auditor's remuneration in financial statements.
- No requirement for financial statements to give a "true and fair view"
- Subsidiary undertakings may be excluded from consolidated financial statements in accordance with applicable accounting standards.
- No requirement to include business review in directors' report.
- No requirement for auditor to express a "true and fair view" opinion on the financial statements.

Exemptions from Business review in directors' report:-

- In order to give more information to the members for the purpose of assess how the directors have performed their duty, a director should prepare business review in the directors report except the companies qualified for simplified reporting

The business review includes:-

- A fair review of the company's business;
- A description of the principal risks and uncertainties facing the company;
- Particulars of important events affecting the company that have occurred since the end of the financial year;
- An indication of likely future development in the company's business;
- An analysis using financial key performance indicators;
- A discussion on the company's environmental policies performance and the company's compliance with the relevant laws and regulations that have a significant impact on the company;
- An account of the company's key relationships with its employees, customers and suppliers and others that have a significant impact on the company on which the company's success depends.

In short, an SM NGO or a group thereof which has annual revenue or aggregate annual revenue of less than HK$25 million are exempted from full financial reporting and exemption from providing business review in its directors' report section of audited financial statements. Hence much less disclosures are required as well as no requirement on "true and fair view" of the financial statements.

For those certain SM NGOs, who hold significant public funds or public interest, the directors should consider whether to take up or not these exemptions but to give more disclosure and transparency to stakeholders to an extent according to their situation.

5.5 Charity Status under Section 88 of the Inland Revenue Ordinance

5.5.1 Being an NGO does not mean that there is no tax payable. The Hong Kong Inland Revenue Department ("IRD") must designate the organization as Section 88 tax exempt under the IRO to have status as a charitable organization ("S.88 charity NGO"). IRD had issued A Tax Guide for Charitable Institutions and Trusts of a Public Character ("Tax Guide") relating to tax exemption to companies granted with charity status under Section 88 of the Hong Kong Inland Revenue Ordinance.

The Tax Guide makes it clear that, a charity must be established for any of the following exclusive charitable purposes to benefit the public or a class of the public, namely:

(1) relief of poverty

(2) advancement of education

(3) advancement of religion, and

(4) other purposes of a charitable nature beneficial to the community not falling under any of the preceding heads.

The purposes under the first three heads may be in relation to activities carried on in any part of the world, but those under the fourth head will only be regarded as charitable if they are of benefit to the Hong Kong community. The tax advantages accorded to charities are that donors will get tax deductions on their donations to the charities, the NGOs profits are also exempt from tax if earned in the course of carrying out the charitable objects (like selling religious sound tracks for religions) substantially for use in Hong Kong. There would also be exemption from business registration fees.

5.5.2 Public Benefit

The charity must be for public benefit which encompasses two aspects-public and benefit. The benefit must be to the public in general or to a sufficient section of the public. In the latter case, the opportunity to benefit must not be unreasonably restricted by geographical or other restrictions.

5.5.3 Governing Instrument

An appendix of the Tax Guide requires clauses which should be in the governing instrument, namely the Articles of Association of the Company, to

- prohibit distribution of income or properties to its members

- members of governing body may not receive remuneration (although may be relaxed in certain circumstances.)

With respect to the requirement of no remuneration to be paid to the governing body, in the 1978 report of Charity Commissioners of England and Wales UK, the principle is set out as follows:

"Shortly, our views are that a power to remunerate trustees is not acceptable, although where there are special reasons remuneration on a reasonable scale within limits set by the governing instrument may be permissible if (i) the trustee or trustees have special for its more effective administration. This, however, must always in our opinion be subject to the overriding provision that (ii) the number of remunerated trustees must be less than a

majority of the quorum (of meeting of trustees) and that (iii) any such trustees must be absent from meetings and discussions concerning their own appointment conditions of services and remuneration and must not vote thereon."

5.5.4 Application for Exemption

The Tax Guide also set out the documents required for an application for Section 88 exemption. Changes in circumstances e.g. alteration of its governing instrument, cessation of operation, dissolution or winding up, must be notified to the IRD within one month. A charity which becomes chargeable to tax must inform IRD within 4 months after the end of the basis period of the relevant basis period of the year of assessment.

5.6 Money Laundering / Terrorist Financing

5.6.1 As charities solicit donations or hold public funds there is a risk that charities be exploited by terrorists and terrorist organizations to raise and move funds, provide logistical support, encourage terrorist recruitment or otherwise support terrorist organizations and operations. There have also been cases where terrorists create "sham" charities or engage in fraudulent fundraising. Such misuses not only facilitate terrorist activities but also undermine donor confidence and jeopardise the integrity of charities. Most charities in Hong Kong are highly localised, i.e. they only receive government grants or local donations, or only serve the local community. The risk of them being abused for terrorist purposes is low as assessed in the Hong Kong Money Laundering and Terrorist Financing Risk Assessment Report published by the Government in April 2018. However, it is now the global trend to combat anti-money laundering and terrorist financing. In this respect, Hong Kong has been placing emphasis of importance in anti money laundering and combat terrorist financing risks and compliance especially in view of the complex environment nowadays in Hong Kong, and globally, and need to be aware mitigate any possible threat and to preserve the integrity of our charity sector.

5.6.2 Statutory Requirements

Money Laundering

Under Section 25(1) of the Drug Trafficking (Recovering of Proceeds) Ordinance (Cap 405) ("DTROPO") and Section 25(1) of the Organized and Serious Crimes Ordinance (Cap 455) ("OSCO"), money laundering is an offence for a person who, knowing or having reasonable grounds to believe that any property which, in whole or in part, directly or indirectly represents any person's proceeds of drug trafficking or indictable offence, deals with that property. Maximum penalty is a HK$5,000,000 fine and 14-year imprisonment.

Terrorist Financing

Under the following provisions of United Nations (Anti-Terrorism Measures) Ordinance (Cap 575) ("UNATMO") which has been fully implemented since 1 January 2011, it is an offence if a person:

- provides or collects property, with intention or knowing or having reasonable grounds to believe, that the property, in whole or in part be used or will be used to commit one or more terrorist acts (section 7)

- makes property or financial services available to or for the benefit of terrorist or terrorist association. (section 8)

  (eg. Funds from a trust used to finance bomb making for terrorists could constitute such an offence)

- Maximum penalty is a fine and 14-year imprisonment.

The Anti-Money Laundering and Counter-Terrorist Financing Ordinance (Cap 615), amended since 1 March 2018, requires financial institutions and designated non-financial businesses and professions to conduct customer risk assessment including charities and other "not for profit" organizations which are not subject to monitoring or supervision (especially those operating on a "cross-border" basis)

The Narcotics Division, Security Bureau of the HK Government has issued "An Advisory Guideline on Preventing the Misuse of Charities for Terrorist Financing" which was first drawn up in 2007, and updated in 2018 taking into account of latest recommendation 8 and other relevant information of the Financial Action Task Force ("FATF"), setting out the legal requirements and practical guidance against terrorist financing. FATF was established in the G7 Summit, Paris 1989. Its current membership comprises 35 member jurisdictions and two regional organizations. Hong Kong has been an FATF member since 1991. FATF is an inter-governmental body that sets international standards on combating money laundering ("ML") and terrorist financing ("TF"). FATF has developed 40 Recommendations based on which the international community has been strengthening efforts to combat ML and TF and members are obliged to follow.

Of FATF's 40 Recommendations, Recommendation 8, as set out below, requires FATF members to protect non-profit organizations ("NPOs"), including charities, from TF abuse –

Countries should review the adequacy of laws and regulations that relate to non-profit organizations which the country has identified as being vulnerable to terrorist financing abuse. Countries should apply focused and proportionate measures, in line with the risk-based approach, to such non-profit organizations to protect them from terrorist financing abuse, including:

(a) by terrorist organizations posing as legitimate entities;

(b) by exploiting legitimate entities as conduits for terrorist financing, including for the purpose of escaping asset freezing measures; and

(c) by concealing or obscuring the clandestine diversion of funds intended for legitimate purposes to terrorist organizations.

Reporting: Section 25A of the DTROPO and section 25A of the OSCO make it an offence if a person fails to disclose to an authorized officer (i.e. Joint Financial Intelligence Unit "JFIU"3) where the person knows or suspects that property represents the proceeds of drug trafficking or of an indictable offence. JFIU is jointly operated by Hong Kong Police Force and the Customs and Excise Department at the Police Headquarters. It was set up in 1989 to receive and analyse STRs, and disseminate the same to relevant units for investigation.

The UNATMO implements relevant Resolutions of the Security Council of the United Nations in preventing and suppressing TF and criminalising the wilful provision or collection of funds for terrorism. Section 12 of UNATMO requires any person who knows or suspects that any property is terrorist property such as any donations or funds received or handled by your charity, to disclose to the JFIU the information as soon as is practicable after that information or other matter comes to the person's attention by filing a "suspicious transaction report" ("STR") to JFIU through the system "STREAM". The maximum penalty for failure to file an STR is a level 5 fine (HKD 50,000) and imprisonment for three months.

It should be noted that once an STR is filed, "tipping-off" is an offence. Under section 12(5) of UNATMO, a person commits an offence if knowing or suspecting that an STR has been filed, the person discloses to another person any information or other matter which is likely to prejudice any investigation which might be concluded following that STR. The maximum penalty is a fine (without cap) and three years' imprisonment.

6. Survey

6.1 Objective The objective of the survey is to obtain some empirical date from SM NGOs on their board practice on corporate governance, ie, the extent of which they have been adopting, and the areas that they need to adopt or improve/enhance in order to have a more proper corporate governance framework. *As a contribution to the society, the researcher will afterwards separately compile a simple guide for corporate governance and send to those surveyed SM NGOs who have given contact address, for their reference for adoption or enhancement in corporate governance, and thereby for the betterment of the society.*

6.2 Target The target is SM NGOs in Hong Kong. In this paper, the researcher will define those NGO with total annual revenue under HK$500,000 as small NGOs; those HK$500,000 to HK$25 million is medium size NGOs. This is in line with the HKCO's definition of a "small guarantee company" which has annual revenue not exceeding HK$25 million. Also the researcher follows he definition from the Hong Kong Council of Social Service ("HKCSS") to define those as small NGO if the number of full time employees is less than 5 persons, and as medium NGOs if number of employees is less than 10.

6.3 Survey Methodology Survey questionnaire through an on-line survey platform with a cover email consists of 28 questions covering NGOs background and various areas of their board practice was sent to 205 SM NGOs by email and some 25 by WhatsApp for on-line completion. The NGOs names were picked up through the search on the HK Companies Registry's search system for companies registered as limited by guarantee, and picked up from the "List of charitable institutions and trusts of a public character, which are exempt from tax under section 88 of the Inland Revenue Ordinance" on the HK Inland Revenue Department website. The researcher then searched their email addresses on their websites or other means through internet. The questionnaire is set in Chinese for better understanding by NGOs. The survey lasted for three months from early July to early September 2020. Regular reminders were sent during this period. Some telephone follow ups were made.

Out of the 205 emails sent, 27 were bounced back. There were 192 (including 178 emails and 14 whatsapps) NGO visited the online platform but only 71 respondents who did the questionnaire. The reply rate is therefore 36.9%. However there was one NGO whose annual revenue is greater than HK$25 million which is a bigger size organization and not this survey's target and therefore not a valid respondent for this research purpose. All the respondents have less than 10 full time employees and so meet the SM NGO definition. The final number of valid respondents is hence 70, although out of which some SM NGOs skipped some questions.

The HKCSS led by a team of survey personnel has conducted a HK NGO Governance Health Survey 2018 which covered all NGOs in Hong Kong, including large and international and small. (Their approach is different that their survey is on taking index on the level of satisfaction that the HK NGOs assess themselves. For this paper, the survey is on areas that the SM NGOs boards have done and have not done). The HKCSS got 77 NGO respondents including large and small NGO. This paper's survey got 70 NGO respondents excluding large NGO, and therefore the samples size should be of good enough representation.

6.4 Empirical Survey Results The respondent SM NGOs come from various sectors including churches, organizations promoting Chinese medical treatment and Chinese medicine, clan associations, care of minorities, alumni associations, project management, commerce associations, youth service, youth entrepreneurship support, elderly services, school organization body, life & death education and care, education support, acupuncture promotion, drug addict healing, profession promotion, religious body, women service and training, caring organization, drug addict rehabilitation support, animal protection, global economic and peace promotion, care of the needy, music therapist association, logistic industry association, environmental promotion, construction safety promotion, heritage association, charitable foundations, financial consultants association, overseas education establishment, cultural exchange promotion, country relations promotion, commercial industrial relation, various charity services association. The first 4 questions relate to nature of the SM NGO, total annual revenue, number of full time employees, and whether they are S.88 charity NGO or not. Around half 52.8% are S.88 charity NGOs and half 47.1% are just ordinary SM NGOs. The results to the remaining 24 questions (translated to English) are as follows:

Table 2 – Empirical Survey Results of SM NGOs Corporate governance practice

Question	Greatest % of answer	Percentage of other answers	Observation on corporate governance inadequacy for most of the SM NGOs
5. Do you issue appointment certificate/appointment letter to directors?	None 63.6%	- Only appointment certificate 14.5% - Only appointment letter 12.7% - Have both certificate & letter 9%	*Duties, legal obligations, commitment as a director not laid out to directors.*
6. Do you brief directors on the organization's constitution, legal & fiduciary duties?	None 45.2%	- Only brief constitution 26.4% - Only brief constitution & legal duties 13.2% - Brief all duties 7.5% - No professional to brief 7.5%	*Not understanding the legal structure, the basic legal requirements especially under the HKCO, nor the fiduciary duties as a director.*
7. How is director's orientation("O")/development("D")?	None 44.2%	- Only when need arises 38.4% - No O, but has D 11.5% - Has O, but no D 3.8% - Has both O & D 1.9%	*No formal organized briefing and development of directors.*
8. What is the term of appointment?	2 years 36.3%	- No fixed term 32.7% - 1 year 16.3% - 3 years 14.5%	*2 years is acceptable, however no fixed term also is of a higher percentage which may lead to stagnation & less chance to expand the board.*
9. How do you arrange directors' declaration of interest?	Not arranged 47.1%	- Declare only on appointment 22.6% - Declare only when there is need 16.9% - Once a year 13.2%	*Independent and unbiased decision not emphasized. Also there may have unknown conflict of interest.*
10. Do you have division of duties between board & management?	No 70.3%	- Yes 29.6%	*No division of duties is not that important especially in small NGOs as the management duties may be divided among directors.*
11. Do you think you have a good mix of directors in respect of gender, age, profession?	Agree 53.7%	- Very agree 22.2% - Not quite 12.9% - Not agree 11.1%	*This is a good practice for board diversity, since "agree" and "very agree" add up to 75.9%*
12. Do you have board retreat?	No 70.3%	- Only when need arises 18.5% - Once a year 9.2% - Once in two years 1.8%	*Lack of opportunities for reflection and forward looking strategic discussions.*

13. Does your salaried chief executive also act as a director?	We have no salaried chief executive 56.6%	- No 37.7% - Yes 5.6%	*SM NGOs not always have a chief executive due to budget constraint. Also if tax exempt charity, it is generally restricted for the salaried executive to act as director by law (subject to certain relaxation).*
14. Who will perform appraisal on the chief executive?	Not applicable 71.1%	- by the Board, once for a while 13.4% - not by the Board 9.7% - by the Board, once a year 5.7%	*Most SM NGOs do not have Chief executive which may due to budget constraint.*
15. Do you have self-assessment on the Board as a whole?	No 87%	- Yes, once a year 11.1% - Yes, once for a while 1.8%	*No opportunity to self-reflection and judge own collective performance.*
16. Do you have self-assessment on individual director?	No 87%	- Yes, once a year 9.2% - Yes, once for a while 3.7%	*No opportunity to assess individual director's performance.*
17. How often you hold board meeting?	Not fixed 38.8%	- once in two months 18.5% - once a month 16.6% - quarterly 12.9% - once a year 7.4% - once in half year 5.5%	*Most SM NGOs do not hold regular board meetings. Board meetings are important for exercising governance.*
18. What is quorum for your board meeting?	Not less than ½ of total directors 53.7%	- Not less than 1/3 of total directors 25.9% -Not less than 1/5 of total directors 9.2% - Not less than 1/4 of total directors 5.5%	*Good practice since more directors for quorum.*
19. Can a director vote on transactions which he has conflict of interest?	No 84.6%	- Yes can. 11.5% - Yes but under specified conditions 3.8%	*Good practice as directors should act for the best interest of the NGO.*
20. What is attendance rate of directors?	Over 75 percent of directors 61.1%	- 50%-74% attendance rate 31.4% - 25%-49% attendance rate 3.7% - less than 25%, 3.7%	*Although majority 61.1% of SM NGOs have high board attendance rate, but still there are nearly 40% of SM NGO which does not have. Need improvement.*
21. How often does your board approve update financial report?	Once a year 44.4%	- At every board meeting 31.4% - Once for a while 22.2% - Others 1.8%	*No financial control*

22. Will you prepare annual budget?	No 62.9%	- Yes, and will obtain board approval 24% - Yes, but not regularly to board for approval 9.2% - Others 3.7%	*Not good financial management*
23. Will you conduct dur diligence on significant donors?	No, since already familiar with these donors	- Not applicable (no significant donors) 35.8% - No 15% - Yes, only for over certain amount 5.6%	*This is acceptable as most SM NGO have already know significant donor.*
24. Do you conduct risk assessment exercise?	No 64.8%	- will do if need arises 27.7% - yes, once a year 7.4%	*Risk assessment should be conducted to avoid fraud, corruption etc. to enhance internal control. It is now more important in view of global trend of anti-money laundering/terrorist financing.*
25. Do you have sub-committees?	None 44.4%	- Executive Committee 33.3% - Nomination Committee 24% - Other types of Committee 24% - Audit Committee 7.4% - Remuneration Committee 1.8%	*Setting up of sub-committee or task force will be beneficial to specialize on special task and invite other expertise.*
26. Do you have policies of **- internal control** **- risk assessment** **- conflict of interest** **- procurement** **- investment** **- whistleblower** **- stakeholder communication**	None 48.9%	- has internal control policy 34.6% - has stakeholders policy 22.4% - has conflict of interest policy 20.4% - has risk assessment policy 18.3% - has procurement policy 18.3% - has investment policy 16.3% - has whistleblower policy 12.2%	*Most NGOs do not have any policy for controlling purpose. For those who have, they mainly have the internal control policy.*
27. Do you have succession plan?	Not specified 74%	- Yes 25.9%	*Succession planning is a recommended practice to ensure the smooth development of an NGO.*

28. Do you buy directors and management liability insurance?	Not necessary 87%	- Yes, also annual review 11.1% - Others 1.8%	*Most surveyed SM NGOs do not buy insurance which may due to financial resources and may also due to lack of awareness of their legal and fiduciary duties' liabilities.*

The survey result shows most of the surveyed SM NGOs have not adopted most of the, ie 21 out of 24, proper or recommended board governance practices set out in the survey questionnaire as outlined in the above table. The only 3 areas that they have good practice relates to board meeting quorum, board diversity, and prohibition from voting when there is conflict of interest in a transaction.

Fig 1 – Some key areas of corporate governance extracted from the survey result and presented as follows:-

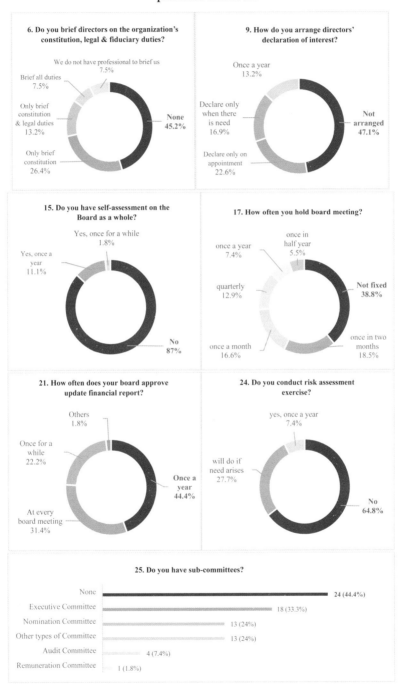

6. Do you brief directors on the organization's constitution, legal & fiduciary duties?

- We do not have professional to brief us 7.5%
- Brief all duties 7.5%
- Only brief constitution & legal duties 13.2%
- Only brief constitution 26.4%
- None 45.2%

9. How do you arrange directors' declaration of interest?

- Once a year 13.2%
- Declare only when there is need 16.9%
- Declare only on appointment 22.6%
- Not arranged 47.1%

15. Do you have self-assessment on the Board as a whole?

- Yes, once for a while 1.8%
- Yes, once a year 11.1%
- No 87%

17. How often you hold board meeting?

- once a year 7.4%
- once in half year 5.5%
- quarterly 12.9%
- Not fixed 38.8%
- once a month 16.6%
- once in two months 18.5%

21. How often does your board approve update financial report?

- Others 1.8%
- Once for a while 22.2%
- At every board meeting 31.4%
- Once a year 44.4%

24. Do you conduct risk assessment exercise?

- yes, once a year 7.4%
- will do if need arises 27.7%
- No 64.8%

25. Do you have sub-committees?

- None 24 (44.4%)
- Executive Committee 18 (33.3%)
- Nomination Committee 13 (24%)
- Other types of Committee 13 (24%)
- Audit Committee 4 (7.4%)
- Remuneration Committee 1 (1.8%)

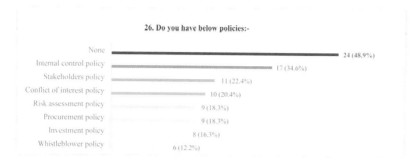

26. Do you have below policies:-

None	24 (48.9%)
Internal control policy	17 (34.6%)
Stakeholders policy	11 (22.4%)
Conflict of interest policy	10 (20.4%)
Risk assessment policy	9 (18.3%)
Procurement policy	9 (18.3%)
Investment policy	8 (16.3%)
Whistleblower policy	6 (12.2%)

7. Case Studies

There are two cases in recent years for study to illustrate the important of corporate governance.

7.1 The Sports Federation and Olympic Committee of Hong Kong

Recently the Audit Commission ("Commission") of the HK Government released its Audit Report No.74 ("Report") on the Sports Federation and Olympic Committee of Hong Kong ("Olympic Committee") in April 2020. The 141 page Audit Report revealed a number of corporate governance problems of the Olympic Committee, which receives fundings from the HK Government.

The major corporate governance failures or issues includes the following: –

(1) *Cost Control (overspending)*

The Commission reported that: "Throughout the period 2014-15 to 2018-19, the committee secretariat has recorded soaring operating deficits, increasing from HK$33,000 [US$4,300] in 2014-15 to HK$588,000 [US$72,000] in 2018-19."

(2) *Procurement*

The Report found that procurement rules were not properly followed. There were 47 cases of goods and services procurement, involving about HK$6.6 million [US$852,000], between 2016-17 and 2018-19. In 20 of them, only a single quotation had been obtained.

24 of the procurements were in fact reimbursements of expenses. 2 of the procurements did not conduct tendering as required by the laid down rules.

The Olympic Committee spends HK$20 million a year of public money which has been abused.

(3) *Lack of sub-committee meetings*

14 out of the 29 sub-committees had not held any meetings between March 30, 2017 and December 31, 2019.

The Commission said that *"could be a cause for concern"* because *"meetings are an important forum where ideas can be exchanged and issues can be discussed in an interactive manner"*. It also said *"Audit considers that [the Olympic Committee] needs to review the frequency of meetings of individual committees to ensure that the functions of the [board and subcommittees] are effectively carried out,"*

(4) *Transparency issue in athlete selection criteria*

At a public hearing of the Hong Kong Legislative Council's Public Accounts Committee following up on the Audit's Report No 74, members slammed the governance of the Olympic Committee on their selection of swimmers for the 2018 Jakarta Asian Games.

The Olympic Committee was criticized that They [seem] to have a different set of standards when choosing swimmers, e.g. Those who come from a big swimming club were given preference and this was not fair.

A member of the Public Accounts Committee, Shiu Ka-fai, questioned the rationale of sending an athlete who had undergone surgery before the Games. *"The athlete's result in Jakarta proved he was not competitive enough,"* he said. *"If he was not able to regain his best form in time, the selection committee should have given his place to another athlete."* It was cited the case of a swimmer who was the fastest in Hong Kong in his event but was not selected, and instead someone below his standard took his place at the Jakarta Games.

The Commission noted a lack of clear criteria for selecting athletes to take part in international games. The Report cited a case in 2018 when 11 of the 17 athletes were chosen for the Asian Games, on grounds that were not stated in a circular distributed previously to sports associations. They included four selected because they had won sports scholarships in Hong Kong. It was criticised as being unfair that athletes were not informed the selection criteria included that being a scholarship athlete at the Sports Institute would gain marks.

(5) *Appeal mechanism*

It was also surprised to learn that any review of the selection was also done by the same selection panel. There had never been any appeal made by national sports associations, including the swimmer's case at the 2018 Asian Games although there is appeal mechanism stated in the constitution.

(6) *The Independent Commission Against Corruption's "Best Practice"*

"The Independent Commission Against Corruption has drawn up a 'Best Practice Reference for Governance of National Sports Associations – Towards Excellence in Sports Professional Development' in 2011 for many years but it seems there has been little progress,"

(7) *No Declaration of interest*

The Olympic Committee introduced the procedure for declaration of interest. However up to January 2020, only 5 out of 29 sub-committees have implemented.

Professor Chung Pak-kwong, of Baptist University's department of sport and physical education, called for an overhaul of the Olympic Committee's leadership. He said *"It has grown into an empire and transparency and accountability are not in their dictionary,"* said Chung, who is the former chief executive of Hong Kong Sports Institute. *"The top guys in the committee's board are just too used to relying on money from the government. They seldom need to explore ways to improve the efficiency of the committee."*

This case shows that poor corporate governance leads to the fact that not the best athletes are chosen thereby affecting Hong Kong's overall performance in international competition; also it leads to the fact that public money was abused with the result that most of the public money is wasted. *(Some news clippings are shown in the Appendix of this paper.)*

7.2 Hing Tak School Scandal

This is a case of the principal Ms Chan Cheung-ping of a school in Hong Kong which receives funding from HK Government, who had various mis-management and fraud. It is a case of failure of corporate governance of an NGO. The failures are as follows:-

(1) Principal Chan forced the teachers to go to Shenzhen China and HK China border to distribute promotion leaflets to solicit potential China mainland students. This may infringe the China mainland law.

(2) Principal Chan installed various cad-cam video camera in various spots of the school premises. This infringed the Personal Data (Privacy) Ordinance (Cap 486).

(3) Hing Tak School is accused of flaking by inflating student numbers, and keeping the names of over 20 mainland pupils who had been absent for up to two years on the roster so as to keep those "Phantom students" or "shadow students" to deceit and to avoid the cut from Hong Kong Government's funding or even closure if student number declines.

But Stephen Lui, who was appointed by the Education Bureau to the school board, said according to his experience as a principal of three different schools, any long absence when a student can not be contacted usually leads to the pupil being deemed to have left the school.

(4) Violation of hiring and promotion procedures: Hing Tak School said that on one occasion, a job interview took place one day after the hiring notice was posted – a violation of recruitment procedures. The school suspects that the sole applicant to the post, has close ties to principal Chan.

School authorities said that, according to regular procedures, roles should first be advertised within the institution so that qualified teachers may opt for an interview. If existing teachers are deemed unsuitable, schools may contact the Incorporated Management Committee (the governing body) ("IMC") for approval to initiate external recruitment and for approval of the chosen candidate.

(5) Resignation of teachers raised from 2014's 17% to 2016's 30%. Some students have already 5 teacher-in charge replaced for them. Some teachers had their promotion every year for 3 consecutive years.

(6) Fake Document: Principal Chan submitted a fake form of job interview record conducted in 2016, to Education Bureau, Hong Kong.

(7) Bribery: teacher who applied sick leave had to pay 6 pieces of cake coupon or $300 cash for each day of sick leave.

After the sandal came to the surface, Hing Tak School finally passed a motion to fire the principal, Chan Cheung-ping. Apart from breach of Education Ordinance, the motion was based on Section 9 of the Employment Ordinance, which states that an employer may terminate a contract of employment without notice or payment if the employee wilfully disobeys a lawful and reasonable order, engages in misconduct, is guilty of fraud or dishonesty, or is habitually neglectful in their duties.

This case, shows the mis-management and fraud of the ex-principal and, above all, a corporate governance failure of the Governing Body, the Incorporated Management Committee ("IMC") of the school. The IMC had not done its role of supervising the Principal (ie. the management head of the school)

There were the issues of transparency, accountability to stakeholders, bribery fraud etc. It was also learnt that most of the members of the governing board IMC had no knowledge on school management and only relied on and believed the Principal on matters of the school, and received no training. (News clippings is shown in the Appendix of this paper.)

8. Conclusion

As NGOs are entrusted, inter alia, by the stakeholders and entrusted with public funds and/ or for public interest, proper corporate governance should be enhanced also to small to medium size NGOs, and not just with attention drawn on commercial organizations by the regulators in Hong Kong, or on big NGOs governed by their international constitutional codes or by various ordinances or by the Government etc. With the growing stakeholder expectations around corporate accountability and the growing global concern on anti-money laundering and terrorist financing, it is time to address this issue and integrate into their organizations with the benefit of adding value potential to these SM NGOs.

9. Recommendation

9.1 Corporate governance guide

In order to make it easy to follow by SM NGO to improve or to enhance their corporate governance, it is recommended to write up a simple user friendly corporate governance manual or guide for these SM NGOs.

A corporate governance guide or manual can help an NGO make clear to its Board and management the legal obligations, structures, processes and principles that it should use to achieve the NGOs objectives and to monitor the NGOs performance and it is a helpful tool for any NGO.

There are best practices that should be adopted regardless of an organisaiton's size, resources or nature. However, for SM NGOs which usually do not have the resources to undertake formal and full flown corporate governance, they can adopt to an extent suitable to their needs and scale under allowable resources or take up those suggestions recommended here which are applicable to their circumstances.

9.2 The corporate governance guide/manual is proposed to be as follows, which as said, should be tailored to individual SM NGOs circumstances & needs:-

Table 3 – Proposed corporate governance guide or manual

1. THE ORGANIZATION	- Background, Legal Status, Services
	- Vision, Mission and Value
	- Strategic Direction
2. GOVERNANCE STRUCTURE	- Board of Directors
	- Composition of the Board, Terms of Office
3. DUTIES AND RESPONSIBILITIES OF THE BOARD	- legal compliance
	- Fiduciary Duties of directors
	- Delegation to Committees
	- Board orientation, development, review and renewal/ succession
4. DIVISION OF BOARD AND MANAGEMENT ROLES	- Role of the Board
	- Role of the Management
	- Division of Board & Management roles
5. BOARD MEETINGS	- Articles of Association
	- Protocols, Procedures and Rules for Effective Board Meetings
	- Minutes
6. CONTROL & RISK MANAGEMENT	- Budget, Internal Reporting
	- Internal control and Risk Management
	- Audit of financial statements
	- Directors & Officers Liability Insurance
7. CODE OF ETHICS	- No remuneration to directors
	- No distribution of income to members
	- Conflicts of Interest
	- Transparency and Accountability
	- Non-Discrimination and Equal Opportunities

9.3 Guidelines and explanations

The followings are the guidelines and explanations in developing the guide or manual:

9.3.1 Section 1 will introduce the background and nature of the organization. It states the legal status which in this paper referred as a company limited by guarantee incorporated under the HKCO. If the SM NGO is given a charity status under section 88 of the IRO, then states so to give a higher confidence to stakeholders and to alert that potential donations are tax deductible for donors' advantage.

It then briefly describes the services it is embarking. It states the vision, mission and value and any statement slogan thereof. Strategic direction is also set.

9.3.2 Section 2 will state out that the governing structure is the board of directors or other equivalent name such as "Council" or "Executive Committee" (but means same as the Board under the HKCO). Any office bearer positions should be listed out and one of the directors shall be the chairman. When determining the composition of the board, consideration should be given to what qualities and perspective are required of the board members. Diversity of board members, such as gender, age, expertise, equal opportunity, culture inclusion should be considered. The Board should determine its optional size based on its needs.

Directors should have a term limit. It is common for the term to be 1-3 years to avoid stagnation, offers opportunity to expand circle of contacts and expertise, provides a smooth channel to remove unproductive board members. On the other hand, there should be mechanism for re-appointment to allow flexibility and to keep valuable board members.

9.3.3 Section 3 Directors shall bear the overall compliance responsibility of the organization under the HKCO. The basic legal requirements under the HKCO, IRO, and BRO can be listed out in the manual here as follows:-

9.3.3.1 Legal Compliance Under HKCO:

- maintain a local registered address in Hong Kong
- must maintain at least two directors, and all of its directors must be natural persons, one of the directors can be the same person as the secretary of the company
- appoint a resident company secretary (individual or body corporate)
- have at least one member (individual or body corporate)
- notify CR of any change of directors and secretary
  - resignation/cessation
  - new appointment
  - particulars (e.g. address) change
- notify the CR of any changes in the company's other particulars recorded in the CR
- renew its annual business registration
- maintain proper accounting records which are required to be kept for 7 years and prepare annual financial statements

- appoint a practising accountant to audit the annual financial statements
- hold an annual general meeting the first of it within 18 months of incorporation and not more than 15 months interval between two AGM's at which the company's audited financial accounts and directors' report are adopted, directors and auditors are appointed/reappointed, and auditors remuneration fixed
- submit a profits tax return together the annual audited accounts to the IRD
- submit annually employer's return on employees' remuneration to IRD
- submit an annual return to the CR together with certified true copy of the relevant audited financial statements, directors' report and auditor's report within 42 days after the annual return date, which is nine months after the end of the company's accounting reference period (ie. financial year end), and
- need not pay the business registration fee or file a profits tax return with the IRD (if it has been recognised as a S.88 charity NGO.

9.3.3.2 Fiduciary duties: Here can list out the general fiduciary duties of directors of duty of care, duty of loyalty, duty of obedience. Directors are referred to a Guide on Director's Duties issued by the Companies Registry in March 2014, or can list out the said duties of the Guide here.

9.3.3.3 Committees The establishment of committees will depend on the need of the Board or the operation nature and size of the SM NGO. The types of any committees formed under the Board can be listed in this section. Examples of committees can be Executive Committee, Audit Committee, Board Nomination Committee, Remuneration Committee, Human Resources Committee, Internal Control Committee, Procurement Committee, Investment Committee, Risk Management Committee, and even Governance Committee etc. Other Committees can be set up to steer specific initiatives or programs. It is recommended that an NGO can have the basic audit, nomination and remuneration committee to demonstrate commitment to corporate governance to donors and thirty party sponsors (including Government). If the size or the resources of the organization does not warrant formal committees, ad hoc task force can be formed to take up specific requirements at specific time and dispensed with after the task requirements have been met. Terms of reference (including functions, responsibilities and limits of authority) should be defined. It needs to be borne in mind that despite of any committees set up, the Board remains with the ultimate responsibility of corporate governance of the NGO.

9.3.3.4 Board orientation, development, review and succession: It is recommended that orientation to new or existing board member (to refresh orientation) be conducted to brief them on the organization, their legal, fiduciary and governance responsibilities, and any expectations on them. Site visit to the organization's premises or facilities is suggested to give a "physical touch" of the organization's work. It is a good practice for a director to complete and sign upon appointment, in a form declaring he is not an undischarged bankrupt nor has he had criminal record and also make declaration of any interest with the SM NGO. It is also a good practice to present a certificate of appointment and/or letter of appointment to formalize and honor their appointment. For more formal NGOs, the appointment letter can lay out job description reflecting collective governance of the Board as well as individual expectations and obligations.

Directors are encouraged to attend periodic trainings or development to keep updated their knowledge with the organization's services or expertise. Board learning and sharing activities outside meetings on strategic discussions can be considered.

The Board should undertake regular review or appraisal to assess its strengths and weaknesses and develop strategies to address any limitations. Even minor improvements to a Board's performance can have significant impact. Review on individual Board member can be on his attendance, any disruptive behavior, whether he acts in the interest of the NGO etc.

The process of Board renewal or succession planning should be open and transparent in accordance with the Articles of Association of the NGO. The task can be delegated to a nomination committee or a panel. The Board should approve the specification (including the knowledge, experience, skills, attitudes required etc) of the role as recommended by the nomination committee. Interview can be conducted and appointment on merit. Committee assignments can be rotated to give Board members experience and opportunity to lead, as a part of succession planning.

9.3.4 Section 4:

9.3.4.1 Role of the Board

The Board leads the NGO and are collectively responsible for ensuring the NGO delivers its objective, apart from meeting with legal and fiduciary compliance. Primary roles includes:-

- Strategic direction: The Board set the direction for the organization, approves and monitor the NGOs core strategies, and make sure financial and human resources are in place.

- Management supervision: The Board is responsible for recruiting, support, appraise, and compensate the management head, and ensure the management team is operating under the Board's direction. Annual assessment on the performance of the management head can be conducted.

- Risk management & internal control: The Board is responsible for ensuring the integrity of the NGOs financials, internal control and risk management are in place and followed. This will be more elaborated in Section 6 below.

- Accountability: The Board is accountable to the NGOs stakeholders such as the NGOs members, donors, the Government (especially those NGOs receiving Government funding support), the service users, the public etc. The Board therefore should act, communicate and make information available in a transparent manner. This will be more elaborated in Section 7.

9.3.4.2 Role of the Management

The head of management leads the management team and reports to the Board. This management head is responsible for the daily operation including setting up an efficient operational structure, executing the strategic policies adopted by the Board, achieving the performance targets set by the Board, and provide operational inputs to the Board for decision making.

9.3.4.3 Division of Board and management roles

The below table shows the division of Board and management role. The division is especially meaningful for those small NGOs with the directors also at the same time involve in the operational matters, so as to distinguish themselves between strategic and operational roles.

Table 4 – Proposed Division of Roles Between the Board and Management Team

TASKS	RESPONSIBLE PARTY	
	BOARD	MANAGEMENT
A. PLANNING		
- Direct the planning process	✓	✓
- Provide input to long term strategy and goals	✓	✓
- Approve long term strategy and goals	✓	
- Formulate annual plans or objectives		✓
- Approve annual plans or objectives	✓	
- Prepare performance reports to track achievement of strategy and goals		✓
- Track achievement of strategy and goals	✓	✓
B. FINANCIAL MANAGEMENT		
- Prepare preliminary budget		✓
- Finalise and approve budget	✓	
- Monitor that expenditure is within authorised budget		✓
- Approve expenditures (if outside authorised budget)	✓	
- Prepare financial statements		✓
- Approve financial statements	✓	
- Draft financial management procedures and policies		✓
- Approve financial management procedures and policies	✓	
- Finalise and approve funding arrangements with the government		✓
- Ensure annual audit of accounts	✓	

Adopted from the HK Social Welfare Department's "Leading Your NGO Corporate Governance A Reference Guide for NGO Boards" June 2002

9.3.5 Section 5: Board meetings:

Board meetings are when the Board exercise their governance authority and should be held on a regular basis. The procedures/rules should be set out according to the Articles of Association relating to convening meetings (including meetings via electronic means, quorum, voting, managing of conflict of interest in transactions, as well as attendance requirement). It is good practice to prepare consent agenda, send the agenda together with the material pack good in advance of board meeting to allow board members to review beforehand. The purpose of consent agenda is to liberate board meetings from administrative routines thereby allowing time to focus on meaningful discussions of real important issues. A copy of the Articles of Association (ie the constitutional document) should be given to the directors upon appointment in an orientation pack. Minutes of Board meeting and all resolutions passed by directors without a meeting have to be documented, approved at the subsequent meeting, and kept for 10 years as required by HKCO. Notify the change of location for minutes record keeping to the Companies Registry within 15 days of the change.

9.3.6 Section 6: Control and risk management:

9.3.6.1 Boards of NGOs, regardless of size, must ensure funds are used properly, assets are protected as they are stewards. Fraudent transactions are blocked and detected early such as under the Prevention of Bribery Ordinance, the Theft Ordinance. Apart from the legal obligations explained above, the NGO can set out other control compliance requirements under other applicable laws of the NGO to which the NGO is subject to, e.g. under the Hong Kong Education Ordinance.

Controls can include internal control, internal reporting, external reporting. Internal controls can include internal authorizations for bank signatories, payment, and other authorizations, separation of key functions.

9.3.6.2 Internal reporting include formalizing and set out the lines of reporting, close reporting relation between each line. For control on financials, the Board is ultimately responsible for the financial reports although it can delegate to the management to prepare and audit committee, if any, to review. It is good practice to prepare an annual budget for approval, and management shall report to the Board at no less frequently than quarterly and explain the variations between budgeted and actual. Financial statements using the appropriate accounting standards should at the same time be prepared and reviewed by the Board. Audit committee comprising of certain directors and/or independent external parties, or financial sub-committee/task force for small NGOs can be considered to be set up to review, monitor the integrity of the NGOs financial statements, internal controls, risk management systems, and recommendation on the appointment of auditor and remuneration thereof.

9.3.6.3 Risk management: Risk management should be on risk based approach instead of rule based, that is, assessment is on the level of risk, namely to implement enhanced risk assessment on those with higher risk. Written policies can be drawn up for areas of significance to the NGO, such as procurement policy which is typically an area that can be highly vulnerable to abuse and bribery. The Board should ensure also policies are in place to protect the stakeholders and reputation of the NGOs. In view of the global effort to combat money laundering and terrorist financing, policies can include due diligence on potential working partners, significant donors, etc. other policy can include investment policy to protect assets of the NGO. There can be also whistleblowing policy and communication policy with stakeholders. It is good practice to review annually the risk management system is in good order, and also review any risk potential and mitigation plans made by management.

9.3.6.4 External reporting/audit: Audit of financial statements: A standard paragraph in a typical SM NGOs auditor's report in the audited financial statements states that *"The directors are responsible for the preparation of the financial statements in accordance with the SME-FRS issued by the HKICPA and the Hong Kong Companies Ordinance, and for such internal control as the directors determine is necessary to enable the preparation of financial statements that are free from material misstatement, whether due to fraud or error. Those charged with governance are responsible for overseeing the Association's financial reporting process."* Therefore even through where the Board has delegated the audit function to an audit committee, the Board is ultimately responsible for the financial statements to be presented fairly. Although the HKCO offers exemption reporting/simplified reporting on the financial statements and exemption from providing business review from directors report for an SM NGO whose annual revenue is less than HK$25 million, if an NGO receives significant public funds or support or of public interest, these exemptions can be considered not to be adopted but enhanced disclosure should be made regarding finance and the NGOs development or even does not take the simplified reporting but take up full disclosure reporting and director's review depending on the circumstances of the NGO.

There are very limited indemnities and ratification rights under the HKCO for director's wrongdoings (Sections 465-473). Directors therefore need to consider purchasing directors and officers (D&O) liability insurance to cover possible exposures to liabilities.

9.3.7 Section 7: Code of Ethics

9.3.7.1 As the input of the head of management in board meeting deliberation is instrumental and valuable for informed decision making, he will attend board meetings "in attendance". That is he is not a director so as to separate the line between oversight and execution. Exception is where applicable law requires for example the Education Ordinance requires the principal of a school to be also a member of the governing board.

9.3.7.2 For a S.88 charity NGO, the Guide can state that directors do not receive remuneration. There is exception under exceptional circumstance, e.g. a church pastor who receives remuneration can also be a director under certain conditions to be complied with and there cannot be distribution of income to members of the NGO.

9.3.7.3 Conflicts of interest: A conflict of interest arises when the "private interest" of a Board member conflict with the interest of the NGO or the Board member's official duties. Board members should remain independent and unbiased and act in the best interest of the organization. The Board should adopt a conflict of interest policy including declaration of actual and perceived interest upon appointment and annually, and maintain a register for that purpose. The policy should also coves that a director cannot vote in a transaction wherein he has interest.

9.3.7.4 Transparency and accountability: The Board needs to be open, transparent, responsive and accountable to stakeholders especially those who provides funding or support. The Board must decide who the stakeholders are and establish procedures that foster effective communication with them. On the other hand, accountability and transparency must be balanced with the Board's confidentiality requirements, such as those relating to data privacy and any confidentiality undertakings under agreements entered into by the NGO.

9.3.7.5 Non discrimination and equal opportunities: The Board could set policy with clear targets (where practicable) to promote and applies equal opportunity and diversity in all areas of its activities.

The researcher fully recognizes the diversity of NGOs in Hong Kong, each with its own historical background, vision and mission, size and organizational structure, etc. This reference guide with the understanding that the practices suggested are not intended to be rigidly followed by every NGO Board. Indeed the size of the NGO and other factors may affect the extent to which some of the suggested practices can realistically be adopted. NGO Boards are however encouraged to test their corporate governance against these suggestions to see if improvements are necessary or beneficial. The researcher hopes that the recommendations here can help SM NGOs to improve/enhance their corporate governance and thereby raise their value and contribute to the society at large.

References

Report of the Committee on the Financial Aspects of Corporate Governance (1992)
Committee on the Financial Aspects of Corporate Governance ("Cadbury Committee"), UK

First Report of the Committee on Standards in Public Life (May 1995)
Committee on Standards in Public Life ("Nolan Committee"), UK
Available at: https://www.gov.uk/government/publications/the-7-principles-of-public-life/the-7-principles-of-public-life--2

Governance in the Public Sector: A Governing Body Perspective, Study 13 (August 2001)
Public Sector Committee, International Federation of Accountants

OECD Principles of Corporate Governance (April 2004)
Organization for Economic Co-operation and Development
Available at: www.oecd.org/daf/ca/Corporate-Governance-Principles-ENG.pdf

Corporate Governance for Public Bodies A Basic Framework (May 2004)
Hong Kong Institute of Certified Public Accountants
Available at: https://www.hkicpa.org.hk/en/About-us/Advocacy-and-representation/Best-practice-guidance/Publications/Public-Bodies-basic-framework

Hong Kong Companies Ordinance (Cap 622), Laws of Hong Kong
Hong Kong SAR Government

Hong Kong Inland Revenue Ordinance (Cap 112), Laws of Hong Kong
Hong Kong SAR Government

Hong Kong Business Registration Ordinance (Cap 310), Laws of Hong Kong
Hong Kong SAR Government

Public Governance Guidance Notes Issue 1 (August 2016), Issue 2 (November 2017), Issue 4 (January 2019), Issue 5 (October 2019).
The Hong Kong Institute of Chartered Secretaries

Belinda Wong. *Hong Kong Company Secretary's Practice Manual 4th Edition 2018*
Wolters Kluwer Hong Kong Limited

Belinda Wong. *Pitfalls in the Companies Ordinance (Cap 622).* Seminar organized by Leader Corporate Services Limited (November 2019)

Mohan Datwani (November 2016). *Hong Kong Incorporated NGOs – Public Governance Standards/Business Review As Limited or Guarantee Companies under NCO.*
Seminar organized by the Hong Kong Institute of Chartered Secretaries

Non-statutory Guidelines on Directors' Duties (January 2004)
Companies Registry, Hong Kong SAR Government
Available at: https://www.cr.gov.hk/en/companies_ordiance/docs/Guide_DirDuties-e.pdf

Kelvin Wong (2014). Highlights of Major Changes in New Companies Ordinance Seminar, organized by S.T. Cheng & Co, Solicitors

A Tax Guide for Charitable Institutions and Trusts of a Public Character (April 2020)
Inland Revenue Department, Hong Kong SAR Government
Available at: https://www.ird.gov.hk/eng/pdf/tax_guide_for_charities.pdf

Susan Lo. *Charitable organizations – Registration of a charity in the form of a company limited by guarantee.* The Journal of The Hong Kong Institute of Chartered Secretaries, December 2019 (pp 32-36)

Mohan Datwani (June 2017). *AML/CFT Risks, Compliance Standards and Tools Potted AML/CFT History.* Seminar organized by the Hong Kong Institute of Chartered Secretaries

Ivan Yiu Tze Leung, Edith Shih (April 2019). *Managing & Sharing on Practical Issues of NGO Governance.* Seminar organized by the Hong Kong Institute of Chartered Secretaries, April 2019

Doug Trafficking (Recovery of Proceeds) Ordinance (Cap 405), Laws of Hong Kong
Hong Kong SAR Government

Organized and Serious Crimes Ordinance (Cap 455), Laws of Hong Kong
Hong Kong SAR Government

United Nations (Anti-Terrorism Measures) Ordinance (Cap 575), Laws of Hong Kong
Hong Kong SAR Government

Anti-Money Laundering and Counter – Terrorist Financing Ordinance (Cap 615), Laws of Hong Kong
Hong Kong SAR Government

An Advisory Guideline on Preventing the Misuse of Charities for Terrorist Financing and related Appendix. Narcotic Division, Security Bureau, Hong Kong SAR Government
Available at: https://www.nd.gov.hk/pdf/guideline_e_20180929.pdf, https://www.nd.gov.hk/pdf/Appendix_e_20180929.pdf

Hong Kong Money Laundering and Terrorist Financing Risk Assessment Report (April 2018). Hong Kong SAR Government
Available at: www.fstb.gov.hk/fsb/aml/en/doc/hk-risk-assessment-report_e.pdf

Good Practice Guide on Charitable Fund-raising
Social Welfare Department, Home Affairs Department, and Food and Environmental Hygiene Department, Hong Kong SAR Government
Available at: https://www.gov.hk/en/theme/fundraising/docs/good_practice_guide.pdf

Best Practice Checklist Management of Charities and Fund-Raising Activities
Independent Commission Against Corruption, Hong Kong SAR Government
Available at: https://cpas.icac.hk/UPloadImages/InfoFile/cate_43/2016/a65540f5-91c5-46a8-8adf-7384e8c694b0.pdf

Best Practice Checklist Governance and Internal Control in Non-Governmental Orgainizations
Independent Commission Against Corruption, Hong Kong SAR Government
Available at: https://cpas.icac.hk/UPloadImages/InfoFile/cate_43/2016/208ec0bd-878e-4dc0-b755-c4a230b39bdd.pdf

Responsibilities of accountants under AMLO.
Seminar organized by Accounting Development Foundation, Hong Kong July 2020

Martin Lim. *Practical Guide to implementing AML/CFT Internal Policies, Procedures and Controls.*
Seminar organized by Ingenique Solutions, April 2019

Hong Kong NGO Governance Health Survey 2018 Summary of Key Findings, Key Insights and Recommendations (June 2019). The Hong Kong Council of Social Service.

Audit Report No.74 April 2020.
Audit Committee, Hong Kong Government
Available at: https://www.aud.gov.hk/pdf_e/e74ch08.pdf

Mayer Brown. Guidelines on How to Develop a Corporate Governance Manual for Hong Kong NGOs. The Hong Kong Council of Social Service. Published in March 2020.

Leading Your NGO – Corporate Governance: A Reference Guide for NGO Boards (June 2002)
Social Welfare Department, Hong Kong SAR Government

National Council of Non-profits, Washington, D.C. USA
Available at: https://www.councilofnonprofits.org/

Board Source, Washington, D.C. USA
Available at: https://boardsource.org/

Rebecca Walker Chan. *Board diversity: best practice tips*
The Journal of The Hong Kong Institute of Chartered Secretaries, July 2018 (pp 24-27)

Appendix

Sports panel head hits back at audit criticism

2020-04-30 HKT 10:25   f Recommend 0   Share this story f 🔽

The Director of Audit investigated the the Sports Federation and Olympic Committee of Hong Kong after it was accused of nepotism over its selection of a swimmer for the 2018 Asian Games. File photo: RTHK

The honorary secretary general of the Sports Federation and Olympic Committee of Hong Kong on Thursday defended its selection criteria and methods, one day after a damning report called for reform at the committee after accusations of nepotism.

The Director of Audit investigated the selection row after the son of a senior member of the Amateur Swimming Association was picked to represent Hong Kong in the 2018 Asian Games - instead of the fastest swimmer.

The minutes of the Sports Federation and Olympic Committee in April 2018, showed that 17 swimmers were shortlisted for the races, including two who had failed to meet the original selection criteria.

The report by the Director of Audit was released on Wednesday and highlighted weaknesses in the way the Committee selected athletes, and called for more transparency and accountability.

It also called on the committee to adopt the best practice used overseas, without directly accusing the committee of nepotism.

The honorary secretary general of the Sports Federation and Olympic Committee, Ronnie Wong, told an RTHK radio programme the report was thoroughly written and admitted there's room for improvement. But he said it was hard to take care of everything when big events are being held back-to-back.

The report also criticised the attendance record of some committee members, but Wong said the committee had never failed to convene a regular meeting because it lacked a quorum, and called the absence of some committee members only "individual cases."

He added that athletes don't have to meet the criteria to take part, because the committee wants athletes who show potential to participate, which justified picking them to represent Hong Kong.

Auditors also criticised the Football Association's practice of giving out an increasing number of complimentary tickets, even though many of them weren't used.

The Football Association's chairman, Pui Kwan-kay, told the RTHK programme that they won't keep giving more complimentary match tickets, but stressed that it would still hand out complimentary tickets to students and grassroots supporters for matches like the Lunar New Year Cup.

He also added that attendance at games is lower because fans can now watch football from around the world on TV.

Hong Kong · Law and Crime

Hong Kong court bars Hing Tak teacher from troubled school

Cheung Kam-fai was charged with 10 counts of remaining in a school without permission

Jasmine Siu

Hing Tak School in Tuen Mun. Photo: K Y Cheng

A former teacher was on Wednesday barred from returning to a Hong Kong primary school where he was accused of staying without permission after his teacher's registration had been cancelled.

The case was the first prosecution arising from a police investigation into the management of Hing Tak School, which attracted headlines last month when it emerged that 21 pupils at the government-subsidised institution were absent for up to two years, but had remained on the student roster.

Cheung Kam-fai, now unemployed, was charged with 10 counts of remaining in a school without permission.

School board in Hong Kong votes to dismiss principal after scandal involving permanently absent pupils

The charges alleged that he had remained in the school grounds from March 30 to April 11 without permission from the permanent secretary for education, despite his registration as a teacher being cancelled.

The offence is punishable by a HK$100,000 fine and two years' imprisonment, under the Education Ordinance.

The 43-year-old was not required to take plea on his first court appearance on Wednesday before acting Principal Magistrate Ivy Chui Yee-mui at the Tuen Mun Court.

The case was adjourned to November 1, with Cheung granted bail on condition that he refrain from school grounds.

Acknowlededgement

The researcher would like to convey thankfulness to the following professors for their guidance and advice:

Prof. Ir Philip K.I. Chan, Professor of my post-doctoral research

Prof. Eric Tao, California State University, Monterey Bay, USA

Researcher's Biography

Dr. TSANG Wai Chun Marianna is an Adjunct Professor (Corporate Governance) of Sabi University. She is a member of the International Postdoctoral Association, a Chartered Governance Professional of The Chartered Governance Institute in UK and a Chartered Secretary of the Hong Kong Institute of Chartered Secretaries, a Chartered Member of Chartered Institute of Personnel UK. She is also a certified tax adviser and a member of the Taxation Institute of Hong Kong, the Society of Registered Financial Planners, the Chartered Institute of Arbitrators, the Institute of Financial Accountants, the Institute of Public Accountants in Australia, the Association of International Accountants in UK and holds other various professional qualifications. Dr. Tsang was a member of the Board of Review (Inland Revenue Ordinance) HK Government from 2010 to 2015. She is a member of the Regulatory Committee of The Hong Kong Independent Non-Executive Director Association. Ms Tsang has been serving chairman of Audit Committee of listed company for over 10 years. She has over 35 years working experience covering corporate governance and related experience in major commercial corporations and in professional firms.

Dr Tsang was appointed as a Special Researcher of the Committee of Experts of National Academy of Governance Government Economics Research Centre Beijing, China and an arbitrator of Quangzhou Arbitration Commission, China in 2019 and a Chief Arbitrator in a recent arbitration case.

Dr. Tsang has in depth experience of over 30 years of voluntary services in NGOs in Hong Kong, both small and big and international, across a wide spectrum, either as Chairman, President, Director, Advisor or Member at different time, including University Court member, various University committees, Management Committee (ie Board) member of schools, Lion's Club International, Y's Men International, Boy's Scott, alumni associations, anti-drug abuse and rehabilitation, the poor and needy, the handicapped, industry associations, Honorary auditor, Honorary Financial Advisor of associations etc. She also has been panel judge chairman or member of the NGO award program and University award programs. Dr. Tsang has therefore a very good knowledge of the situation of corporate governance in NGOs in Hong Kong. Dr. Tsang has received various awards including "Hong Kong Most Valuable Companies", "Outstanding Business Award", "Greater Bay Area Outstanding Leader", "The First Hong Kong Baptist University Distinguished Alumni" etc.

Postdoc WONG Kwok Keung keith
黃國強 博士後

　　筆者於中華人民共和國（中國）廣東省四會市出生，小學畢業後移居香港特別行政區（香港），並在香港理工大學高級文憑課程畢業後，投身屋宇設備工程行業。在工程界奮鬥的 20 餘年裡，筆者一直不忘自我增值的理念，於日間工作外，晚上持續進修工程相關的專業課程。在持續進修的道路上，分別獲得：英國中央蘭開夏大學屋宇裝備工程學的（榮譽）學士、香港公開大學工程管理工程學碩士、法國北歐大學的法學博士及工程學榮譽博士，其後更成為：美國斯坦福／伯克萊加州／加州州立大學蒙特利灣分校的博士後及中國公共經濟研究會的經濟學博士後。

　　筆者目前就職於香港某大型中央企業集團，專注於粵港澳大灣區的房地產行業，主要負責住宅和商業建築的屋宇裝備系統設計和施工計畫審查。除此以外，筆者還是中國南京仲裁院的經貿專家及仲裁員、國際博士後協會的院士、英國特許水務學會資深會員及香港一級持牌水喉匠。筆者致力於研究香港的水利工程對法律及經濟的影響，以幫助節約用水保護世界食水資源。

　　筆者近年熱心公益事業，積極參與志願服務，以支援不同地區的需要者。在社區服務中分享自己的知識和經驗，以鼓勵年輕人永不停止學習，並幫助少數族裔、新來港人仕和弱勢群體融入社區。2019 年新中國成立 70 周年，筆者有幸成為「共築中國夢」慶祝中華人民共和國成立七十周年紀念郵票珍藏冊（海內外 70 位入選華人楷模）內的一員，擁有個人的郵票。多年來，筆者一直心存感恩，對之前學習和工作的付出獲得肯定而感到十分榮幸。

The author was born in the Sihui City, Guangdong Province of People's Republic of China (China). He moved to Hong Kong Special Administration Region (Hong Kong) with his family after graduated from primary school, and continue his studies by completing a high diploma education from The Hong Kong Polytechnic University before joining the Building Services industry. In his next 20 years of working in the Engineering industry, his ambition has led the path of continuous self-enhancement by studying professional courses at night time despite daily work. Throughout this path, he has attained a bachelor's degree with honour in Building Services Engineering from the University of Central Lancashire (United Kingdom); a degree of Master of Engineering in Engineering Management from the Open University of Hong Kong (Hong Kong); the honorary Doctorate in Engineering and the degree of Doctor of Laws from the SABI University (France); a postdoc at Stanford University/California State University Monterey Bay/ University of California at Berkeley (United States of America) and a postdoc of Economics at China Public Economic Research Association (China).

The author is currently employed by a State-owned enterprise group, which focused on real estate industry in the Guangdong-Hong Kong-Macao Greater Bay Area. His duties specialize in building services system design and approving method statement for residential and commercial property development. Also, he is an economic and trade experts and arbitrators of Nanjing Court of Arbitration, academicians of the International Postdoctoral Association, fellow member of the Chartered Institute of Plumbing and Heating Engineering (CIPHE) and a Grade 1 licensed plumber in Hong Kong. He is committed to studying the legal and economic impact of Hong Kong's water conservancy projects to help save water and protect the world's fresh water resources.

In recent years, the author was continuously devoted into community services by participate in volunteering works and providing support to local poverty. Through his sharing of knowledge and experience during community service, he encourages young people to never stop improving themselves. Nevertheless, he also provides assistance to ethnic minorities, new arrivals from other place and the disadvantaged to integrate into the community. In Year 2019 during the 70th anniversary of the founding of the People's Republic of China, the author was included as 1 of 70 Prestige Overseas Chinese in "Building a Chinese Dream Together" Commemorative Stamp Collection to own a personal set of collection stamp. This achievement was both memorable and thankful as it was delightful to be recognized after all these years of contribution in academic and career.

香港可持續發展水資源

黃國強 [1,2,3,4*] 和陳勤業教授、博士後導師 [1,2,3,4,5]

[1] 中國公共經濟研究會，北京市海澱區長春橋路 6 號，中共中央黨校（國家行政學院）
[2] 美國加州蒙特里加州州立大學，美國加州，CA 93955-8001
[3] 美國伯克來加州大學，美國加州，CA 94720-1234
[4] 美國斯坦福大學，美國加州，CA 94305
[5] 國家行政學院、蒙特里加州州立大學、伯克來加州大學、斯坦福大學，博士後導師
* wongkwokkeungkeith@gmail.com

摘要

香港特別行政區（香港）位於我國的南部，屬亞熱帶氣候，雖然每年有充沛的降雨量，但是香港集水設施的集水量並不足以應付香港的食水需求，僅佔香港食水量 30%。另一方面，由於香港每年降雨量並不穩定，波幅極大，以致香港集水量相差達 2 億立方米。為解決降雨量不足、不穩的挑戰，香港自 1961 年起輸入東江水，現在每年用 47.8 億元向東江購買 8.2 億立方米水，佔香港食水量 80%，以滿足香港用水的需求。

本文的目的提供概念，說明香港政府可以清晰找出食水不足的原因。香港食水的水源由 2 個增至 5 個，以幫助解決香港長期依賴東江水，並以創新去探索，及時性，開源及節流。本文包含比較新舊水資源的調查，包括在成本、保護水資源方面。

主要開源建議：

1) 擴大收集雨水設施範圍，善用香港渠務處防洪系統的蓄洪池收集香港的雨水。
2) 使用海水化淡設備，轉換海水為淡水，然後直接供水到供水系統。
3) 香港使用海水來作沖廁用途，減少依賴食水。

主要節流建議：

1) 更換老化水管，原有喉管已超出使用年限，不斷出現漏水及爆喉事故，更換新喉管可大幅度減低滲漏的機會。

2) 香港供水系統的操作壓力為 30 米壓力差，若降低供水喉管的操作壓力至 20 米壓力差，可以減低 30% 的滲漏機會。

研究結果表明，通過創新的想法、系統的提升、市民的節水意識來減少食水的使用，從而開源節流，使香港減小對東江水的依賴程度，額外出來的水資源可以轉供給粵港澳大灣區其他城市，帶動粵港澳大灣區城市的經濟發展，而香港的成功案例亦可複製到其他地方，提供借鑒作用。

關鍵字：創新，提升，雨水系統，海水化炎，蓄洪池

引言

　　預計於 2025 年全球有三分之二人口會面臨中度至高度的水資源供應不足壓力，而香港亦都面對相同的問題。香港地處我國的南部，屬於亞熱帶氣候，每年都有充足的雨水量。但香港近海，地底是花崗岩質，儲蓄水的能力比較差，加上沒有足夠河流和湖泊來儲水，香港集水設施的集水量不足以應付香港的食水需求，只佔香港食水總用水量 9.87 億立方米的三分之一，約 3.58 億立方米。由於香港每年降雨量並不穩定，波幅極大，以致香港集水量相差達 2 億立方米。為解決降雨量不足及不穩的挑戰，香港自 1961 年起由我國輸入食水，水量由開始 2270 萬立方米提升到 2017 年的 8.2 億立方米，價格每立方米由 HK$0.05 升至 HK$5.83。在 2017 年香港花費 HK$47 億來購買食水。東江水供水採用「統包總額」的方式運作，香港每年在廣東粵港供水有限公司有 11 億立方米的購買水權，使香港有可靠和靈活食水供應，並確保 99% 供水穩定性。縱使香港遇到百年一遇的旱季，仍能有足夠食水供應。食水在香港是珍貴的資源，以創新的方法使用現有的及拓展新的水資源，可大大減低環境的影響。

　　本文目的是提供概念，幫助香港政府可以清晰找出食水不足的原因。香港食水的水源由 2 個增加到 5 個水資源，以 5 管的供水的結構。該結構由二個來源，一本地收集雨水，東江水供應，及 3 個新水源，回收雨水，淡化海水及使用海水沖廁構成，以幫助解決香港長期依賴東江水。以開源及節流的方法提升香港供水保障及穩定性。

開源

增加收集雨水的範圍

　　主要開源建議：

　　1. 擴大收集雨水設施範圍，香港現有 17 個水塘作為集水區，佔香港土地面積的三分之一，集水區收集雨水佔總食水量約 36%。香港地少人多，大部份平地已蓋房，只可以在位置偏遠的地方設立集水區和水塘，這樣可以收集到年降雨量的 30% 雨水，而有 70% 的雨水將流往大海。

　　為了擴大收集雨水設施範圍，可於市區範圍內利用香港渠務署防洪系統的蓄洪地收集香港雨水。香港的渠務系統設計分開「污水渠」及「雨水渠」兩個獨立系統。污水渠會經過污水處理廠的處理後排出大海，而雨水系統則由收集地面上所有雨水後統一排出海。

香港的主要供水系統及現有收集雨水設施的範圍

2012 年至 2016 年全年供水量

　　蓄洪池是渠務署一個管制雨水溢流的管理方法。蓄洪池普遍設計於人煙稠密的市區內。即使遇上十年一遇的大雨，雨水超出下游雨水渠的排水量，雨水在儲存蓄洪池內短時間停留，限制在下游的排水系統用量之內，可以防止市區出現水浸的情況，減少對市民的不便和降低對經濟損失的影響。蓄洪池建於地底，能騰出地面的空間建造政府的大型球場等康樂設施，香港可以在各區建造蓄洪池來收集雨水為各區的大型球場、公園和高

九龍大坑東蓄洪池計畫示意圖

爾夫球場進行灌溉，這樣便可以節省大量食用水資源。目前全香港有 3 個蓄洪池，分別是香港上環蓄洪池（容量達 9 千立方米）、香港跑馬地地下蓄洪池（容量達 6 萬立方米）及九龍大坑東蓄洪池（容量達 10 萬立方米），如果將市區內的雨水渠接駁蓄洪池，並採用可調式溢流堰來控制，在雨水期間截取大量雨水，待超出處理量後降低溢流堰，排走多餘的雨水，以達到儲水的效能。

　　將收集的雨水進行除污淨化後，便可以供應臨時淡水沖廁、工業用水、建築用水、清洗街道、車輛及消防系統救火用途、灌溉公園、高爾夫球場、運動場等非飲用水所用。市區進行都市化時已鋪設雨水排放系統，在舊區的雨水渠過了使用期限後，不少地方出現破

裂及淤塞情況。更改雨水渠時要面對空間不足、地下公共設施須要改位等困難,很難全部使用創新方法。但可考慮修改渠務處的雨水排放隧道的辦法,以截取及輸送雨水去蓄洪池。香港一共有四條雨水排水隧道,總長度 21.6 公里,借著隧道輸送系統把大量雨水瞬間沿著山上建築物和斜坡收集流入隧道。在隧道出口築建蓄洪池,把高地收集的雨水改道到蓄洪池。經過處理後,清除垃圾及除去有害物質,再泵到水務署的集水區進入供水系統。每年經原有的集水設施收集的雨水有 2 至 4 億立方米,經雨水排放隧道及蓄洪池收集的雨水有 2 至 4 億立方米,這樣,每年可收集高達 4 至 8 億立方米的雨水。2016 年香港非飲用水需求量是 4.04 億立方米,而非飲用水的主要使用地區接近蓄洪池,可以利用這一點使用處理後的雨水,解決非飲用水的供應問題。

九龍大坑東蓄洪池

雨水排水隧道

海水化淡

2. 香港鄰近海洋,有豐富的海水資源。1971 年香港建造及營運年產量只可以滿足當時 48 天用水量(6643 萬立方米水)。但因當時科技落後及石油燃料費的格價比較昂貴,增加海水化淡的成本,當時的海水淡化廠營運三年後就停止運作及拆卸。但是 2000 年後化淡海水的科技有大幅的進步,加上國際上

上世紀的海水化淡廠

石油價格大約大幅下降到 60 美元一桶。2018 年底開始建造的海水化淡廠產量每日達 27 萬立方米。可以滿足本港 10%(1 億立方米)食水需求,生產食水的成本為每立方米 10.2 元(和美國相同)。增加建造海水化淡廠規模,可以減低運作成本。如香港所有食水都採用海水化淡技術供應,僅需要 6 座海水化淡廠就足夠應付香港日常用水的需要。

減少使用食水作臨時淡水沖廁

香港使用海水作沖廁用途，現在 80% 人口用海水沖廁，即有 20% 使用食水作臨時淡水沖廁，佔全年用水量 7.9%（0.75 億立方米）。香港政府可以繼續擴大推動使用海水沖廁，提升到 95% 人口使用海水沖廁，但香港北部地方遠離海岸邊，使海水沖廁不符合經濟原則，但可以使用蓄洪池的設計，收集雨水以供沖廁及清潔用水。這樣臨時淡水沖廁使用食水減少到佔全年食水用水量 1.5%（15 萬立方米水）。

節流

主要節流的建議是更換老化的喉管，香港開埠以來大部份的食水管埋藏於地下。當時設計的使用年期為 30 年，原有喉管早已超出使用年限，不斷出現漏水和爆裂情況。2014 年，香港共發生爆裂 173 次，滲漏 9831 次，滲漏率高達 16%（1.52 億立方米水），更換新喉管可以減低滲漏機會，香港分 15 年完成更換 3000 公里老化水管。佔全部水管 5700 公里的 53%，估計費用 236 億。可以採用內喉緊貼法，用聚乙烯喉，拉長及縮細，進入舊的水喉內，安放好了用充氣的方法把聚乙烯喉回復原來尺寸，亦緊貼舊有喉內壁，可以減少開挖路面，影響市民的日常生活，減輕對交通造成的影響。

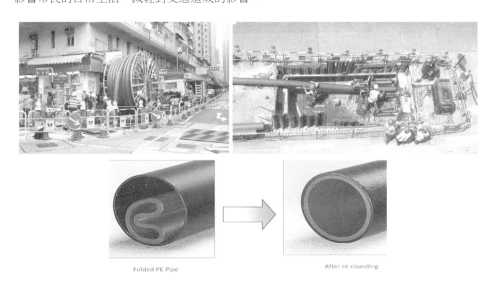

Folded PE Pipe   After re-rounding

2. 香港供水系統的操作壓力為 30 米壓力差，更改新建築物的設計，減低工作壓力為 20 米壓力差後，可以減低滲漏機會達 30%。

香港水費相當低廉，沖廁用的海水是免費使用，而食水則按政補貼的收費機制收費。香港食水收水準較其他發達國家城市低。東江供港水價錢在 2015 至 2017 年的分別為每立方米 HK$5.15、5.48 及 5.83 港元，遠低過海水化淡的成本價格 10.2 港元。當然原水價格是未經處理的，水處理後大約為 8 港元是／每立方米，價錢依然比海水化淡少 20%。

經過香港可持續發展水資源之問卷調查

經過香港可持續發展水資源之問卷調查 500 份，成功回覆 425 份，訪問的人員主要有：持牌水喉匠 100 人，水喉從業員 75 人，工程從業員 50 人，工程師 50 人，教師及教授 35 人，普通市民 115 人，問卷調查方式主要為面對面調查，或通過電郵完成問卷。

統計調查後發現香港人經過上世紀因沒有足夠食水供應而需要停水的人群，會非常渴望有充足的食水供應，市民可以接受購買東江水來補充雨水量不足，亦都接受比現有價錢 3 倍去使用海水化淡技術，但是不接受東江水供應不足時，使用經雨水渠系統收集的雨水作飲用水，可以接受利用收集雨水作灌溉，洗地之用。訪問中市民提到購買東江水的方法需要更改。

香港可持續發展水資源之問卷調查的內容	是	否	其他
是否願意支付較多金錢 (現在水價 3 倍) 使用海水化淡食水？	120	300	5
是否願意使用再造水用作灌溉、洗地等非飲用水？	400	23	2
是否願意支付金錢作購買東江水作飲用水？	325	95	5
是否接受使用經雨水渠系統收集的雨水作飲用水？	20	400	5
東江水供水不足時 , 是否接受使用經雨水渠系統收集的雨水作飲用水？	50	350	25
是否接受使用經雨水渠系統收集雨水灌溉、洗地等非飲用水？	300	120	5

每年供水統一按 8.2 億立方米定價，無論供水量多少都收取這個價錢。根據統計數字，近 5 年內有 2 億立方米的食水不需要供水公司提供。但是香港政府需要一個保障，以免在枯水期間沒有充足食水供應，香港政府在意的水權而不是買水。

香港近 5 年的用水量由 9.33 至 9.87 億立方米，保持相對平穩的狀態。但枯水年和豐水年的降雨水量相差可達 2.26 至 3.58 億立方米。未來是豐水年，還是枯水年，我們都很難預測，也很難去計算。例如：以枯水年的降雨水量計算，每年使用食水量大約 10 億立方米，收集雨水量 2 億立方米，東江水需要供應香港大約 8 億立方米食水。從 2006 年開始，供水公司保障每年可以提供 11 億立方米的供水權。(代表香港全年收集雨水量是 0 立方米，也不會有食水供應問題)。

根據統計 2012 年至 2016 按用水類別劃分的食水量，香港用於飲用水大約 5.37 億立方米，其他 4.5 億立方米用水大部份都是非飲用水，可以建議使用經雨水渠系統收集的水。經分析後每年由需要供水公司提供食水，豐水年的 0 立方米到枯水年的 2.5 億方米。

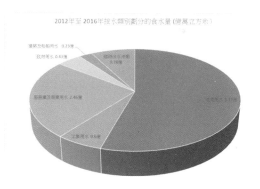

水的來源	2015 年	2016 年
全年用水量	9.73 億立方米	9.87 億立方米
香港現在集水區收集雨水	2.07 億立方米	3.58 億立方米
經率蓄洪池雨水用現在經及排水隧道收集雨水	2.07 億立方米	3.58 億立方米
臨時淡水沖廁 (0.75-0.15) 億立方米	0.6 億立方米	0.6 億立方米
海水化淡	0.987 億立方米	0.987 億立方米
漏水率 16%	1.55 億立方米	1.57 億立方米
	2.453 億立方米（欠淡水）	0.45 億立方米（有盈餘）

供水權

東江供水系統能力達到 24.23 億方米水，對香港的供水配額為 11 億立方米水，對深圳的供水配額為 8.73 億立方米，對東莞的供水配額為 4 億立方米。

香港經過開源節流後可以節省大量的食水資源，估計購買 2 億方米的東江水就足夠應付日常需要。節省下來的食水可以轉供給粵港澳大灣區其他城市，帶動粵港澳大灣區內經濟的快速發展。從 1961 年東江供水給香港以來，香港經歷過幾次旱災，已傷及民生和經濟，為了保障供應香港食水水質安全。東江水流經的江西，廣東兩省一直重視東江的沿岸的環境保護，頒佈和制定了一系列的保護東江水措施。

相關的水質資源保護條例，規章，和檔合計 35 個，內容十分詳細，涵蓋面非常大，反映廣東省政府對東江水的重視。以前為保障香港居民用水安全，充足用水，廣東及江西兩省犧牲沿岸各地的經濟發展，禁止發展有機會影響水安全的行業進駐當地，使到當地人民的生活不富裕。例如東江水沿岸尋烏縣境內，2014 年村民人均收入 2400 元人民幣，為了保障食水安全，全縣嚴格控制開發，只可以種植果樹，影響經濟發展；廣東的新豐縣，為了保障食水安全關閉為政府財政帶來巨大收益的造紙廠，拒接了幾十個涉及方面的專案，關停 200 間企業；廣東河源拒接年 30 億元的產值和 6 億元的稅收的紙漿廠，等於當地全年工業總產值。香港減少東江水的供應，東江沿岸城市可以大力發展經濟，配合國家大力發展粵港澳大灣區的發展戰略，東江沿岸的地區可以直追其他灣區內的城市。

總結

以創新去探索及研究香港的水資源，開源及節流的方法，科學性，有系統管理水資源。經過深入調查及研究，表明了用創新想法供水系統的提升，市民的教育來減少使用食水。然後再使用雨水回收的方式，減少長期依賴東江水，節省的水資源可以提供給粵港澳大灣區其他城市，可以使到他們高速發展經濟。然而香港的成功開源及節流，可以在其他地方使用同一的方法，為全球節省水資源出一分力。

鳴謝

陳勤業教授 博士後導師

余非女仕

香港水務署

香港渠務署

參考文獻

1). 余非〈東江水一本通〉2018

2). 何耀光〈香港的城市防洪排澇策略及活化水體意念〉2015

3). 香港渠務署〈渠務署可持續告 2017-2018，活化河道，上善若水〉2018

4). 香港水務署〈供水管理 2016-2017 年報〉2018

香港可持續發展水資源之問卷調查

Questionnaire of sustainable development of water resources in HK

Name 姓名 :_____   Sex 性別 :_____

Age 年齡 :_____   Trade 行業 :_____

Education 學歷：

☐ Secondary school or below 中學或以下   ☐ College to University 大專至到大學

☐ University or above 大學或以上   ☐ Others 其他 :_____

感謝您撥空填寫這一份問卷，本問卷目的是在探討您對香港可持續發展水資源意見。您在問卷中所提供的資訊，僅作為學術研究之用，不會提供其他單位，敬請安心填寫。您的意見對我們非常重要，衷心期盼您依自己的實際感受填答。感謝您的熱情支持與協助！敬祝平安快樂，萬事如意！

Thank you for completing this questionnaire. The purpose of this questionnaire is to discuss your views on sustainable water resources in Hong Kong. The information you provide in the questionnaire is for academic research purposes only and will not provide to other units. Please feel free to fill out. Your opinion is very important to us, and we sincerely hope that you will answer it according to your actual experience. Thank you for your support and assistance! Wishing you peace and happiness, good luck!

1. 是否願意支付較多金錢（現在水價 3 倍）使用海水化淡食水？

   Are you willing to pay more money (currently 3 times the price of water) using seawater desalinated water?

   ☐ Yes 是
   ☐ No 否
   ☐ Others 其他 :_____

2. 是否願意使用再造水用作灌溉、洗地等非飲用水？

Are you willing to use recycled water for non-potable water purpose such as irrigation and washing?

☐ Yes 是

☐ No 否

☐ Others 其他 :_____

3. 是否願意支付金錢作購買東江水作飲用水？

Are you willing to pay for the purchase of Dongjiang water for drinking water?

☐ Yes 是

☐ No 否

☐ Others 其他 :_____

4. 是否接受使用經雨水渠系統收集的雨水作飲用水？

Do you accept to use the water collected by rainwater collecting system for drinking water?

☐ Yes 是

☐ No 否

☐ Others 其他 :_____

5. 東江水供水不足時，是否接受使用經雨水渠系統收集的雨水作飲用水？

Do you accept to use the water collected by rainwater collecting system for drinking water, when there is a shortage of DJ water supply?

☐ Yes 是

☐ No 否

☐ Others 其他 :_____

6. 是否接受使用經雨水渠系統收集雨水灌溉、洗地等非飲用水？

Is it acceptable to use rainwater collection systems to collect rainwater for irrigation, washing and other non-potable water purpose?

☐ Yes 是

☐ No 否

☐ Others 其他 :_____

Sustainable Development of Water Resources in Hong Kong Special Administrative Region, China

Wong Kwok Keung, Keith [1,2,3*] and Philip K.I. Chan[1,2,3,4]

[1] Stanford University, Stanford, California, USA 94305
[2] University of California, Berkeley, Berkeley, CA, USA 94720-1234
[3] California State University, Monterey Bay, 100 Campus Center, Seaside, CA, USA 93955-8001
[4] Stanford, Berkeley, Monterey Postdoc Advisor
* wongkwokkeungkeith@gmail.com,

Hong Kong Special Administrative Region (HK) a coastal city located at South of People's Republic of China (China). HK's climate is of subtropical with occasional rainfall. However, only 30% of annual water consumption is collected by the extensive water gathering grounds due to a massive demand of water supply and unreliable rainfall pattern. Difference in quantities of local net yield by rainwater collection system could be more than 200 million cubic meter among years. To cope with this inadequate and unstable challenge, HK started procurement of fresh water from Dongjiang (DJ) River of Guangdong Province in China since 1965. Now, near 80% of annual water consumption (0.82 billion cubic meters) in HK is purchased under this scheme to support the enormous water demand in HK, pricing at 4.78 billion HKD.

This paper aims at planning and assisting the HK government to find out reason for the lack of water resources. To explore ways to increase the total number of water resources in HK from 2 to 5, to help solving the current liability on DJ Water Supply Scheme by using an innovative and timeliness way to explore how to expand supply and reduce demand. A comparison will also be shown regarding water resources from the old times to modern years, including cost and protection method.

A. Recommendations to Expand Supply

A1) To expand the total surface area of rainwater collection by addition of facilities, make use of existing flood storage ponds by the Drainage Services Department (DSD) to collect rainwater all over HK.

A2) Develop reverse osmosis technology for desalination to convert seawater into drinking water and connect to existing system.

A3) Continue to rely on seawater instead of freshwater for flushing purpose, and to eliminate facilities with temporary fresh water flushing supply.

B. Recommendations to Reduce Demand

B1) Replace and upgrade aged water mains which approach the end of their service life to minimize water leakage and risk of main pipe bursting.

B2) Reducing current residual pressure of 30 meter head in most existing fresh water supply zones to become 20 meter head to lower the extent of water leakage by 30%.

Our research indicates that by innovative ideas, upgrade of existing systems and education to the public, water resources can be saved by expanding supply and reducing demand. The reduction of DJ water usage can be reverted to other major cities within in the Guangdong-Hong Kong-Macao Greater Bay Area in China to speed up their economic development. These key success factors can also be copied towards those cities as references.

Keyword: innovative, upgrade, rainwater collection system, seawater desalination, flood storage for stormwater

Introduction

It is expected that by Year 2025, two-thirds of the world's population will face a moderate to high degree of water shortage pressure, while half of the world's population will suffer from actual water resource shortage. By the time, HK will also face the same problem.

HK is located in the southern part of our home country – China. It has a subtropical climate and abundant rainfall every year. However, the whole city does not have enough rivers and lakes, and the underground is granitic in the offshore area of HK, resulting a relatively poor ability to store water by natural means. Also, the water harvesting capacity of HK's catchment facilities is insufficient to meet its water demand, with this method only accounts for 358 million cubic meters out of its total water consumption of 987 million cubic meters. Due to this unstable annual rainfall in Hong Kong, the fluctuation is so great that the difference of water catchment in HK can be up to 200 million cubic meters. To address the challenge of insufficient and unstable rainfall, HK has imported drinking water from China since 1961, from 22.7 million cubic meters in the first year to 820 million cubic meters in 2017. The price per cubic meter has risen from HK $0.05 to HK $5.83 nowadays. In 2017, HK $4.7 billion was spent to purchase drinking water. With Dongjiang Water Supply adopts the mode of <total contracted amount>, HK has an annual water purchase right of 1.1 billion cubic meters in Guangdong, through Guangdong and HK Water Supply Co., Ltd., which enables HK to have reliable and flexible water supply and ensures 99% reliability. Even in the 100-year-a-time dry season, there is still enough water supply. Even with this method, fresh water is a precious resource in HK, the use of existing and new water resources in innovative ways has greatly reduced the impact of the environment.

This paper aims at planning and assisting the HK government to find out reason for the lack of water resources. To explore ways to increase the total number of water resources in HK from 2 to 5, to help solving the current liability on DJ Water Supply Scheme, using an innovative and timeliness way to explore how to expand supply and reduce demand. A comparison will also be shown regarding water resources from the old times to modern years, including cost and protection method.

A. Recommendations to Expand Supply

A1) Increase the scope of rainwater harvesting

To expand the total surface area of rainwater collection by addition of facilities

Nearly 30% of the Territory are catchment areas and there are 17 impounding reservoirs across HK. Local yield generally accounts for about 36% of our total fresh water consumption, meaning that the other 64% are being wasted. In order to expand the scope of rainwater harvesting facilities, rainwater can be collected in urban areas and rainwater can be collected throughout the city using flood storage ponds of the Drainage Services Department's (DSD) flood control system. The drainage system in the area is designed separately from the sewerage and rainwater drainage systems. Sewage canals will pass through sewage treatment plants, in which the water is being treated and discharged into the sea. Rainwater systems are arranged by collecting rainwater on the ground and gathering them together.

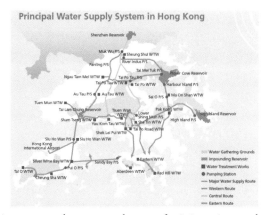

Hong Kong's main water supply system and scope of existing rainwater harvesting facilities

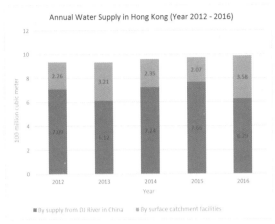

Annual Water Supply in Hong Kong (Year 2012 - 2016)

By supply from DJ River in China ■ By surface catchment facilities

Annual water supply in Hong Kong from 2012 to 2016

In HK for the past few decades, underground stormwater storage tank serve as a way to deal with rainwater flooding by DSD of the government bodies. These design of flood storage pools are usually used in urban and densely populated areas. When rainfall exceeds the drainage capacity of downstream rainwater canals, which in cycle happens once in 10 years, water will normally flood on road surface. After constructing this facility, the rainwater can instead be stored in flood storage ponds temporarily, buffering the amount of rainwater towards the downstream drainage system, thus can prevent urban road flooding to reduce inconvenience towards the

Tai Hang Tung Stormwater Storage Scheme main components layout plan in Kowloon Peninsula, HK

public and reduce economic loss. Flood storage pools are built under the ground to make room for construction of large government sports ground, car parking facilities and large-scale parks. Within a small city like HK, constructing these flood storage pools in all districts to collect rainwater would be the best option. Moreover, large stadiums, parks and golf courses require a lot of fresh water to irrigate plants, which they can utilize the stored fresh water to save water usage. There is a total of three flood storage pools in the whole city, located in Hong Kong Island and Kowloon Peninsula; the smallest with a capacity of 9,000 cubic meters located in Hong Kong Island's Sheung Wan; 60,000 cubic meters in Hong Kong Island's Happy Valley and 100,000 cubic meters in Kowloon's Tai Hang Tung. Connecting rainwater canals in urban areas to flood storage pools, adjustable overflow weirs are used to control the interception of rainwater. If the overflow weirs exceed the treatment capacity, the overflow weirs will be reduced and the excess rainwater will be discharged.

To increase the scope of rainwater harvesting, all rainwater shall be collected and treated in disposal of solid waste. After further treatment and regular inspection to ensure removal of harmful substances, the water can be supplied for use as temporary freshwater toilet flushing, industrial water, building water, street cleaning, fire-fighting use of vehicles and fire-fighting systems, irrigation parks, golf courses, stadiums and other non-drinking water source. Rainwater drainage system has been laid during urbanization in urban areas, while rainwater pipe in old areas has passed its service life, and are in the brink of break-down and silt up and should be replaced. During replacement work of rainwater pipe, challenges such as shortage of space, re-location of underground public facilities need to be faced. It would be a difficult task to completely use innovative methods to replace all rainwater drainage tunnel of the DSD to intercept pipe which is in use and transfer rainwater to temporary flood storage facilities.

There are four stormwater drainage tunnel facilities in HK, summing up a total length of 21.6 kilometers. Through the tunnel system, large amount of stormwater can be collected into the tunnel instantaneously from upland areas for discharge directly into the sea. At the exit of tunnels, flood storage pools are usually constructed to divert stormwater collected from upland area to these flood storage ponds. After treatment, garbage and harmful substances are removed and shall be pumped into the water supply system through the catchment area of the Water Supply Department. Every year, there are 200 to 400 million cubic meters of rainwater collected by the original catchment facilities, while the other 200 to 400 million cubic meters of rainwater are collected by rainwater drainage tunnels and flood storage ponds, in total about 400 to 800 million cubic meters of rainwater can be collected. In year 2016, 404 million cubic meters of non-drinking water were needed in HK. The main non-drinking water use area is close to the flood storage facilities, which can facilitate the use of treated rainwater.

Tai Hang Tung flood storage ponds, located in the Kowloon Peninsula of HK

Drainage tunnel system of DSD in HK

A2) Seawater Desalination

HK is adjacent to the sea and has abundant marine resources, including the advantage of applying seawater desalination which once started in HK before in 1971. However, due to the high price of petroleum fuel, cost of desalination of seawater was expensive. Also, for the low technology used at that time, annual yield of the seawater desalination plant was only enough to supply water for 48 days (66.43 million cubic meters). Finally, its operation was soon found failed and the plant was discontinued and dismantled after three years.

Since then, the technology of desalinating seawater has made great progress since entering the 21st century. Also, the petroleum oil price has dropped to about $60 a barrel. At the end of 2018, a typical seawater desalination plant was able to provide a daily output of 270,000 cubic meters. This amount of output can cover 10% (100 million cubic meters) of water demand, while the cost of producing water is HK $10.2 per cubic meter (the same as in the United States). This operation cost can be further reduced by increasing the scale of seawater desalination plant. It was estimated that daily drinking water demand can be fully satisfied only by desalination of sea water if a total of six desalination plants are built in HK.

Record photo of last century's Seawater Desalination Plant located in HK

A3) Reduce the use of drinking water for temporary freshwater flushing toilets

Continue to rely on seawater instead of freshwater for flushing purpose, and to eliminate facilities with temporary fresh water flushing supply

HK has a long history of using seawater instead of fresh water for flushing purposes. Currently, 80% of the HK population use seawater to flush toilets. In the other way around, up to the rest of 20% citizen still use fresh water for temporary fresh water flushing due to unavailable supply of seawater by the government, accounting for 7.9% (75 million cubic meters) of water use throughout the year. This is due to the northern part of HK being too far away from the coast, thus resulting the use of seawater flushing is not economical. To decrease this amount of fresh water usage, the government is suggested to explore ways to expand the use of seawater flushing, with an aim to raise the population to 95% from existing 80 % to adopt seawater for flushing. To help solving this, design of the flood storage facilities can be used to collect rainwater for flushing and also for cleansing water. If the final goal can be achieved, use of fresh water for flushing water can be reduced to 1.5% of the annual water consumption (150,000 cubic meters of water).

B. Recommendations to Reduce Demand

B1) Replace and upgrade aged water mains

The major water saving proposal is to replace aging water supply pipe network. A major part of these water pipes have been installed underground since the early stage of 20th century. At that time, those pipe's design usage period was around 30 years only. The original pipe had now exceeded its service life, resulting frequent leaks and bursts. Statistical finding reveals in the year 2014, a record of 173 times broken pipe, leakage of 9831 times was found; and the leakage rate was as high as 16% to 152 million cubic meters of water.

Through replacement of new water supply pipes, leakage frequency can be largely reduced. Up to this moment, HK completed the replacement of 3,000 kilometers of aging pipes in the past 15 years. This accounts for 53% of the total water supply main pipe, with a length of around 5,700 kilometers and an estimated cost of HK $23.6 billion.

It is suggested to use an innovative "close-fit lining" technique to "rehabilitate" the old pipe, instead of replacing them. This method is done by using a continuous string of polyethylene (PE) pipe, firstly folded into a 'c' shape along its length. This kind of compact shape is created during manufacturing by folding the pipe while it is still hot. It is completed from the manufacturing plant wound on a drum. With its folded shape, the PE pipe will become substantially smaller in terms of cross sectional area, and then allowing it to be easily inserted into the existing old pipeline inner wall. After inserting, the polyethylene pipe can be re-rounded into its original size by heating up, pump air to cause inflation to make the plastic pipe expand to a size close to the old pipe's inner wall. The advantage of this method is that it can reduce the extent of road surface excavation, minimizing the affect towards daily life of the citizens and also reduce the impact on traffic.

Folded PE Pipe   After re-rounding

"Close-fit lining" technique adopted and used in HK

B2) Reducing system residual pressure

The operating pressure of the HK's water main supply system is 30 meters head. If the design of new buildings are changed by reducing the working pressure to 20 meters head, this can reduce water leakage opportunity by 30%.

Seawater used for flushing is free of charge for users, while the fee charged towards fresh water usage in HK are relatively low, with the fresh water is being charged after deducting the government's own subsidy charging mechanism. In the whole, HK's water charging scale is far lower than other developed cities. Comparing the price of various water supply source, the water price of buying DJ water supply from China to HK was HK$ 5.83 per cubic meter in 2017. This cost is comparatively far lower than that of seawater desalination at HK$ 10.2. Of course, the raw water price is zero when untreated. After water treatment, the cost is estimated at about HK$ 8, still 20% less than desalinated seawater.

Questionnaire survey on sustainable water resources in HK

Recently, a questionnaire on sustainable water resources in HK was conducted towards 500 people, with successful respondent numbered at 425. These conductors include 100 licensed plumbers, 75 plumbers, 50 engineering practitioners, 50 engineers and 35 teachers and professors. Remaining 115 ordinary people are conducted and responded through face-to-face or by email.

The survey result indicates that HK people tends to be very eager to have ample water supply, mainly due to their past experience of water supply shortage happened in the last century in which citizens did not have enough water supply. Respondents are openly acceptable to the purchase of DJ water from China to supplement lack of rainfall during rain seasons. Acceptable range to use desalination technology comparing to purchase price of DJ water is up to 3 times. However, they tend not to accept to utilize rainwater collected by the rainwater canal system be used as drinking water for collecting irrigation water for irrigation and land washing, even though the water supply to the DJ River has shortage. During the survey, they mentioned that the method of purchasing DJ water needs to be changed, such as how to negotiate the price of DJ Water.

Currently, the HK government has agreement to Guangdong Potable Water Supply Co., Ltd. with a fixed amount of water supply package. The annual water supply is fixed at 820 million cubic meters, which means that a fixed price is charged regardless of the actual amount of water supplied. According to government statistics, up to 200 million cubic meters of drinking water in the past five years are actually not required by Hong Kong from China's water company. However, this action is still required because the Hong Kong Government needs this as a protection to avoid the lack of water supply during unforeseeable dry season. What HK Government is concerned about the water rights rather than buying water itself.

Result of questionnaire of sustainable development of water resources in HK

Questionnaire on sustainable development of water resources in HK	Yes	No	Others
Are you willing to pay more money (currently 3 times the price of water) using seawater desalinated water?	120	300	5
Are you willing to use recycled water for non-potable water purpose such as irrigation and washing?	400	23	2
Are you willing to pay for the purchase of Dongjiang water for drinking water?	325	95	5
Do you accept to use the water collected by rainwater collecting system for drinking water?	20	400	5
Do you accept to use the water collected by rainwater collecting system for drinking water, when there is a shortage of DJ water supply?	50	350	25
Is it acceptable to use rainwater collection systems to collect rainwater for irrigation, washing and other non-potable water purpose?	300	120	5

HK's water consumption in the past five years has been quite stable, ranging from 933 to 987 million cubic meters. However, in dry years, the amount of rainfall when compared to wet years can varies from 226 to 358 million cubic meters. That means, it is in fact quite challenging to predict if the next year would be wet or dry years, therefore it would be best to have buffer. Based on the amount of rainfall in dry years, the annual water consumption is about 1 billion cubic meters, with annual rainfall is 200 million cubic meters. After calculating, it was estimated that DJ water needs to supply about 800 million cubic meters of water to HK every year. Since 2006, the water company from China has guaranteed a water supply of 1.1 billion cubic meters per year. (Meaning that there can be enough water supply to HK even though the amount of rainwater collected in the whole year is 0 cubic meters)

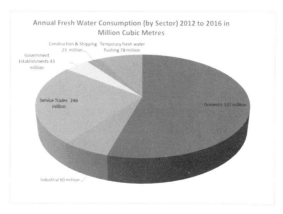

Compare two years of Sources of water

Source of water	2015	2016
Annual water consumption	973 million cubic meters	987 million cubic meters
Rainwater collection from existing catchment areas	207 million cubic meters	358 million cubic meters
Rainwater collection from existing flood storage ponds and rainwater collection tunnels	207 million cubic meters	358 million cubic meters
Temporary fresh water for flushing (0.75-0.15) billion cubic meters	60 million cubic meters	60 million cubic meters
Seawater Desalination	98.7 million cubic meters	98.7 million cubic meters
Leakage rate 16%	155 million cubic meters	157 million cubic meters
Conclusion	245.3 million cubic meters (insufficient fresh water)	45 million cubic meters (sufficient fresh water)

According to the statistics of water consumption by water category from 2012 to 2016, HK uses about 537 million cubic meters of drinking water. However, most of the other 450 million cubic meters of water are non-potable water. It is recommended to use water collected through the rainwater collection system. After analysis, the water supply company needs to provide water every year, from 0 cubic meters in the year of water to 250 million cubic meters in dry water year.

Compare water price by different methods of water collection

Year 2017	Each cubic meter HK$
Rainwater collection though flood storage ponds	1
Dongjiang water supply	5.83
Seawater Desalination	10.2
Replacement and rehabilitation of aged water mains	236
Temporary fresh water flushing	4.58
Reducing the working pressure	2

According to our previous survey, people tends not to accept drinking the water collected by flood storage ponds. However, the amount of rainwater collected in the flood storage ponds is very large and should not be wasted. Seeing this result, it is suggested to utilize this water source for non-potable use such as road washing, construction site use and irrigation purpose. Conversely, they also suggested that the large amount of money was wasted to buy non-used Dongjiang water, which wastes the earth's natural resources. Therefore, in some day when water resources are scarce, rainwater collected by flood storage ponds is still regarded as a drinking water source in the long run.

Water supply rights

Before the start of supplying Dongjiang River water to HK in 1961, HK has experienced several water source shortage which affected citizen's livelihood and the economy. In order to ensure the stability of water supply in HK, Jiangxi Province, which runs through the Dong-jiang River, has always attached importance to the protecting Dongjiang's living environment to ensure quality of Dong-jiang water source. Nowadays, the Dong-jiang water system has a capacity to supply 2.423 billion cubic meters of fresh and clean water to nearby province. The water supply quota for HK is 1.1 billion cubic meters of water. The water supply quota for Shenzhen is 873 million cubic meters, while the water supply quota for Dongguan is 400 million cubic meters.

As discussed, HK is capable of saving lots of water resources after implementing new innovative technology and construction of new facilities. Also, it is estimated that 200 million cubic meters of Dongjiang water is sufficient to meet daily needs. Therefore, the redundant water source can be reserved for other cities in the Greater Bay Area, together the economy can be developed at a higher speed.

The water resources protection regulations in China includes rules and guidelines of a total number of 35. The content inside is very detailed with large coverage, reflecting the Guangdong government's emphasis on preserving the quality of Dongjiang water. As a result, other cities in the Guangdong-Hong Kong-Macao Greater Bay Area are able to increase the demand of water source, thus rapidly develop their economy by actively promoting industries to the cities with high water consumption, including papermaking and textiles, and to increase investment in various places.

The past story to forbid industries affecting water quality to protect HK's water source in Guangdong and Jiangxi provinces have vanished. Now, these cities can easily develop their economies along the coast. In the past, in the territory of Xunwu County along the coast of Dongjiang River, the per capita income of villagers in 2014 was 2,400 yuan. In order to ensure the safety of drinking water, the county strictly controls development and can only grow fruit trees and affect economic development. In Xinfeng County, Guangdong Province, paper mills that have brought huge profits to the government's finances in order to ensure the safe closure of water have rejected dozens of water-related projects. The pulp mill that shut down 200 enterprises in Guangdong Heyuan refused to receive an annual output value of 3 billion yuan and a tax of 600 million yuan is equivalent to the local total industrial output value. HK can reduce the supply of Dongjiang water. The Dongjiang River can develop its economy vigorously. Together with the country's vigorous promotion of the development of Dawan, the areas along the Dongjiang River can directly catch up with other cities in the Guangdong-Hong Kong-Macao Greater Bay Area.

Conclusion and Future Research

Innovate to explore the timeliness of HK's water resources, open source and throttling methods, scientific, systematic management of water resources. After in-depth investigation and research, it shows that the use of innovative ideas system, the public's education to reduce the use of water. By using rainwater recycling, the long-term dependence on Dongjiang water can be reduced, and the water resources saved can be provided to other cities in the Guangdong-Hong Kong-Macao Greater Bay Area, enabling them to develop their economy at a high speed. However, HK's successful open source and thrift can use the same method elsewhere to contribute to global water conservation.

Acknowledgement

This study was supported by my parents, my wife and professors from my Post-Doctoral research study. Also, with support from the HK Government Departments and including the following persons:

Prof. Philip K.I. Chan

Ms. Yu Fei

HKSAR Drainage Services Department

HKSAR Water Supply Department.

I really thank you for them deeply from my bottom heart.

References

1). 余非〈東江水一本通〉2018

Yu Fei (Dongjiang Water One Pass) 2018

2). 何耀光〈香港的城市防洪排澇策略及活化水體意念〉2015

He Yaoguang (Hong Kong's urban flood control strategy and activated water concept) 2015

3). 香港渠務署〈渠務署可持續告 2017-2018，活化河道，上善若水〉2018

DSD (DSD Sustainability Report 2017-2018: River Revitalisation for the Good of Water) 2018

4). 香港水務署〈供水管理 2016-2017 年報〉2018

WSD (Water Supply Management 2016-2017) 2018

5). Philip Kan Ip Chan (Evaluation of the Practice of Quality Assurance in Greater China Construction Industry) 2012. UQAM of Quebec at Montreal, Canada

香港可持續發展水資源之問卷調查

Questionnaire of sustainable development of water resources in Hong Kong

Name 姓名 :_____   Sex 性別 :_____

Age 年齡 :_____   Trade 行業 :_____

Education 學歷：

☐ Secondary school or below 中學或以下   ☐ College to University 大專至到大學

☐ University or above 大學或以上   ☐ Others 其他 :_____

感謝您撥空填寫這一份問卷，本問卷目的是在探討您對香港可持續發展水資源意見。您在問卷中所提供的資訊，僅作為學術研究之用，不會提供其他單位，敬請安心填寫。您的意見對我們非常重要，衷心期盼您依自己的實際感受填答。感謝您的熱情支持與協助！敬祝平安快樂，萬事如意！

Thank you for completing this questionnaire. The purpose of this questionnaire is to discuss your views on sustainable water resources in Hong Kong. The information you provide in the questionnaire is for academic research purposes only and will not provide to other units. Please feel free to fill out. Your opinion is very important to us, and we sincerely hope that you will answer it according to your actual experience. Thank you for your support and assistance! Wishing you peace and happiness, good luck!

7. 是否願意支付較多金錢 (現在水價 3 倍) 使用海水化淡食水？

   Are you willing to pay more money (currently 3 times the price of water) using seawater desalinated water?

☐ Yes 是

☐ No 否

☐ Others 其他 :_____

8. 是否願意使用再造水用作灌溉、洗地等非飲用水？

Are you willing to use recycled water for non-potable water purpose such as irrigation and washing?

☐ Yes 是

☐ No 否

☐ Others 其他 :＿＿＿＿＿＿＿＿＿＿＿＿

9. 是否願意支付金錢作購買東江水作飲用水？

Are you willing to pay for the purchase of Dongjiang water for drinking water?

☐ Yes 是

☐ No 否

☐ Others 其他 :＿＿＿＿＿＿＿＿＿＿＿＿

10. 是否接受使用經雨水渠系統收集的雨水作飲用水？

Do you accept to use the water collected by rainwater collecting system for drinking water?

☐ Yes 是

☐ No 否

☐ Others 其他 :＿＿＿＿＿＿＿＿＿＿＿＿

11. 東江水供水不足時，是否接受使用經雨水渠系統收集的雨水作飲用水？

Do you accept to use the water collected by rainwater collecting system for drinking water, when there is a shortage of DJ water supply?

☐ Yes 是

☐ No 否

☐ Others 其他 :＿＿＿＿＿＿＿＿＿＿＿＿

12. 是否接受使用經雨水渠系統收集雨水灌溉、洗地等非飲用水？

Is it acceptable to use rainwater collection systems to collect rainwater for irrigation, washing and other non-potable water purpose?

☐ Yes 是

☐ No 否

☐ Others 其他 :＿＿＿＿＿＿＿＿＿＿＿＿

CURRICULUM VITAE

(Prof. Dr. Po Shing Liu)

Occupation	Community Volunteers; Engineering Technologies; Technological and educational promotion; Economic and Quality Promotion; Volunteers Promotion; Support Charity Promotion;	Position	(i) Hong Kong Chairman (ii) Chairman HKIIE
Company/Organization	(i) The International Postdoctor Association (IPostdocA) (ii) Hong Kong Institution of Incorporated Engineers (HKIIE)		
Email	eddieliu88@gmail.com psliu88@gmail.com		

University Education & Academic Learning

Post-doctoral Education:

Year	Degrees / Qualifications Awarded
2015 – 2016	Post Doctoral Fellowship (awarded by California State University at Monterney Bay USA)
2015 – 2016	Post Doctoral Fellowship (awarded by the joint universities of Stanford, UCB, CSUMB (USA))
2013 – 2015	Postdoctoral (technology and management) (awarded by China National School of Administration's Government Economics Research Center, Chinese Academy of Governance)
2012 – 2014	Post Doctoral Fellowship (technology and management) (awarded by the University of Quebec at Montreal Canada)

Doctoral Education:

Year	Degrees / Qualifications Awarded
2020 – Present	PhD In Public Administration (study pursueing in Central School via higher education programme)
2017 – 2020	Doctor of Law (LLD) (awarded by Sabi University, France)
2012 – 2015	Doctor of Business Administration (DBA) (awarded by Tarlac State University)
2007 – 2012	PhD In Engineering Management (awarded by Nueva Ecija University of Science and Technology)

Masters Education:

Year	Degrees /Qualifications Awarded
2000 – 2003	MSc(Eng) In Communications Engineering (awarded by the University of Hong Kong)
1998 – 2000	MSc In Electronic Engineering (awarded by the Chinese University of Hong Kong)
1993 – 1997	MSc In Engineering Business Management (awarded by the University of Warwick, U.K.)

Under-graduate Education:

Year	Degrees / Qualifications Awarded
1994 – 1998	BEng(Hons) In Electronic Engineering (awarded by the City University of Hong Kong)
1992 – 1994	Post-Experience Certificate In Manufacturing Engineering (awarded by the Hong Kong Polytechnic University)
1983 – 1985	Higher Diploma In Electronic Engineering (awarded by the Hong Kong Polytechnic University)
1980 – 1983	Diploma In Electronic Engineering (awarded by the Hong Kong Polytechnic University)

Knights and Honours Bestowed

2021 – Present	Bestowed with the ennobled title of Grand Knight KGCO (bestowed by H.M. the Chief (King) of Sefwi Obeng-Min, the Royal House of Obeng-Min, Western North Region of Ghana)
2021 – Present	Bestowed with the ennobled title of Knight KGCG (bestowed by H.M. the Chief (King) of Sefwi Obeng-Min, the Royal House of Obeng-Min, Western North Region of Ghana)

Professorship & Engineering Fellowship & Think Tank & Expert Member

2020 – Present	Guest Professor in Engineering Technology Management (appointed by Peking University's PKU Yan Yuan Ren He Education Organization, China)
2019 – Present	Adjunct Professor in Engineering and Technology (confered by the Sabi University, France)
2016 – Present	Engineering Fellow (FEng) (awarded by the Hong Kong Professional Validation Council for Industry (PVCHK))
2014 – Present	Think Tank of Dashun Foundation (invited and admitted by Dashun Foundation, Hong Kong)
2015 – 2018	Experts Committee Expert Member (SZAQ) (admitted and appointed by Shenzhen Association for Quality, China)

Arbitration Commission Listed Arbitrator

2020 – Present	首席法務師（Chief Legal Office） (admitted by 全國工商聯人才交流服務中心 , China)
2020 – Present	Arbitrator (NJAC) (admitted and listed by Nanjing Arbitration Commission, China)
2017 – Present	Arbitrator (QZAC) (admitted and listed by Quanzhou Arbitration Commission, China)
2014 – Present	Arbitrator (GZAC) (admitted and listed by Guangzhou Arbitration Commission, China)

Professional Institution Fellowships

2020 – Present	Fellow of IPostdocA (FIPostdocA) (admitted by the International Postdoctor Association (IPostdocA))
2011 – Present	Fellow of HKIoD (FHKIoD) (admitted by the Hong Kong Institute of Directors (HKIoD))
2011 – Present	Honorary Fellow of HKIIE (Hon.FHKIIE) (admitted by Hong Kong Institution of Incorporated Engineers (HKIIE))
2007 – Present	Fellow of HKIIE (FHKIIE) (admitted by Hong Kong Institution of Incorporated Engineers (HKIIE))
2006 – Present	Fellow of IET (FIET) (admitted by the Institution of Engineering and Technology (IET), UK)

Experience in national and international meetings, conferences

●	Liaison Officer of the Organizing Committee of the IEEE International Symposium on Product Compliance Engineering-Asia (IEEE 2019 ISPCE-CN); Date: 23-26 October 2019; Venue: Open University of Hong Kong (OUHK), Ho Man Tin, Kowloon, Hong Kong S.A.R.
●	Elected Committee Member of the Planning and Organizing Committees of the Asia Pacific Radio Spectrum Conference 2018 (APRSC2018); Tentative Date: April 2018; Venue: (Hong Kong Special Administrative Region, HKSAR).
●	Publication Chair; Track Co-Chair; Conference Coordinator; of the Organizing Committee of the Asia Pacific Radio Spectrum Conference 2016 (APRSC2016); Date: 09-10 March 2016; Venue: Regal Hong Kong Hotel, Hong Kong S.A.R.
●	General Chair of the Organizing Committee of the IEEE 2015 International Conference on Consumer Electronics – Shenzhen China (IEEE 2015 ICCE – Shenzhen China); Date: 9-11 April 2015; Venue: Shenzhen Conference and Exhibition Center (SZCEC) and Harbin Institute of Technology (Shenzhen) Research Center (HITSZ).
●	Exhibit Chair of the Organizing Committee; and Technical Papers Oral Presentation Session judge for the IEEE 2014 International Conference on Consumer Electronics – Shenzhen China (IEEE 2014 ICCE – Shenzhen China); Date: 9-13 April 2014; Venue: Shenzhen Conference and Exhibition Center (SZCEC).
●	Honorary Secretary of the Organizing Committee of the Asia Pacific Radio Spectrum Conference 2012 (APRSC2012); Date: 12-13 April 2012; Venue: Hotel Shangri Hotel, Tsim Sha Tsui East, Hong Kong S.A.R.
●	Honorary Secretary of the Organizing Committee of the Asia Pacific Radio Spectrum Conference 2010 (APRSC2010); Date: 8-9 April 2010; Venue: Shangri Hotel, Tsim Sha Tsui, Hong Kong S.A.R.
●	Honorary Secretary; Technical Paper Committee Co-Chair; and Local Arrangement Committee Co-Chair of the Organizing Committee of the Broadband World Forum Asia 2009 (BBWFA2009) – Technical Session: IET Next Generation Wireless Broadband Communication Conference; Date: 20-22 May 2009; Venue: Hong Kong Convention and Exhibition Centre (HKCEC).
●	Local Arrangement Chair of the Organizing Committee of the Asia Pacific Spectrum Management Policy Symposium 2009 (APSMPS2009); Date: 03 April 2009; Venue: Mira Hong Kong Hotel, Hong Kong S.A.R.
●	Honorary Secretary; Liaison Committee Chair; Local Arrangement Committee Member; and Papers Poster judge of the Organizing Committee of the Broadband World Asia Forum 2008 (BBWAF2008); Date: 16-17 July 2008; Venue: Hong Kong Convention and Exhibition Centre (HKCEC).

Publications

no	Paper Title	Congress / Conference	Author	Published Date	Conference / Proceedings
1	A Study on Innovation and e-Service Quality for Developing e-Retailing Mass Entrepreneurship	Asian National Quality Congress 2016 Vladivostok, (Far Eastern Federal University)	1st author (Po Shing Liu)	2016.09. 21-22	International ANQ Congress 2016 Official Proceedings (HK-06) (Far Eastern Federal University, Vladivostok, Russia)
2	Consumer Electronics Retail Entrepreneurship: Opportunities and Development	IEEE International Conference on Consumer Electronics (Shenzhen) China 2015	1st author (Po Shing Liu)	2015.04.10 p.m.	IEEE International Conference on Consumer Electronics (Shenzhen) China 2015 official proceedings
3	A Study on e-Service Quality of Consumer Electronics Retail Business	IEEE International Conference on Consumer Electronics (Shenzhen) China 2015	1st author (Po Shing Liu)	2015.04.10 p.m.	IEEE International Conference on Consumer Electronics (Shenzhen) China 2015 official proceedings

Continuous development plan / Professional institutes' committees / Meetings / Events

2013 – Present	Hong Kong Chairman The International Postdoctor Association (http://www.IPostdocA.org)
2010 – 2012	ECS Chairman and Committee Chair Electronics and Communications Section (ECS) & working committees The Institution of Engineering and Technology (IET) Hong Kong Branch
2012 – 2014	ECS Immediate Past Chairman Electronics and Communications Section (ECS) The Institution of Engineering and Technology (IET) Hong Kong Branch
2014 – 2020	ECS Committee Member (Ordinary Member; Helper; Co-opted member; Technical Visits; Seminars; Symposiums; Conferences;) Electronics and Communications Section (ECS) & working committees The Institution of Engineering and Technology (IET) Hong Kong Branch
2009 – 2010	ECS Honorary Secretary Electronics and Communications Section (ECS) & working committees The Institution of Engineering and Technology (IET) Hong Kong Branch.
2007 – 2009	ECS Committee Member (Technical Visits Coordinator) Electronics and Communications Section (ECS) & working committees The Institution of Engineering and Technology (IET) Hong Kong Branch.
2006 – 2007	ECS Committee Member (Publicity Officer) Electronics and Communications Section (ECS) & working committees The Institution of Engineering and Technology (IET) Hong Kong Branch.
2018 – 2021	Chairman HKIIE Electronics and Communications Section (ECS) & working committees The Institution of Engineering and Technology (IET) Hong Kong Branch.
2018 – 2021	Honorary Secretary & Registrar (for RPEM – Registered Professional Engineering Managers) Engineering Managers Registration Council (EMRC) Hong Kong Institution of Incorporated Engineers (http://www.HKIIE.org.hk)
2007 – 2018	Vice Chairman HKIIE Council Hong Kong Institution of Incorporated Engineers (http://www.HKIIE.org.hk)
2013 – 2018	Chairman The Chinese Institue of Electronics (Hong Kong Electronics Branch)

A Study on Innovation and e-Service Quality for Developing e-Retailing Mass Entrepreneurship

Po-Shing LIU [1,2,3,*] and Philip K.I. CHAN [1,2,3,4]

[1] Stanford University, Stanford, California, CA 94305, U.S.A.
[2] University of Calfornia at Berkeley, California, CA 94720-1234, U.S.A.
[3] California State University at Monterey Bay, 100 Campus Center, Seaside, CA 93955-8001, U.S.A.
[4] Stanford, Berkeley, Monterey Postdoc Advisor
*Email: eddieliu88@gmail.com, *(852) 61333567

Abstract

In this twenty-first century the world is undergoing a trend with a number of significant changes in customer buying behaviors and business doing ways electronically. These changes as in ancient times of human beings were namely the activity of exchanges or the trading of valuables, and as in nowadays they are referring to as doing business activities electronically in e-commerce e-business platform in the local and global business world environment. These changes of customer buying behaviors electronically have an impact on the performance of business doing activities, business activity processing time, trading transactions, payments, etc, making business transactions more easily, more efficient, more effective, speed faster. This trend together with innovation, creativity, curiosity, passion, fashionable liked quality products, e-service quality, and e-platform execution can lead to a smart development of customer e-business e-retailing mass entrepreneurship, driving a society's business activities with more economic growth and the various related supporting technology and industrial manufacturing sectors a bit forwards.

In short it is noticeable that: Firstly, there is an emerging trend in the shift of proportion on the act of buying behaviors from traditional ways, physical outlets shifting towards online e-commerce e-platform purchases at relatively faster trade transactions and shorter time electronically. Secondly, with the maturity of high speed internet, wireless and mobile technology, sophisticated and delicate powerful functional working algorithms and the represented by the popularity of high speed smart and intelligent phones internet as trading platform and the concept of global village, the new form of marketplace online is now solidly gaining its share as the major channel of retail business. Thirdly, companies have to be pragmatic in maintaining their current competence, as well as developing new competence and hook these up with performance and profitability in order to achieve long term development and sustainable profitability.

As a result, e-commerce entrepreneurship in e-retailing is in an emerging current and the trend is now becoming the blue ocean of established business enterprises and SME as well as the biggest opportunities in new business ventures. This paper outlines the value-added elements of innovation, creativity, curiosity, passion, fashionable liked quality products, e-service quality, e-platform execution such as e-Bay, in driving and developing e-commerce e-business consumer-to-consumer e-retailing and mass entrepreneurship. This paper also shares with practical experiences and offers alternatives for success based on a practical company example in Hong Kong.

Keywords: Innovation, e-Service Quality, e-Business, e-Retailing, Mass Entrepreneurship.

1. Introduction

E-commerce has been taken up tremendously since 1990s. It started up with selling airplane tickets online. Then, Amazon Book Company made itself one of the top three book sellers in the early 1990's through online order of books. Since then, many companies built their web sites for global access and networking of businesses. Then, eBay, Yahoo, and Google has been very successful in selling consumer products through internet. Lately, even business-to-business products are now available through web sites such as Alibaba and Made-in-China in China. The increasing number of people from the general public using e-banking and e-shopping has not only resulted in a booming of electronic business transactions at a rate of 10% rise per year, but also caused a change of buying habits globally. The popularity of smart phones and its related applications of e-transactions is a typical example.

This paper outlines and addresses the importance of the key value-added elements of innovation, creativity, curiosity, innovation strategy, applying in e-business platforms such as e-Bay in driving and developing e-commerce e-business consumer-to-consumer e-retailing and mass entrepreneurship.

This paper also serves to use a Hong Kong example of a consumer-to-consumer electronic business (e-business) e-retailing company's operation as a platform to demonstrate the opportunities, challenges, learning and insights of developing e-service quality in retail business via the internet and mobile wireless network platform. It will cover issues such as the e-business operating environments, e-retailing e-commerce opportunities and challenges, strategic alternatives and how e-service quality development gives the winning edge in e-commerce business mass entrepreneurship.

This study is significant to a number of stakeholders as stated in the following:

Firstly, entrepreneurs stakeholders who want to start up e-commerce with a proven e-business model; Secondly, e-retailers stakeholders look for effective strategies to raise the bar of their e-businesses to next level; Thirdly, investors and bankers stakeholders who provide funds for E-commerce business projects; and Fourthly, people who want to know more about e-commerce e-business e-retailing and wholesale mass entrepreneurship and how to put the theories into practices in the practical e-business environment. Fifthly, innovators stakeholders who design, invent and innovate efficient systematic e-business new models.

More importantly, there was limited literature about hands-on operational strategies in developing service quality of a Hong Kong based e-commerce retail setup with its main revenue generated from the eBay marketplace. This action research will serve as an effort to fill the knowledge gap.

As the scope of this paper is confined to being an action research on the specific case of e-commerce firm in Hong Kong, it will not cover concerns on the wider scope of e-commerce beyond consumer-to-consumer retail like electronic banking services or other electronic form of professional services.

The discussion in this paper takes an exploratory approach. Fact-finding and action spirals span around a qualitative study of a Hong Kong consumer-to-consumer e-business e-retailing private company. The author would like to acknowledge the contribution of the business owner of the subject company under study in providing valuable business information and trade secrets to facilitate analysis and generating insights.

2. Literature Review

2.1 Parent Literature

E-commerce is a comparatively young knowledge stream, but it is gaining popular attention and with a wide spectrum in what it refers to. From the 1990s onwards, electronic commerce would additionally include Enterprise Resource Planning systems (ERP), data mining and data warehousing. An early example of many-to-many electronic commerce in physical goods was the Boston Computer Exchange, a marketplace for used computers launched in 1982. An early online information marketplace, including online consulting, was the American Information Exchange, another pre Internet online system introduced in 1991.

In 1990 Tim Berneers-Lee created the first World Wide Web server and browser (Palmer, 2007). It opened for commercial use in 1991. In 1994 other advances took place, such as online banking and the opening of an online pizza shop by Pizza Hut (Palmer, 2007). During that same year, Netscape introduced SSL encryption of data transferred online, which has become essential for secure online shopping. Also in 1994 the German company Intershop introduced its first online shopping system. In 1995 Amazon launched its online shopping site, and in 1996 eBay appeared (Palmer, 2007).

At first, the main users of online shopping were young men with a high level of income and a university education (Bigne, 2003). This profile is changing. For example, in USA in the early years of Internet there were very few women users, but by 2001 women were 52.8% of the online population.

Sociocultural pressure has made men generally more independent in their purchase decisions, while women place greater value on personal contact and social relations.

Why does electronic shopping exist? For customers it is not only because of the high level of convenience, but also because of the broader selection; competitive pricing and greater access to information (Jarvenpaa, 1997) (Peterson, 1997). For organizations it increases their customer value and the building of sustainable capabilities, next to the increased profits (Peterson, 1997).

The main idea of online shopping is not in having a good looking website that could be listed in a lot of search engines and it is not about the art behind the site (Falk, 2005). It also is not only just about disseminating information, because it is all about building relationships and making money (Falk, 2005). Mostly, organizations try to adopt techniques of online shopping without understanding these techniques and/or without a sound business model (Falk, 2005). Rather than supporting the organization's culture and brand name, the website should satisfy consumer's expectations. (Falk, 2005) A majority of consumers choose online shopping for faster and more efficient shopping experience. Many researchers notify that the uniqueness of the web has dissolved and the need for the design, which will be user centered, is very important. (Falk, 2005) Companies should always remember that there are certain things, such as understanding the customer's wants and needs, living up to promises, never go out of style, because they give reason to come back (Falk, 2005). And the reason will stay if consumers always get what they expect. McDonaldization theory can be used in terms of online shopping, because online shopping is becoming more and more popular and website that wants to gain more shoppers will use four major principles of McDonaldization: efficiency, calculability, predictability and control.

Organizations, which want people to shop more online for them, should consume extensive amounts of time and money to define, design, develop, test, implement, and maintain website (Falk, 2005). Also if company wants their website to be popular among online shoppers it should leave the user with a positive impression about the organization, so consumers can get an impression that the company cares about them (Falk, 2005). The organization that wants to be acceptable in online shopping needs to remember, that it is easier to lose a customer then to gain one. (Falk, 2005) Lots of researchers state that even when site was a "top-rated", it would go nowhere if the organization failed to live up to common etiquette, such as returning e-mails in a timely fashion, notifying customers of problems, being honest, and being good stewards of the customers' data (Falk, 2005). Organizations that want to keep their customers or gain new ones try to get rid of all mistakes and be more appealing to be more desirable for online shoppers. And this is why many designers of webshops considered research outcomes concerning consumer expectations. Research conducted by Elliot and Fowell (2000) revealed satisfactory and unsatisfactory customer experiences (Falk, 2005).

What is innovation? The word "Innovation" in dictionary comes from the Latin root "innovates" and has a meaning that it is "to renew something" or "to change something". Innovation is generated from a combination of ideas, visions, creativity, intelligence, inspiration, passion, solution, execution. Innovation is interpreted as the creativity and/or invention plus technology to result in with meaningful value. It launches with new product, new service, new process, etc into marketplace via commercialization, entrepreneurship, Innovation is usually economic motive and situation and environment induced. Innovation can sum up to invention plus commercial exploitation. (Kao, 2016).

Why we require innovation? Innovation can improve performance, raise productivity, increase yield and output, and create growth, resulting in economic money and wealth. Innovation can reduce waste, minimize environment damages. Innovation acts as a driver force to and has a driving effect that can increase competition; have greater availability of potentially useful technologies coupled with a need to exceed the competition in these technologies; reduce product life cycles; improve the financial pressure to reduce costs, increase efficiency, do more with less, etc; give industry and community needs for sustainable development; increase demand for accountability; improve demographic, social and market changes; raise customer expectations regarding service and quality; improve economy and wealth (Kao, 2016).

How we apply innovation? ......

e-Service quality is the strategy that is gaining momentum for online business operators to position themselves more effectively in the marketplace (Parasuraman, Zeithaml, and Berry 1988). However, the strategic implementation of service quality requires going over the hurdle of defining and measuring service quality. Service quality has been defined as the comparison of expectations with performance. In essence, it is between what the customers feels should be offered and what is provided. e-Service quality can be assessed by measuring customer's expectation and perceptions of performance level on a number of key service attributes (Parasuraman, Zeithaml, and Berry 1985, 1988).

It is important to take the country and customers into account. For example, in Japan privacy is very important and emotional involvement is more important on a pension's site than on a shopping site (Stephen, 2004). Next to that, there is a difference in experience: experienced users focus more on the variables that directly influence the task, while novice users are focusing more on understanding the information (Kumar, 2006).

E-commerce product sales totaled $146.4 billion in the United States in 2006, representing about 6% of retail product sales in the country. The $18.3 billion worth of clothes sold online represented about 10% of the domestic market (CNN, 2007).

Although the benefits of online shopping are considerable, the lack of full disclosure with regards to the total cost of purchase is one of the concerns of online shopping. While it may be easy to compare the base price of an item online, it may not be easy to see the total cost up front as additional fees such as shipping are often not be visible until the final step in the checkout process. Some services such as the Canadian based Wishabi attempts to include estimates of these additional cost (E-commerce News 2010), but nevertheless, the lack of general full cost disclosure remains a concern.

2.2 Immediate Literature

Hong Kong was one of the first places among in Asia to embrace business trading and consumer retailing electronically by means of electronic commerce (e-commerce) platform. All but the smallest shipping and freight-forwarding companies have been using electronic data interchange (EDI) since the early 1980s. More recently, Hong Kong traders have established their presence in consumer-to-consumer e-business e-retailing via the eBay platform. Generally, these consumer-to-consumer traders are using seven adaptable generic strategies in meeting the market forces (Porter, 2010).

The first strategy is overall cost leadership. This strategy aims at reducing costs at all levels. The company reduces costs through high market share, favorable access to raw materials, well-designed parts for ease in manufacturing, maintaining a wide line of related products to spread costs, serving all major customer groups in order to build volume, and streamlining the production line.

The second strategy is differentiation. This strategy differentiates products or services by creating something that is perceived industry-wide as being unique. Differentiation comes from brand names, unique design, packaging, use of technology, unique features, extraordinary customer service and a wide dealer network.

The third strategy is "focus". This strategy focuses on a particular niche, like buyer group, segment of the product line, segment least vulnerable to substitutes, segment where competitors are the weakest, or different geographic market. It means to focus efforts and resources on a niche. By doing this, a company will be strong enough to compete with more powerful companies.

The fourth strategy is "innovation". This strategy focuses on the systematic approach of particular areas of innovation in technology, product, service, process, organizational, business and marketing strategy;

The fifth strategy is applying "innovation strategy". The strategy focuses on the degree of newness, such as incremental or radical, and focuses on the degree of impact in the range of continuous or disruptive.

The sixth strategy is "e-service quality". This strategy focuses on its five key dimensions, namely, the reliability, responsiveness, assurance, empathy, and tangibles.

The seventh strategy is the "timing" and "time". This strategy focuses on the important issues of the time to market and the concerned products to the consumers.

How do these seven generic strategies interact with the increasing demand of service quality from customers? There are five common criteria that customers use in evaluating e-service quality delivery through online purchase experience. These include information availability, ease of use, privacy and security, graphic style, and fulfillment (Zeithaml, Parasuraman and Malhotra 2002). To get the most out of this exploratory study, an action research system model, which is a three-stage model with double loops feedback, shall apply.

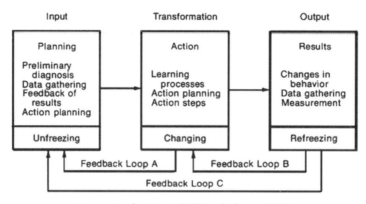

Source: Based on Lewin (1988) and Johnson (1996)

The action research model is shown clearly and precisely in Figure 1 and it closely followed Lewin's (Lewin 1988) repetitive cycle consisting of Input stage, Transformation stage, Output stage. Planning, action, and measuring results. As indicated in the diagram, the input (planning) stage is a period of unfreezing, or problem awareness. The transformation (action) stage is a period of change, namely, trying out new forms of behavior in an effort to understand and cope with the system's problems. The output (results) stage is a period of refreezing, in which new behaviors are tried out on the job and, if successful and reinforcing, become a part of the system's repertoire of problem-solving behavior. This model serves as a guiding principle for making the best use of knowledge discovered to benefit stakeholders.

An attempt has been made in this paper to convert knowledge discovered into effective behaviours in the context of web e-service quality for consumer-to-consumer e-business e-retailing operators and entrepreneurs.

3.   Research Methodology

The research instrument is Action Research such that the researcher went through the Strategic Management Process to develop strategies. The analysis and ideas would be carefully and completely documented in field notes recording forms. A questionnaire of probing questions will also be used in in-depth interviews. Eventually the collected data would be consolidated into a critically reviewed, feasible strategic plan for implementation, evaluation, next stage fact-finding and further action spirals, which are shown in Section 4 and Section 5 below.

The collection of data would be from eBay web site, the subject company, the iOffer website, the Yahoo! Auction site, the competitor's website, as well as other websites and locations giving supportive functions to the broader e-commerce field.

Data processing and statistical treatment shall be qualitative. Data analysis techniques such as content analysis and grounded theory shall be adopted through the whole data collection process to capture the subtle, intrinsic "tacit" knowledge in e-commerce retail and wholesale practices. This shall take the best advantage of case study action research. The codification and personalization of knowledge into a value chain will establish a firm base for the formulation of strategic alternatives in the next stage and as a result contributing to the integral effort of strategic development.

4. Opportunities for and Challenges to Consumer-to-Consumer e-Business e-Retailing Entrepreneurship

In the following, a company named Charter Lift Limited shall be used as the subject of study to reveal the opportunities and challenges of consumer-to-consumer e-business e-retailing mass entrepreneurship.

Charter Lift Limited, a Hong Kong based company with its main business activities in consumer-to-consumer e-business e-retailing, is now in its ninth year. The company is traded under a number of business names in eBay (the biggest and representative online marketplace), with its flagship store under the name "cqCharter camera-chess". The mission of the company is "to provide lifestyle products to customers with convenience, value, competitive prices and quality services".

The structure of this company under study (hereafter called "cqCharter") is a typical Asian business model, with the enterprise carrying a strong family constituent. The close family tie and the kinship factor ensure commitment, obedience and loyalty, which form the foundation of a unified, efficient steering team. The company started small a few years ago, with an initial capital of just US$5,000 and a reasonable 50% profit margin. Its main products were Chinese Checkers Board Games. As of December 2013, the company is selling more than 100 different types of chess and board games, 140 different types of cameras, 134 combinations of jewelry trees, plus a number of lifestyle items such as automatic watch winders. The annual sales of cqCharter are US$400,000 and a profit margin ranging from 50% to 400%. Amazingly, the cqCharter has fewer than 5 regular staff handling daily routine business activities. By the standard of a small enterprise with little experience in online retail business and a very small startup capital, this is a great achievement.

An analysis of internal environment of the company revealed some astounding facts contributing to its success. The special structure and the tie of kinship in cqCharter ensured a cohesive and highly efficient team during the initial growth stage of the company. This was evidenced by a number of record breaking milestones:

(a) The company was the first ever online business venture in Hong Kong with both retail and wholesale business model on Board Games and Toy Cameras;

(b) It occupied over 90% of the consumer-to-consumer e-business e-retailing market share on the Yahoo! Auction marketplace from June to August 2007. At the time the company was in its first year of business;

(c) The company obtained the "powerseller" status in a record-breaking three months' period after it entered the eBay marketplace in 2008. The "powerseller" status represent the top 5% of online sellers in the eBay marketplace in terms of overall ratings in sales and service quality; (note: the company is currently ranked "top-rated" seller, which denoted the top 1% of online sellers in the eBay marketplace);

(d) The company entered the consumer-to-consumer e-business e-retailing wholesale market in Indonesia in the 4th quarter of 2009. By the 4th quarter of 2010, it was the largest overseas online wholesale supplier in the Indonesian market, with its customers being the major consumer-to-consumer e-business e-retailers in the city of Jakarta, Bandung, and Batem;

(e) The company owner-manager, as the company representative, was invited by eBay, the world class online marketplace, as "sales professional". The role of "sales professional" is to act as mentor and advisor for company and individuals who will develop online e-commerce business model in eBay. This is an unparalleled recognition of the achievements and contribution on e-commerce retail and wholesale business.

There is, however, also a downside for family based business. The heavy reliance on loyalty and kinship relations is at the same time a hindrance to the establishment of systems and more formal structure to the company, a necessity for further development and growth to breakthrough the bottleneck of the work capacity and the span of direct supervision by the top management members. The company is currently in the outermost boundary of a relatively small, self-sufficient and profitable family operation. A further step will be the transformation of the company into a bigger size and a world-class e-commerce e-business e-retailing company with renewed vision, mission, values and culture. This will be a giant leap from the current traditional operational state. Although it is a common goal of the company top management members to see this happened, there are concerns among the management members towards such a major transformation. The fear of change, the stress on the accompanying risk, and the lack of knowledge on how to work out the change have become mental blocks among some of the senior managers. While the whole management team trusts the owner-manager and respect him very much, there are hesitations as the owner-manager suggested to move the company ahead in the direction of formalizing systems and structures, investing in technology and expanding business operations beyond the initial span of control of the senior managers.

As a result, the company's growth has been slowed down over the past 6 months as the company is approaching the maximum operational strength under the current family style operational structure. Sourcing and customer development has been piled up, while the warehouse inventory record update has experienced delay. These are signs that the future sales and service level are deteriorating. cqCharter is in need of a "revived" strategic development program to continue the momentum of the company's long term growth and sustainable profitability.

A favourable external environment complemented the advantages of internal environment for consumer-to-consumer e-business opportunities. The availability of several popular consumer-to-consumer marketplaces at very low cost means that retailers are able to open a store at a very low "rental" cost. This is an unparalleled competitive advantage over retailers operating traditional stores in high traffic locations in the streets. The specialization of these online marketplaces, e.g. eBay for everyone, Yahoo! Action site for young people in Hong Kong and Taiwan, IOffer as an eBay alternative, Taobao for people in mainland China, Facebook for social networkers, Uwants and 3boys2girls for Hong Kong teens all open up exchange platforms for a vast variety of products and services offerings. This facilitates cost effective market segmentation on buyers of generation Y, parents, individual consumers, business buyers and resellers. With the smart use of free photo storage websites to store product photos and linked them to advertisement pages, and to set a "right" price (such as not exceeding USD60), consumer-to-consumer e-retailers may easily attract buyers to buy from impulse or loyalty.

To expand the pie, cqCharter made an effort to pair up cultural product preferences by listing different product illustrations in different countries, prepare wholesale catalogue for wholesalers, and apply different sales focus in different online marketplaces. The company established the following nine performance standards to cater for the challenges of the external environment:

Product Design and Quality: The product should be such a way that it is not easy to find from the local retail shops. The product design should be unique and attractive. The quality has to be good because the buyer could complain at eBay. The seller has to deal with enquiries and complaints promptly and properly so as to prevent losing credit and reputation.

Price: Price preferably at USD 60 or under so that eBay consumers could easily buy it at impulse. This is the price the buyer can take a risk and do not bother a loss of it. It is because the buy does not have a chance to feel and touch the product like seeing it in a retail shop.

Service: eBay requests excellent service, such as prompt and polite response to all buyer's request. Quick shipments and guarantees are important too. eBay allows customers to write about their complaints or appreciations on the service. Therefore complaints will hurt the business greatly. Excellent operation management is needed for eBay business. For example, giving daily reply to enquiries and registering the tracking number on the transaction record are typical daily chores and should be complete on time.

Reliability: Customers want a reliable seller. They find information about the seller's past sales record from eBay. If the seller received adverse feedback or negative comments on its reliability in previous sales, then prospective buyers are likely will not buy from the seller.

Tangibles (services): The services have to be tangible. For example, good control of delivery date and the communications on shipment. Also return with refund and no questions asked would be a good tangible service.

Security (on deliveries): The seller has to promise on the arrival date and be secured on the date, in particular for seasonal gifts and birthdays.

Responsiveness in answering requests: eBay has template on how to answer customer's requests politely and detail so that the buyer would not feel frustrated.

Assurance on delivery: The seller has to assure when the goods be arrived and the intermittent communications to keep the buyer informed about the progress.

Empathy: This is to be patient with the buyer in their dissatisfaction, and to be prompt in actions to solve their problems.

Innovation:

This is to be very important as the innovation and design can contribute to add innovation value as shown in the smiling curve. Innovation can be incremental innovation and breakthrough innovation. In incremental innovation, it refers to the analytical problem solving approach. For example, the DMAIC model, namely, define problems, measurement data, analyse the data and the problem situation, improve the situation and parameters, control the improvement processes, etc. In the breakthrough innovation, it refers to the creative problem solving approach, in which creativity, curiosity, passion, design thinking, etc are of significant importance.

Out of these ten performance standard, six of them (reliability, tangibles, security, responsiveness, assurance, and empathy) are e-service quality standards. Innovation is linked to (creativity, curiosity, passion, design thinking, execution) are innovation standards. Generally, cqCharter has consistently performed well in all the above nine aspects.

A SWOT (Strength, Weakness, Opportunity, Threat) chart is summarized below for a detailed view of opportunities and challenges for consumer-to-consumer e-business entrepreneurship.

Figure 2: SWOT Analysis for Consumer-to-Consumer e-Retailing Entrepreneurs

Source: Qualitative Data obtained from after Taking In-depth Interview of owner-manager, Charter Lift Limited (2015)

Applying the SWOT Analysis – The Strengths (S) concerned issues
1. Innovation and trending.
2. Creativity and iterative execution.
3. Curiosity and passion.
4. Reputation given by highly rated seller status at eBay.
5. Growing profitability ensures cashflow.
6. Good communication skills in writing gain attractions and credits.
7. The operation team is strong in quality deliveries.
8. Cohesive team with passion.
9. Sourcing advantages, customer loyalty.

Applying the SWOT Analysis – The Weaknesses (W) concerned issues
1. Lack of management and leadership knowledge for long term growth.
2. Mental blocks about change.
3. Rules from the online marketplace (e.g. eBay) are changing fast. The current advantages and buyer pool may not last for long.
4. Lack of vision and marketing analytics data.
5. Product life cycle is fast for most retail items. Bestseller items succession requires good planning and an intuition of the market.

Applying the SWOT Analysis – The Opportunities (O) concerned issues
1. Expand the business with internal fund.
2. Bring about transformation through adopting one or more strategic alternatives.
3. Develop overseas wholesale clientele.
4. Develop local corporate clients.
5. Popularity of high speed 4G / 5G smart phones, wireless communications network, and the maturity of ICT and IoT technologies continually increase Consumer-to-Consumer customer base.
6. Low entry barrier.
7. Small capital requirements for maintaining business.
8. Improvement in trade policy of online marketplace boosts confidence.

Applying the SWOT Analysis – The Threats (T) concerned issues
1. Competition is increasing.
2. Internet and wireless network security is always an issue.
3. Increasing postage costs causing decreasing profit margin.
4. Challenges from offline retailers via Business-to-Consumer e-retailing model.

Like all businesses, challenges always come with the opportunities. Competition among Hong Kong sellers is more intense than the competition between Hong Kong sellers and other sellers from different countries. This is partly because sellers from Hong Kong share the same advantages of the low tax environment and the close proximity of sourcing from mainland China. There are about six hundred professional Hong Kong retailers selling at eBay. Therefore, the key to success is to maximize uniqueness and to minimize direct competition. The selection of chess products in 2007, underwater camera products in 2009, jewellery tree products in 2013 and automatic watch winder products in 2014 by cqCharter are typical examples of going around competitions in the company's favor. Still, the company has to face new challenges of increasing postage costs and the growth of B2C retailing by traditional retailers.

5. Strategic Options for Consumer-to-Consumer e-Retailing Entrepreneurship

The following strategic options have been engineered with reference to the opportunities, challenges and comparative advantages of cqCharter.

5.1 e-Retailing Entrepreneurship – Strategic Option One

The owner-manager adopts a "do nothing" strategy with his company. He will give more time to his senior managers so that they will see the need for change in a few months' time. By that time, it will be easy to reach consensus for actions.

This is a conservative move. It could maximize rapport in the management team. However, the "do nothing" strategy is postponing the problem to a later time. This strategy lacks a long-term vision to steer the company to the next level of development, and is the least preferable strategy. In the real case of cqCharter, this strategy had not been adopted.

5.2 e-Retailing Entrepreneurship – Strategic Option Two

The owner-manager keeps on running the current operations, and continues his individual, own effort to break through the bottleneck of growth. He keeps on looking for "star" merchandise to help his business go through the present bottleneck of growth.

The success of this alternative depends on personal strength, time and energy resources, as well as luck to capitalize on the "right" opportunities. The positive message in this alternative is that it is always a good quality of a modern manager to keep on searching and observing around. The negative aspect is that it was giving up teamwork. Lacking synergy, the efficiency of personal effort will likely be low.

In reality, the owner-manager of the subject company identified and researched on the following new product categories: flash memory drive with client logo to be printed, scenic candles, Chinese opera masks, silk embroidery, and i-yo-yo.

5.2.1 Client Logo Printed product – Flash Memory Drive

The prices of 32GB USB flash drive varied in the eBay market. Most of the prices shown were less than those at the retail store. However, there were no competitors who would sell flash drive with printing logo. The owner-manager anyway made an attempt to test market.

The deal was to sell at USD 2,500 for a lot size of 200 with the client's logo and free shipping. This had been put up on the owner-manager's eBay store for 60 days. There were about 80 visitors and no request. The lot size was then reduced to 50, with the product price reduced to only USD 550, which is more acceptable in eBay's price range. Finally, the item was pulled off from the shelf after two months. No sale had been made.

5.2.2 Scenic LED Candles

The owner-manager found that the sources of scenic candles from China were just simple decorative candles for restaurants. Competitors on eBay, however, were selling large size candles with paintings on it for special occasions such as Christmas. It was found out that the supply source of sellers on eBay were mainly local homemade candles. The owner-manager simply could not find Chinese suppliers of homemade candles for the western markets. This project was suspended.

5.2.3 Chinese opera masks

The owner-manager had found a few suppliers from Taobao, the famous Chinese consumer-to-consumer marketplace. However, they did not have many choices nor their products look good. This project was suspended.

5.2.4 Silk embroidery

This is an idea that certain embroidery could be liked by western ladies. The owner-manager has not picked any specific item for considerations. More researches are needed.

5.2.5 i-Yo-Yo

This product was first noticed by the owner-manager in the 2010 Gifts and Premium Items Expo in Hong Kong. The same product range reappeared in 2011 Expo in Hong Kong. Though the unit price of this item is low, the minimum order quantity is high (5,000 pieces per model per color). Taobao marketplace did have some yo-yo models on sale for a small quantity order but the available choices were unattractive. This project was finally suspended.

To sum up, one of the most important competences to be successful to sell at online marketplaces is the knowledge of a suitable product, and the owner-manager of cqCharter possessed this competence. However, it took a fair amount of energy and time resource to find a good product. Working individually without team support put any smart manager in a difficult position.

5.3 e-Retailing Entrepreneurship – Strategic Option Three

The researcher advised the owner-manager of the captioned company that he may convince his senior managers on the concept of "business units". This is to use the company available reserved fund in starting up the second, the third and the fourth e-commerce retail stores, with each store possessing a general product mix as well as a specialized line of commodity. With the anticipated success of the second e-retail store, change shall prevail. This will become the much needed pushing force for getting on the right track of sustainable development.

This alternative requires the support of other operations managers in the team. If this can be done, this alternative is likely to work. The owner-manager has kept his team people in the picture by communicating the progress of Strategic Option Two to them during the regular company meetings. This serves two purposes. First, it will give the operations managers a sense of urgency for the change. Second, it will remind the managers the available option of starting up a second e-commerce retail outlet. If they "buy" the plan and are getting confident with the change, it will be the best way to "soft sell" the owner-manager's team into this development path.

In the end, Strategic Option Three was gradually gaining weight over Strategic Option Two during the three-year transformation period of cqCharter from 2010 to 2013. By December 2013, the main store of cqCharter has a positive rating score of 6086, as compared to the score of 1494 in December 2010. This represented a 50% increase in the growth rate of past three years when compared to the early years of the business. Currently, the subject company has four consumer-to-consumer stores in operations, while the fifth one will go online in January 2014.

6. Learning and Insights

Based on the owner-manager's valuable sharing of his experiences and examples in the consumer-to-consumer e-retailing business, the critical success factors (CSF) lie on: Consumers' living ways, life style, quality of life, service quality, inspiration, innovation, creativity, curiosity, design thinking, execution. Innovation is very important; Innovation can be expressed as the product effect of the Creativity and Execution; Curiosity can be expressed in terms of the Curiosity Quotient.

Product: It is important that the seller should have a good knowledge of the product and also a comparative advantage in acquiring and selling such products. First of all, the buyers are usually very knowledgeable on the product. Therefore, if the seller does not know his product, he is deemed to fail. On the other hand, the seller has to find an entry point, such as a general supplier who can support him to get started. Then, he should be able to find a distributor or a manufacturer who can give him a more competitive price.

Service: The services have to be excellent because eBay allows buyers to comment on the service. A poor buyer comment could discourage a buyer. Some examples on good services are: reply client's question with solutions and actions within one day, make sure customers always feel the seller on their side, use standardized letters with a courteous tone in addressing receipt of fund upon the close of a deal, use a further letter in informing shipment.

Guarantee: A good guarantee is important because a buyer would be scare off on losing money. Some of the guarantees are: 60 day "no question asked" return policy, guaranteed responsibilities for postal problem, communications.

Product photos should be taken from real products and preferably in a real life setting. Also photos should illustrate the size and different angles of the product. The unambiguous photo description, together with the content description, will help to reduce the gap between buyer's visualization and his/her impression when he/she receives the real item.

Product Design and Quality: The product should be such a way that it is not easy to find from the local retail shops. The product design should be unique and attractive. The quality has to be good because the buyer could complain at eBay. The seller has to deal with enquiries and complaints promptly and properly so as to prevent losing credit and reputation.

Price: Price preferably at USD 60 or under so that eBay consumers could easily buy it at impulse. This is the price the buyer can take a risk and do not bother a loss of it. It is because the buy does not have a chance to feel and touch the product like seeing it in a retail shop.

The 5 Key Dimensions of e-Service Quality:

The five key dimensions for the e-Service quality to perform are namely, reliability, responsiveness, assurance, empathy, and tangibles.

Reliability: Customers want a reliable seller. They find information about the seller's past sales record from eBay. If the seller received adverse feedback on its reliability in previous sales, prospective buyers will not buy from the seller. The concerned issues are dependability, delivering on promises, accuracy, consistency.

Responsiveness in addressing requests: eBay has template on how to answer customer requests politely and detail so that the buyer would not feel frustrated. The concerned issues are promptness, helpfulness.

Assurance on delivery: The seller has to assure the buyer when the goods would be arrived and carries out intermittent communications to keep the buyer informed about the progress. For seasonal gifts and birthdays, the seller has to keep promise on the arrival date. The concerned issues are competence, courtesy, credibility, security.

Empathy: This is to be patient with the buyers when they are communicating their dissatisfaction, and to be prompt in actions to solve their problems. The concerned issues are physical evidence.

Tangibles: The services have to be tangible. For example, good control of delivery date and the communications on shipment. Also return with refund and no questions asked would be a good tangible service.

The Importance of the e-Service Quality:

E-Service Quality: eBay requests excellent service, such as prompt and polite response to all buyer's request. Quick shipments and guarantees are important too. eBay allows customers to write about their complaints or appreciations on the service. Therefore complaints will hurt the business greatly. Excellent operation management is needed for eBay business. For example, giving daily reply to enquiries and registering the tracking number on the transaction record are typical daily chores and should be complete on time.

It is equally important to bridge the e-service gaps. Zeithaml et al. (2002) have identified information gap, design gap, communication gap and fulfillment gap as the four major gaps in improving e-service quality (Figure 3). These gaps can be closed by learning what customers expect, putting customer requirements into website design, and improved internal communication between website designers and marketers. In the end, customers will find great online purchase experience with an easy-to-use website and problem solving with one click of the button, which are the e-service quality they expect.

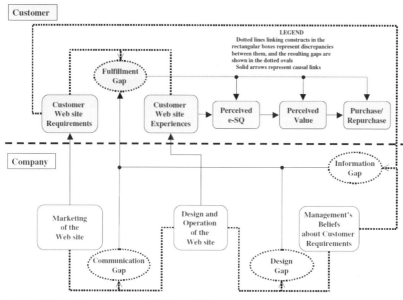

Figure 3: Schematic for e-Service Quality Upgrading and Improving

Source: Zeithaml, Parasuraman and Malhotra 2012

Applications of Analytics:

Last but not least, never ignore statistical tools and analytics. eBay offers quite a number of electronic applications to help you read your business performance. Most of these tools are free, and sometimes because of this their values are undermined.

As a failure example, the owner-manager revealed that he had chosen to ignore these numbers and charts and rankings in his business operations until the very recent 12 months. This was because his operational staffs were not familiar to read those numbers, and business had been at a comfortable good level to allow doing nothing against these statistics.

This attitude changed over the past 12 months as business went down, and it was proved the analytic tools were extremely useful in helping the consumer-to-consumer sellers to maximize their efforts in sales, promotion and in offering package deals with the right product category to the right customer segments and in the right place (country).

This effectively contributed to a turnaround in business volumes and trends over the final quarter of 2013. Two examples of Listing Analytics page is demonstrated in Figure 4 and Figure 5.

Figure 4: eBay Web Info of Analytics Listing on Camera Ranking

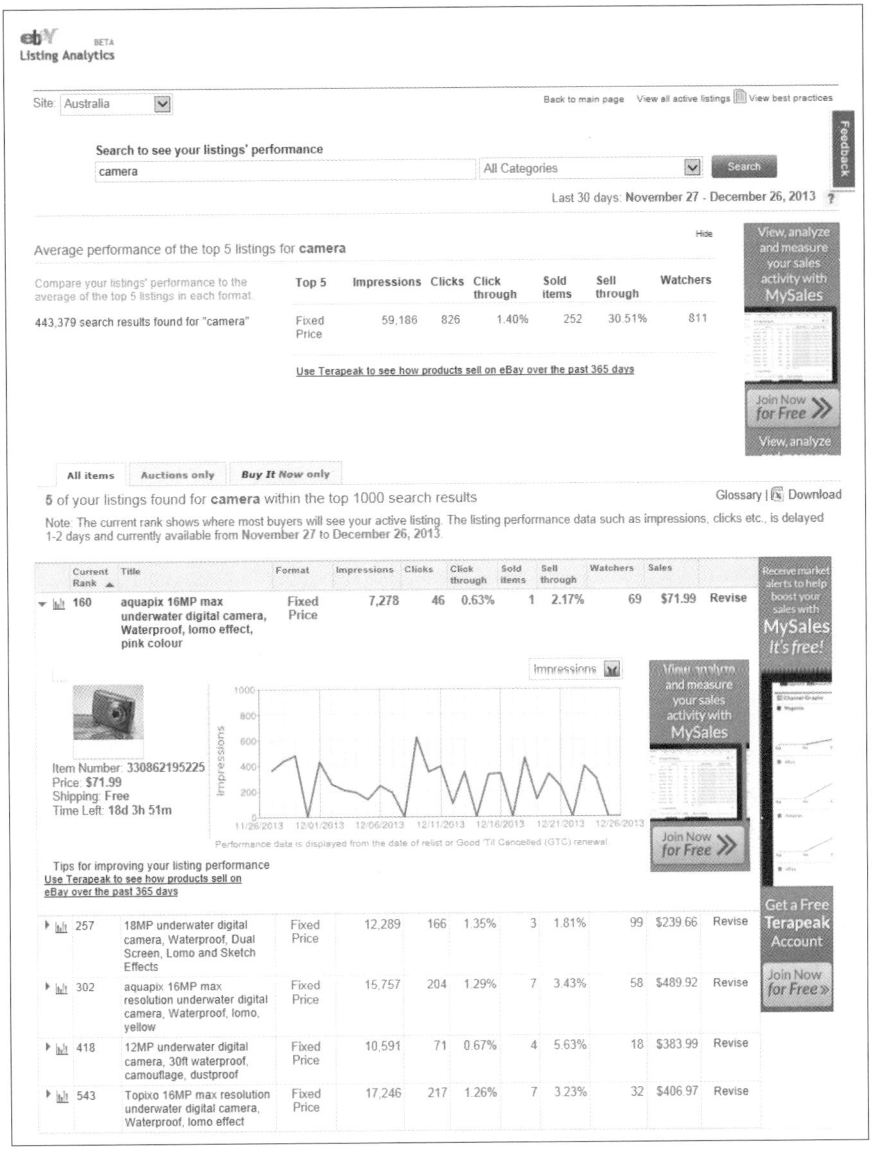

Source: eBay Listing Analytics page, accessed 30 December 2013 (compliments of Charter Lift Limited)

Figure 5: eBay Analytics Listing Data on Camera Ranking

Source: eBay Listing Analytics page, accessed 30 December 2013 (compliments of Charter Lift Limited)

7. Conclusion

This case study action research, together with the exclusive business information released by the owner-manager's company, are representative of the current consumer-to-consumer e-commerce e-retailing sector. It filled the gaps of the consumer-to-consumer e-commerce e-retailing field in a number of strategic and operational issues, and sheds light on how e-commerce business can be performed in retailing and wholesale under the current macro environment for mass entrepreneurship.

The information developed in this research will be helpful for anyone who wants to know more about e-business in online marketplaces.

Consumer-to-consumer e-retailing on e-Commerce platform is easy to start and go in for its low entry barrier, but it is harder to excel as most people would think.

Firstly, the seller has to possess very good product knowledge as well as a good sense of the market trend. This will enable him or her to find the saleable, fashionable products for the smart and knowledgeable buyers.

Secondly, the seller needs to have good promotional and service knowledges, skills, and techniques. It is because competition is global, severe, and keen. For instance, eBay allows previous customers to comment on the seller's service. Any customer's negative feedback could have a significant impact on the e-business e-retailing sales. It can even force a seller out of business. This means that excellent services and e-service have to be maintained at all times, to all customers, and under all circumstances. E-service quality acts as the deciding factor of a company's existence, profitability, expansibility, and sustainable growth.

Thirdly, the issues on inspiration, innovation, quality, life style, and design thinking are of significant to e-business e-retailing entrepreneurship. Inspiration enlights innovation; Innovation promotes quality; Quality changes life style and life; Life touches inspiration. Design thinking is important to empathize, define, ideate, prototype, and test.

An entrepreneur needs to keep on identifying his next set of opportunities available and the changes in external environment, and selects the most suitable ones for implementation. This research serves to ignite the insights and imagination of current and potential entrepreneurs, and it is hoped that it can be referenced to contribute in small measures on their future e-business and e-retailing mass entrepreneurship developments.

Acknowledgment

I would like to express my deepest gratitude to the owners and managers of Charter Lift Limited, for their time energy, and selfless contribution to this research.

I would like to thank Professor Dr. Eric Y. TAO for his advice and guidance, as well as his time and efforts in peer reviewed this paper. Professor Dr. TAO is a Full Professor and Founder Head teaching and conducting research in computer science, data analysis, innovation and entrepreneurship, e-learning and globalization, etc at the School of Computing and Design, the California State University at Monterey Bay (CSUMB), U.S.A. Professor TAO is also Director at the Institute for Innovation and Economic Development (iiED) CSUMB, serving the Greater Monterey Bay region by providing a venue for innovative ideas, the commercial transfer of research and technology (R&T), and an investors and mentors network for start-ups and entrepreneurs. Prof TAO is a board member of WASC, CALMAT, and Pacific Grove Museum of Natural History, Leadership Monterey Peninsula. Professor TAO is appointed as a Digital Technology Advisor for the National Health Research Institute of the Republic of China.

Several other friends who declined to be name had helped in a number of practical issues beyond research, patiently proof read my writing and did the final touch for illustrations and graphics.

Special thanks go to the organizing committee of ANQ Congress 2016, the Hong Kong Society for Quality (HKSQ), and the conference organizer host's administration of the Far Eastern Federal University (FEFU), Vladivostok, Russia for their assistance in the administrative work in the submission of this paper. My research would not have been possible without their helps.

References

Altshuller, G. (2008), "The Innovation Algorithm – TRIZ, Systematic Innovation and Technical Creativity"

Bigne, Enrique.(2005), The Impact of Internet User Shopping Patterns and Demographics on Consumer Mobile Buying

Box, G.E.P. and Woodall, W.H. (2012), "Innovation, Quality, Engineering, and Statistics", Quality Engineering 24, 20-29

CNN, (2007) "Total online sales expected to grow 19% to $174.5B in 2007". CNN. 2007-05-14. http://money.cnn.com/2007/05/14/news/economy/online_retailing/. Retrieved 2008-12-24.

eBay Listing Analytics, (2013). [online]

Available at: http://app.listinganalytics.com/?pg=report&keyword=camera&categoryId=0&globalId=EBAY-AU

[Accessed 30 December 2013].

E-Commerce News, (2010) "Keeping It Real for Cross-Border Online Shoppers". http://www.ecommercetimes.com/story/69105.html. Retrieved 2010-02-10.

Falk, Louis K. et al (2005) "E-Commerce and Consumer's Expectations: What Makes a Website Work." Journal of Website Promotion, 1(1) (65–75)

Jarvenpaa, S. L., & Todd, P. A. (1997). Consumer reactions to electronic shopping on the world wide web. International Journal of Electronic Commerce, 1, 59–88.

Johnson, R.A. et al (1996) Management, Systems, and Society: An Introduction. Calif.: Goodyear

Kao, William (2016), Innovation, Creativity And Invention, University of California at Berkeley

Lewin, Kurt (1988). Group Decision and Social Change. New York: Holt, Rinehart and Winston. pp. 201.

Palmer, Kimberly (2007) News & World Report.

Parasuraman, A., Zeithaml, V.A., & Berry, L.L. (1985). A conceptual model of service quality and its implications for future research. Journal of Marketing, 49, 41-50.

Parasuraman, A., Zeithaml, V.A., & Berry, L.L. (1988). SERVQUAL: A multiple-item scale for measuring consumer perceptions of service quality. Journal of Retailing, 64, 12-40.

Peterson, R. A. et al (1997). Exploring the implications of the Internet for consumer marketing. Journal of the Academy of Marketing Science, 25, 329–346. Porter, Michael (2010), Three Generic Strategies [online]; Available from: http://en.wikipedia.org/wiki/Porter_generic_strategies

Porter, Michael (2010); Five Forces Strategies; Available from: http://hk.search.yahoo.com/search?p=Michael+Porter&fr=FP-tab-web-t&fr2=tab-web&ei=UTF-8&myip=61.10.161.136&tmpl=&meta=rst%3Dhk

Ram L. Kumar et al. (2004), "User interface features influencing overall ease of use and personalization", Information & Management 41 (2004): 289–302.

Steve Elliot and Sue Fowell (2000), "Expectations versus reality: a snapshot of consumer experiences with Internet retailing", International Journal of Information Management 20 (2000): 323–336.

Tsung, Fugee (2016), Smiling Curve of Design and Innovation Values, HKUST Quality And Data Analytics Lab

Zeithaml, V.A., Parasuraman, A., and Malhotra, A., (2002), Service Quality Delivery through Web Sites: A Critical Review of Extant Knowledge, Journal of the Academy of Marketing Science, Vol. 30 No.4, pp. 362-375.

Authors' Biographical Notes:

Po-Shing LIU, Ph.D. DBA, is a UQÀM Postdoctoral Fellow at University of Quebec at Montreal, Canada and is presently a postdoctoral researcher at California, U.S.A. Dr. LIU is a Registered Professional Engineering Manager and is currently the advisory Past Chairman Committee Member at the Electronics and Communications Committee of the Institution of Engineering and Technology Hong Kong (IET-HK), the elected Chairman (Promotion and Public Relations) at the Hong Kong Institution of Incorporated Engineers (HKIIE), the elected Chairman of the Chinese Institute of Electronics (Hong Kong Branch) (CIE Hong Kong), and a Committee Member of the Council of Hong Kong Professional Associations (COPA). Dr. LIU is a Fellow of IET (UK), a fellow of the Hong Kong Institute of Directors (FHKIoD), and a Hon. Fellow of HKIIE. Dr. LIU is a Think Tank of the Dashun Foundation Hong Kong, an quality expert of the Shenzhen Association for Quality P.R. China, and also a senior member of the Institute of Industrial Engineers (USA), a senior member of American Society for Quality (USA), a member of IEEE (USA), a member of ACM (USA), a member of the Society of Motion Picture and Television Engineers (USA), a member of The Optical Society (USA), an elected member of the Hong Kong Professionals and Senior Executives Association (HKPASEA). (Email: eddieliu88@gmail.com)

Philip K.I. CHAN, Ph.D., is a Guest Professor at Peking University, Beijing Normal University, and Tianjin University, P.R. China; is a registered professional engineer of various national engineering institutions; and is a Postdoctoral in Management Sciences and Engineering at Tianjin University, P.R. China and a Postdoctoral Fellow at University of Quebec at Montreal, Canada. Ir Prof. Dr. CHAN is a Chartered Engineer, a Chartered Surveyor, a Chartered IT Professional, a Chartered Marketer, a Chartered Manager as well as an accredited Mediator and an Arbitrator in Hong Kong and is a registered Arbitrator of mainland China's Arbitration Commissions. Ir Prof. Dr. CHAN is a Fellow of several engineering, professional institutions and learned societies (HKIE, IET, SOE, BCS, CMI, CIM, CILT, HKIoD, HKQMA). Ir Prof. Dr. CHAN is elected as a Member of the Academy of Experts (MAE), UK and is appointed as an Advisor at the Institute for Innovation and Economic Development (iiED) CSUMB, U.S.A., and is an Post Doctoral Programs Advisor.

Prof. Dr. LAM Wai Pan, Wilson

Honorary Doctorate in Law, MIPostDocA, Adjunct Professor of SABI University on ADR,

Guest Professor of Peking University, Yuenpei Business School and Academy of Experts, B.A.(A. S.), B.Arch (Dist),

MSc. Const. & Econ., MBA (1st Class Hon), MSc Finance (1st Class Hon), MSc. Marketing (1st Class Hon), LLM (Arb. & DR),

HKIA, RIBA, Registered Architect (HK), AP(Architect), PMP®, MCIArb, FHKIArb, BEAM Pro. (NB, EB & Int.), FAIADR, FICM-MCN,

APMG-Agile P, Prince 2® Practitioner, ITIL® Expert, ITIL MP, CIMP, PECB ISO 31000 Risk Manager, PECB/ISO 27001 Lead Auditor

Cert. Big Data Scientist, Cert. Big Data Consultant, Cert. Blockchain Architect, Cert. Scrum Master, CIC Cert. BIM Manager,

Arbitrator, Adjudicator, Accredited Mediator, CEDR Trained Assessor, Expert Witness, CIC Cert. NEC 4 Project Manager.

Professor Dr. Lam Wai Pan, Wilson is a registered architect and authorized person in Hong Kong. He is a member of the Hong Kong Institute of Architects and corporate member of the Royal Institute of British Architects. He has more than 25 years of experience in architectural practice, building construction and project management. He had worked in both private and public sector to cover a wide range of commercial and residential developments; prestigious offices and commercial developments; urban design and redevelopments; institutional buildings such as major hospital developments; major educational campus developments; hospital A&A works; schools and international schools; infrastructure; as well as public open space/promenade and government complex.

He is a Project Management Professional, BEAM Professional, Prince 2 Practitioner and Agile Practitioner and is experienced in various types of project management strategies. He is now working at the Architectural Services Department. He also served the Hong Kong Institute of Architects (HKIA) in Urban Planning and Design Committee, Building and Lands Committee, Contract and Dispute Resolution Committee (CDRC) and Belt and Road and Great Bay Area Task Group in preceding years. He is a Board Member of Board of Practices and Board of Mainland Affairs of HKIA and Chairman of Contract and Dispute Resolution Committee since 2019. Besides, he had served in the Legal Affairs Committee of the Architects Registration Board (2016-2020) and Committee Member of the Chartered Body, Association of Project Management (Greater Bay Area Branch) from 2018 to 2022.

On internet, information technology services management and big data related developments, he possesses diverse experiences in IT-related projects in Architectural Services Department since 2007. He is humbly a Prince 2 Practitioner, Certified Big Data Science Professional, Big Data Scientist, Big Data Consultant as well as an Information Technology Infrastructure Library (ITIL) Expert and Certified Internet Marketing Practitioner (CIMP). He is a qualified Agile Practitioner and a globally accredited PECB ISO 31000 Risk Manager and PECB ISO/IEC 27001 Lead Auditor. He is also certified globally as Information Technology Infrastructure Managing Professionals (ITIL MP), Certified Blockchain Architect, Certified Scrum Master in 2020. He was accredited as the Construction Industry Council (CIC) Building Information Modelling (BIM) Manager and CIC Certified NEC 4: ECC Project Manager.

He was appointed as Adjunct Professor on Alternative Dispute Resolution by the SABI University and Guest Professor for Project Management by the Peking University – Yuenpei Business School in 2020. He was also appointed as Adjunct Professor by the Academy of International Dispute Resolution and Professional Negotiation in June 2021 for a term of 3 years.

Unlocking the Benefits of Alternative Dispute Resolution in Enhancement of Change Management for Construction Projects in Hong Kong Special Administrative Region from the Project Management's Perspective

Lam Wai Pan, Wilson[1,2,3*] and Philip K.I. Chan[1,2,3,4]

[1]Stanford University, Stanford, California, USA 94305
[2]University of California, Berkeley, Berkley, CA, USA 94720-1234
[3]California State University, Monterey Bay, 100 Campus Center, Seaside, CA, USA 93955-8001
[4]Stanford, Berkeley, Monterey Postdoc Advisor
*wilson.wp.lam@hotmail.com and cw9888@gmail.com (852) 68311488

Abstract

Construction industry contributed 5.2% of Gross Domestic Product ("GDP") of the HKSAR's economy in 2016 that was 73% higher than 2006. The HKSAR Government estimated this sector employed 328,000 people in 2016 with target increase in number in 2018.

Effective project management is the cornerstone to successful construction project delivery. Managing changes due to design, project, site constraints, statutory requirements or defective works etc. are essential in cost, quality and programme controls. Conflicts among project stakeholders may become potential or actual disputes. This paper aims to identify and evaluate the benefits of using a spectrum of Alternative Dispute Resolution (ADR) to enhance change management. It covers conflict resolution via negotiation skills in early stage, mediation, statutory or contractual adjudication and finally arbitration to address various conflicts or disputes due to project changes. The paper investigated the construction industry standards and statutory framework, including HK

Arbitration Ordinance, HK Mediation Ordinance, proposed Security of Payment Legislation and project management mechanisms (e.g. PMBOK). Furthermore, it summarizes the ADR based upon literature review. These benefits are compared against open litigation in courts.

Survey questionnaires were sent to project stakeholders for empirical results. They are supplemented by in-depth interviews to experienced practitioners and case studies. Research findings would be analyzed. Conclusions on the benefits are evaluated on how they enhance effective change management are highlighted. Research limitations (e.g. additional cost of ADR) and practical recommendations to foster collaboration (e.g., Building Information Modelling (BIM) or New Engineering Construction (NEC) Contracts) will be discussed.

Keywords: Conflict Resolution, Change Management, Alternative Dispute Resolution, Arbitration/Mediation/Adjudication, Continuous Improvement.

1. Introduction

1.1 Construction sector as a key driver of Hong Kong's economy

Research Office of Legislative Council Secretariat (2019) estimated that the construction industry contributed 5.2% of Gross Domestic Product ("GDP") in 2016 and was 73% higher than the year 2006. Construction sector had employed 328,000 people in 2016 and would increase to 353,000 people in 2018. There would be 881 and 608 construction sites in the private and public sector respectively in 2018. At nominal prices, gross value of construction works increased from HK$ 99.6 billion in 2008 to HK$ 249.9 billion in 2017.

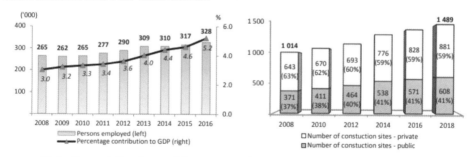

Fig 1 – Economic Significance of Construction Industry; Fig 2 – Number of Construction Sites

Source @ Research Office of the Legislative Secretariat, "Statistical Highlights ISSH07/18-19

The Office of the Government Economist, Financial Secretary of the HKSAR Government (2020) estimated that the construction sector represented 4.5% of H.K.'s GDP in 2018. It is closely related to the "Financing and insurance, real estate, professional and business services" which represented 30.1% of the HK economy's GDP in 2018. Construction sector employment represented 8.6% of the total work force in 2018. Effective project delivery of construction projects is essential to support the supply chain of land, housing, and infrastructure.

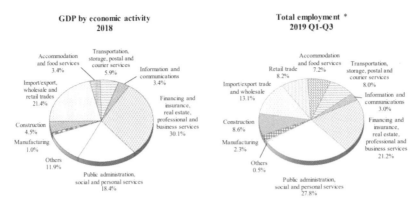

Fig 3 – GDP by Economic Activity 2018; Fig 4 – Total Employment 2019: Q1–Q3

Source @ Office of the Government Economist (2020), "2019 Economic Background and 2020 Prospects"

1.2 Problems statement

Conflicts and disputes are common that affect the performance and key performance indicators of construction projects. At Chapter 2 below, potential conflicts or disputes have resulted in essential problems for the construction industry. These problems include (i) core challenges facing the construction industry on high construction costs, delay and over-expenditure; (ii) complex construction ecosystem has derived different types of conflicts or disputes among stakeholders; (iii) improper change management could result in additional conflicts and potential disputes; and (iv) uncertainties or queries from construction stakeholders regarding the benefits of alternative dispute resolution (ADR) mechanism due to its confidential nature.

2. Literature review – The issues

2.1 Conflicts and disputes resulting at core challenges to the construction industry

In 2019, the Development Bureau (DevB) of the HKSAR Government and KPMG had published a report known as "Construction 2.0 – Time to Change". It highlighted various key challenges including high construction costs, delay and over-expenditure. Firstly, DevB (2019) concluded that average construction costs in Hong Kong was ranked the 3rd highest cities globally following New York and San Francisco. Between 2007 to 2017, construction tender price indexes for building contracts and civil engineering contracts had grown at a compound annual growth rate of 7.23% and 4.62% respectively. Secondly, from 2008 to 2017, additional funding above the approved project estimates were required in approximately 10% of the projects. Thirdly, a number of complex mega-projects had suffered from "underperformance in form of delays, cost overrun and/or quality failures in recent years". A few selected mega-projects are summarized in Table 1 below.

The Hong Kong Construction Industry Council ("CIC") (2015) had commented that the resolution of disputes could be expensive and time-consuming. CIC recommends addressing claims or potential claims as early as practicable for dispute prevention. If disputes are unavoidable, they should be managed constructively and collaboratively to obtain early or effective settlements while arbitration or litigation should be a last resort.

Table 1 – Performance of selected mega-projects in Hong Kong @Source and modified from: Development Bureau (2019), "Construction 2.0 – Time to Change" at pages 13-15.

	Cost		Timeline	
Project	Original APE	Revised APE	Original scheduled completion date	Latest scheduled completion date
Hong Kong-Zhuhai-Macao Bridge	HK$56.5bn	HK$70.8bn	End of 2016	Oct 2018
Guangzhou-Shenzhen-Hong Kong Express Rail Link	HK$69.6bn	HK$89.2bn	2015	Sept 2018
Shatin to Central Link	HK$79.8bn	HK$89.7bn	2018-2020	2019-2022
Liantang/Heung Yuen Wai Boundary Control Point	HK26.7bn	HK$35.4bn	End of 2018	Feb 2020

2.2 Conflicts and disputes unique in the construction industry ecosystem

DevB (2019) highlighted that the construction industry ecosystem "is a complex collection of public, private and other stakeholders' groups that either play a role in, or are impacted by, the day-to-day execution of the industry's activities." Fig. 5 below illustrates the ecosystem involving public or private sector, consultants, contractors, suppliers etc. Census and Statistics Department (2016) recorded that over 75% of construction work activity was less than HK$5 million. The unique construction industry ecosystem and pre-dominance of small to medium scale construction work activities including subcontracts or supplies contracts generate various forms of conflicts or disputes among different stakeholders.

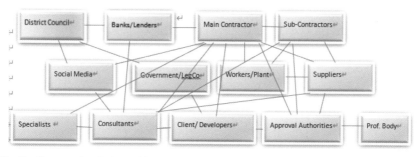

Fig. 5 – Construction Industry Ecosystem @Source and modified from: Development Bureau (2019), "Construction 2.0 – Time to Change" at page 9.

2.3 Improper change management resulting in conflicts and disputes

Grace Ellis (2021) summarized the latest statistics at the Digital Builder of the Autodesk Construction Cloud based upon KPMG International's research[1]. The global study illustrates that improper change management is common in the global context resulting in additional potential conflicts or disputes.

- · 1% reduction in construction costs can save society USD $100 billion globally [KPMG International: Building a technology advantage – Global Construction Survey 2016]
- · Over 50% of engineering and construction professionals report one or more underperforming projects in the previous years. [KPMG International]
- · Just 25% of projects came within 10% of their original deadlines in the past 3 years. [KPMG International]
- · Only 31% of all projects came within 10% of the budget in the past 3 years. [KPMG International]

2.4 Dispute avoidance and resolution promoted by the Construction Industry Council (CIC)

Conflicts and disputes as discussed in Section 2.1, 2.2 and 2.3 of Chapter 2 affect the quality performance of building and construction projects. After the appointment of the Construction Industry Review Committee (CIRC) by the Chief Executive of the HKSAR, CIC (2010) published a set of CIC Guidelines and concluded that the resolution of disputes could be expensive and time-consuming. CIC and CIRC recommended to address claims or potential claims as early as practicable for dispute prevention. If disputes are unavoidable, they should be managed "constructively and collaboratively to obtain early or effective settlements" while arbitration or litigation should be considered as a last resort.

5 types of Alternative Dispute Resolution (ADR) approaches were proposed to include: (i) Arbitration, (ii) Expert Determination, (iii) Adjudication, (iv) Independent Expert Certifier (IEC) and (v) Mediation. The CIC Guidelines paved the roadmaps for Statutory and Contractual Adjudication. However, the advantages or limitations for different forms of Dispute Avoidance and ADR Mechanisms had not been discussed or explained in the CIC Guidelines. The researcher (having more than 25 years of experiences as architect, authorized person and project management professional) observes that majority of construction stakeholders may not have sufficient awareness or knowledge on the merits of ADR. Unlike legal practitioners, they remain as special areas of expertise or confidential issues that construction stakeholders may not have access or knowledge about. Hence, they may often avoid or misunderstand dispute prevention and dispute resolution in construction projects.

---

1 See website: https://constructionblog.autodesk.com/construction-industry-statistics/

3. Objectives

A. This research's primary objective is to identify and evaluate the benefits of using a spectrum of Alternative Dispute Resolution (ADR) approaches to enhance change management for the construction industry in Hong Kong. It covers conflict resolution through negotiation skills for dispute avoidance as well as mediation, statutory or contractual adjudication and finally arbitration for dispute resolution due to project changes for construction projects in Hong Kong from project management perspectives.

B. Upon evaluation of the benefits of ADR approaches, the research further aims to discuss on limitations and recommendations taking into account the research findings, core challenges of the construction industry and the construction industry ecosystem.

4. Research Methodology

The paper firstly investigated the construction industry standards, international project management best practices and standard forms of construction contracts. Then it reviews the project management statutory framework, including HK Arbitration Ordinance, HK Mediation Ordinance, proposed Security of Payment Legislation and major project management mechanisms (e.g. PMBOK). Furthermore, it summarizes the ADR based upon literature review. These benefits are compared against open litigation in courts for project delivery. Survey questionnaires were sent to project managers and practitioners for empirical results. They are supplemented by in-depth interviews to experienced practitioners and case studies. Research findings would be analyzed based upon both quantitative and qualitative research findings.

5. Literature Review – Conflict resolution and negotiation skills for dispute avoidance

5.1 Industry Standards in Hong Kong

5.1.1 Overview

For small-scale building and construction projects (e.g. minor works, renovation works or alteration and addition), conflicts may appear at different project stages. They may be resolved by conflict resolution through negotiation skills. If conflicts cannot be resolved and resulted into disputes, they would normally be resolved by ADR approaches through mediation or arbitration. Where there is no provision for arbitration, disputes would need to be decided in courts by open litigation. For a medium to large-scale project, conflicts or disputes may appear at various work stages among different project stakeholders. In most circumstances, conflict resolution requiring negotiation skills is the essential approach for dispute avoidance and prevention. If such conflicts continues and result into disputes, they would be resolved by different approaches of ADR. The following diagram developed by the researcher aims to illustrate the various approaches to resolve conflicts or disputes.

Fig. 6 Use of Conflicts Resolution and ADR approaches for construction projects in Hong Kong.

5.1.2 CIC Guidelines on Dispute Resolution (September 2010)

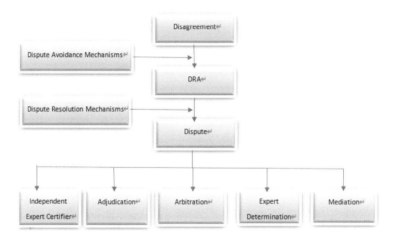

Fig. 7 Flowchart to illustrate different forms of Dispute Avoidance and Dispute Resolution Mechanisms proposed by the Construction Industry Council in 2010

The above diagram summarizes the proposed framework of disputes avoidance and dispute resolution approaches originated from CIC's recommendations in their "Guidelines on Dispute Resolution" published in 2010. CIC proposed a dispute resolution flowchart for Public Works to consider with different ADR methods.

5.1.3 CIC Reference Materials for Alternative Dispute Resolution (August 2015)

CIC further published the "Reference Materials for Application of Dispute Resolution in Construction Contracts" in 2015. CIC (2015) recommended the stakeholders to adopt the Reference Materials and to adhere to such standards or procedures (i.e. including the model rules and procedures). CIC recommended to consider using a "constructive and collaborative approach for ADR" instead of the traditional approach of arbitration or litigation as a final solution to improve project performance. The ADR methods include: (i) mediation; (ii) adjudication; (iii) independent expert certifier review; (iv) expert determination; and (v) short form arbitration.

5.2 International Standards

5.2.1 Project Management Institute (PMI)

For dispute avoidance, the commonly used methods are conflict resolution and negotiation skills. It may be taken up by the Client's project manager, or project consultants (e.g. architect, engineer or surveyor) or Contractor's representative (e.g. construction manager). Throughout this Postdoctoral Research, it is found that the construction industry is generally lack of international standards or references for conflict resolution or negotiation skills for dispute avoidance. Hence, the researcher has referred to the international standards promoted by the PMI that are globally accepted and adopted in other information technology, business or financial sector projects in Hong Kong. Project Management Professional (PMP)® is widely adopted in many countries as the "Gold Standard for Project Management". There are over 910,000 certified PMP worldwide in all major countries. It is considered as the "gold standard" for project management. PMI (2017) recommends three key approaches, namely the Predictive (waterfall), Agile and Hybrid approaches to cover the project leadership skills necessary for project management.

5.2.2 PMBOK and Conflict Resolution

According to **"PMBOK Guide: A Guide to the Project Management Body of Knowledge (6th Edition)"** (referred as **"PMBOK Guide (2017)"**), conflict management is unavoidable in a project environment. Conflicts management and resolution skills are essential qualities and skillsets for leadership in project management:

- Being a visionary (e.g. product innovation, goals, objectives etc.)
- Optimistic and positive.
- Collaborative.
- Managing relations & conflicts – building trust, consensus, negotiation and conflict resolution.
- Communication skills.
- Being respectful, friendly, kind, honest, trust-worthy, loyal and ethical.
- Exhibit integrity and being culturally sensitive, courageous, problem-solving and decisive.
- Being a life-time learner and to be result and actions-oriented.

5.2.3 Conflict Management and Negotiation Skills

PMI (2017) stresses that conflict resolution and negotiation skills as vital skillsets and are useful in the following processes to link up process inputs and outputs:

- <u>Acquire Resources</u> – Negotiation skills are essential techniques and tools.
- <u>Develop Team</u> – Team development has the key benefits of improving team spirt, improved interpersonal skills and competences, motivation and improved project performance.

Conflict management and negotiation are the essential skills for achieving benefits of team development.

- Conduct Procurements – Negotiation is an essential for the process that ensure selection of qualified supplier or seller and to implement effective and legally binding agreement

- Manage Stakeholder Engagement – Conflict management and negotiation skills are essential to the process. It refers to communicating and coordinating with stakeholders to meet their needs, expectations, address to project issues in priority and foster stakeholder involvement

5.3 Change management – International and domestic standards

5.3.1 International standards

PMI (2017) recommended that change management plan is an essential component of the project management plan. Change management plan gives directions for managing the change control process and documents roles and responsibilities of the change control board (CCB). It is an essential input for the processes below:

- Perform Integrated Change Control – PMI (2017) defines as a process to review all changes requests, approval and management of changes to deliverables, project documents, project management plan and communication towards final decision-making. Change requests due to design, project requirements, site constraints, user-requests and statutory requirements should be monitored and managed.

- Control Scope – PMI (2017) defines as a process to monitor project scope, management of change requests according to the scope baseline. For construction projects, any unjustified change of scope shall not be permitted for funding purposes and fiducial duties.

- Control Procurements – PMI (2017) defines as a process for management of procurement stakeholders' relationships; monitor of contract performance; allowance for necessary changes or corrections and finally the effective closing out of contracts. For construction projects, it ensures sellers and buyers deliver outputs effectively according to contract and legal requirements.

- Manage Stakeholder Engagement – PMI (2017) defines as a process to ensure effective communication and management of stakeholders to meet their needs, expectations, concerns, technical issues and foster stakeholder engagement through techniques including conflict management and negotiation skills. Change management plan is important due to the unique construction stakeholders' relationships.

5.3.2 Domestic standards on variations

Variations in construction project may arise due to changes of design, project requirements, site constraints, users-requests as well as statutory requirements. For instance, the Civil Engineering and Development Department (CEDD) requires the following control and procedures for variations under CEDD TC No. 7/2019:

- Measures for avoiding variation – by project team and consultants during design stage.
- Measures for avoiding variation – overall financial implications to be assessed for new requirements.
- Consideration of the approved scope of project, contract sum and approved project estimate etc.
- Consideration before ordering variations – possibility of other more economical or efficient options.
- Consideration before ordering variations – financial implications including contractual claims.
- Authorities for approving variations and increase in contract sum shall be followed.

5.3.3 Domestic standards on standard forms of building/construction contracts

The Hong Kong Institute of Architects (HKIA) had presented to three hundred professionals including architects, surveyors, engineers, project managers and legal practitioners regarding time and cost related claims for building and construction projects. Four experts had investigated from different perspectives including: (i) quantity surveyor, arbitrator and solicitor; (ii) civil engineer and solicitor; (iii) architect and real estate developers; and (iv) architect, authorized person and project management professional. The essential points as supported by the Contract Dispute Resolution Committee of the HKIA are summarized below.

(A) <u>Time and cost related claims</u>

Cheung (2020) summarized three categories of delay that may affect the assessment of extension of time under the HKIA/HKIS/HKICM Standard Form of Building Contract (2005 Edition):

- <u>Employer's fault</u> – Architect's instruction (or Engineer's instruction), late instruction, late nomination of nominated sub-contractor, delay caused by separate contractor or statutory undertaker or utility company, delay in supply of materials or equipment by the Employer, late possession of site, special circumstances, act of prevention, breach of contract or default of Employer.

- <u>Neutral events</u> – Force majeure, inclement weather, typhoon warning signal no. 8, black rainstorm raining, excepted risks (e.g. war, acts of terrorists, riots etc.), special perils (e.g. fire, lighting, explosion).

- <u>Contractor's default</u> – No extension of time.

(B) <u>Impartial role of the contract administrator (architect, engineer or surveyor)</u>

- Contract administrator as independent certifier with duty to act fairly and independently.
- Abdication of responsibility by the Employer's representative.[2]
- Potential liability in tort if the contract administrator fails to discharge his duties.[3]

---

2 *Chun Wo Building Construction Ltd. v Metta Resources Limited* (2016) 2 HKLRD.

3 *John Mowlem & Co. PLC v Eagle Star Insurance Co. Ltd.& Others* (No. 2) [1995] 62 BLR 126.

· HKIA Code of Professional Conduct – A member shall act impartially in all cases in which he is acting between the parties and shall interpret the conditions of a building contract with fairness.[4]

(C) Time and cost related claims relating to 11 types of standard forms of building/construction contracts

Lam (2020) analyzed 4 major types of standard forms of building contracts with different contractual relations among employer, main contractor, nominated sub-contractors and nominated suppliers. These standard forms of contracts would give different provisions in contracts for contract administrators and project managers to consider the adverse impacts of COVID-19. Lo (2020) extended the analysis to 11 types of standard forms of contracts and assess the relevant possibilities of claims on extension of time (E.O.T.) and additional payment due to adverse effects of COVID-19 for construction contracts. They cover all major standard forms of contract in the public and private sector in Hong Kong. They are summarized in Tables 2 and 3 below.

Table 2 & 3 – Summary of E.O.T. and additional payment due to COVID-19 for construction contracts: Source @ Deacons (2020), "Presentation slides by Mr. Stanley Lo for HKIA-CDRC CPD Seminar – Legal and Practical Aspects of Time Related Claims and Unforeseen Risks (including COVID-19)".

Contract	2005 Private Form	1999 Government Form	NEC 4	HK Government NEC 3	FIDIC (Silver, Red, Yellow, MRD, Harmonized)	MTR (SCL x XRL)
Contractual Grounds	Cl. 25.1	Cl. 50(1)	Cl. 60.1(19)	Cl. 60.1(19)	Cl. 19.1	Cl. 22.6 Excepted Risks)
Notice	Cl. 25.1 & 25.2 (28 days)	C. 50(1) (28 days)	Cl. 61.3 (8 weeks)	Cl. 61.3 (8 weeks)	Cl. 19.2 (14 days)	Cl. 68.3 & Cl. 82.1 (28 days for cost & time)
Time	Depend upon if coronavirus is construed as special circumstances		OK	OK	OK	Unlikely since pandemic is not recognized as Excepted Risks
Cost	X	X	OK	X	X	X

---

4 Hong Kong Institute of Architects (2008), "The Code of Professional Conduct", 4th Revision, Rule 1.3 of the Principle 1.

Contract	HKHA	AA	WKCD	CIC Standard Domestic Sub-Contract	HKCA Standard Form of Domestic Sub-Contract
Contractual Grounds	C. 8.4(2((b)(xiii) (special circumstances)	Cl. 20.6 (Excepted Risks) Cl. 74.4 (Special Risks)	Cl. 21.5 (Excepted Risks) Cl. 77 (Special Risks)	Item 34 of Sub-Contract Particulars	Cl. 6.2 Cl. 10
Notice	Cl. 8.4(2)(a) (28 days) for Time	Cl. 44.1 (14 days) for time Cl. 56.1(a) (28 days) for cost	Cl. 45.1 (28 days) for time Cl. 58.1(a) (21 days) for cost	Cl. 4.3(a) (14 days)	Cl. 6.2 (21 days) Cl. 10
Time	Dependent upon if coronavirus is construed as special circumstance	Unlikely as epidemic is not recognized as an Excepted Risk or Special Risk	Unlikely as epidemic is not recognized as an Excepted Risk or Special Risk	OK	Dependent upon if compensation is available under the underlying main contract.
Cost	X	X	X	X	

(D) Employer's perspective

Yeung (2020) analyzed from the real estate developer's perspective. As a Chinese businessman, time and cost related claims for building and contracts can be reviewed from 3 perspectives, including (i) legal perspective; (ii) reasonableness from commercial perspective; and (iii) relationships of the parties and stakeholders.

6. Literature Review – Major Alternative Dispute Resolution (ADR) approaches

6.1 Arbitration

6.1.1 Arbitration Ordinance (Cap. 609)

S.2(1) of the Arbitration Ordinance (Cap. 609) (referred as "HKAO") defines arbitration as "any arbitration, whether or not administered by a permanent arbitral institution". The Hong Kong International Arbitration Centre (HKIAC) defines arbitration as "a consensual dispute resolution process based on the parties' agreement to submit their disputes for resolution to an arbitral tribunal usually composed, of one or three independent arbitrators appointed by or on behalf of the parties." Under the current HKAO, the two distinct domestic and international regimes were abolished and governed by the UNCITRAL Model Law[5] with modifications as incorporated into the HKAO. The HKAO has fourteen parts that correlates to the sequence of arbitration and the Model Law to make it

---

5 Model Law on the International Commercial Arbitration (1985), with amendments as adopted in 2006, of the United Nations Commission on International Trade Law ("UNCITRAL Model Law").

user-friendly and conform to international arbitration practices.

6.1.2 Special provisions for building and construction industry

Schedule 2 of the HKAO was developed to cater for the needs of construction industry in Hong Kong. The "Opt-in provisions" of Schedule 2 in pursuant to Section 99 of the HKAO have the following features for Hong Kong to use the method of "domestic arbitration" commonly adopted by building and construction sector.

- · Para. 1 of Schedule 2 – Dispute to be submitted to a sole arbitrator.
- · Para. 2 of Schedule 2 – Consolidations of arbitrations by the Court of HKSAR.
- · Para. 3 of Schedule 2 – Determination of preliminary points of law by the Court of HKSAR.
- · Para. 4 & 7 of Schedule 2 – Challenges to arbitration awards only permitted on grounds of serious irregularity and those expressly provided under the New York Convention.
- · Para. 5, 6 & 7 of Schedule 2 – Appeals to the Courts of HKSAR on questions of laws in connection of arbitration awards shall be in accordance with Hong Kong Arbitration Ordinance (Cap. 609).

Section 100 of the HKAO allowed for these major "Opt-in provisions" in Schedule 2 to be automatically opt-in for all arbitration agreements or arbitration clauses which expressly provide for "domestic arbitration" instead of "international arbitration" (as default under the UNCITRAL Model Law and the HKAO) which were entered into contract before the date the HKAO came into effect or within 6 years from the date of the Ordinance came into effect (i.e. until 31 May 2017). This was mainly based upon the requests from the construction industry during the drafting and development of the HKAO to replace the old Arbitration Ordinance (Cap. 341).

There is no automatic opt-in by specifying "domestic arbitration" after 1 June 2017. From this date onwards, parties to the arbitration and contract must expressly specify which provisions under Schedule 2 of the HKAO they may wish to opt-in and provides for in their arbitration agreement or arbitration clause. If the Main Contract has entered into domestic arbitration before the date of 31 May 2017, the sub-contracts and nominated sub-contracts would also follow this domestic approach even if they were entered into at a later date. Schedule 2 of the HKAO would apply to these sub-contracts if the Main Contract has applied and expressly provided for.

6.1.3 Advantages of arbitration approach

Construction arbitration is confidential. Throughout the literature review, there was a lack of confirmed or published benefits of construction arbitration including sources from major construction industry stakeholders from "project management's perspectives."

The literature review had identified the following advantages under the HKAO or the UNCITRAL Model Law. The sources of these advantages include the Department of Justice for the HKSAR Government, HKIAC and World Intellectual Property Organization (WIPO). These advantages would be investigated by questionnaire, interviews and case studies as discussed in Chapters 7 and 8.

- Arbitration is consensual – Parties can agree in the formation of contract to use arbitration to resolve disputes by arbitration agreement or arbitration clause.

- Choice of arbitrator or arbitral tribunal – Parties are free to decide on the choice of arbitral tribunal including his or her background, professional qualifications, experience or language etc.

- Seat of arbitration – It determines the procedural law governing the arbitration process and procedures (e.g. Hong Kong law or UK law etc.). It is distinguished from the place of hearing or venue of hearing. If the seat of arbitration is in H.K., it also determines the arbitration award shall be published in H.K.

- Choice of language and venue – Parties are free to choose the language to be used and venue of hearing.

- Arbitrator is neutral, impartial and independent – Arbitrators are considered as private judges to be chosen and agreed by the parties or to be nominated and appointed by arbitration institution.

- Arbitration procedures are flexible – may not need to follow Court procedures.

- Arbitration shall be speedy and cost-effective – as provided in HKAO and UNCITRAL Model Law

- Confidential – Arbitration process is confidential to protect the parties and arbitration procedures. Witness and hearing will not need to be published in comparison with open court litigation. It helps to maintain business or commercial relationships as well as parties' reputation.

- Arbitration awards are final and binding – Arbitration awards shall be final, binding that are only subject to limited exceptional grounds under the HKAO and UNCITRAL Model Law. Unlike court litigation, it avoids the possibilities of appeals that may be time-consuming or costly.

- Arbitration Awards enforceable in H.K. & worldwide – Arbitration awards are legally enforceable in Hong Kong, countries under the New York Convention and Mainland PRC.

6.2 Mediation

6.2.1 Mediation Ordinance (Cap. 620)

Section 3 of the Mediation Ordinance (referred as "HKMO") provides for the objects as: "… To promote, encourage and facilitate the resolution of disputes by mediation" and "to protect the confidential nature of mediation communications." Section 4 of the Ordinance defines mediation as follows:

- "A structured process comprising one or more sessions in which one or more impartial individuals, without adjudicating a dispute or any aspect of it, assist the parties to the disputes to do any or all of the following: -

· (a). identify the issues in dispute.

· (b). explore and generate options.

· (c). communicate with one another.

· (d). reach an agreement regarding the resolution of the whole, or part, of the dispute."

Section 6 provides that the HKMO applies to the Government. Disputes involving Government or Government officers are covered by the HKMO and Government is not exempted. With regards to the essential principles of confidentiality, Section 8(1) provides that: "A person must not disclose a mediation communication except as provided by sub-section (2) or (3)." Confidentiality is one of the most important elements of mediation. It encourages constructive exchange, makes mediation more effective and protects the integrity of the process.

6.2.2 Hong Kong Mediation Code and Practice Direction 31 ("PD 31")

The HKMO should be considered together with the PD 31 that became effective in January 2010 and the Hong Kong Mediation Code that was published by the Department of Justice. Some professional bodies may have their own mediation rules. If there are no such rules provided, the use of Hong Kong Mediation Code would be the recommended approach. According to PD 31, the lawyers or legal advisors should advise the parties to consider the use of mediation as compared to litigation in court. If the party refuses to use mediation without reasonable grounds or causes, the court may consider requesting the party to be responsible for the legal and litigation costs even if the party can win the litigation case.

6.2.3 Advantages of mediation approach

The Hong Kong Mediation Accreditation Association Limited (HKMAAL) stresses the importance for mediator to build trust with integrity of the mediator and good faith or rapport built with the parties to resolve their needs, problems and impasse as well as to resolve their needs, problems or emotional concerns through the mediator's skillsets in "facilitative approach of communication." The HKMAAL and the Judiciary of the HKSAR propose the following advantages of using mediation approach. But these benefits are not confirmed or tested by construction industry stakeholders. They would be investigated by questionnaire, interviews and case studies as discussed in Chapters 7 and 8.

· Creates a supportive and constructive environment.

· Enables the parties to control the outcome of their dispute.

· Promotes communication between the parties.

· Time efficient and cost effective.

· Confidential process.

· Helps to teach the parties an effective way of resolving disputes through co-operative decision making.

· Not an imposed settlement..

6.3 Statutory adjudication and contractual adjudication

6.3.1 Proposed Security of Payment Legislation (SOPL) for Hong Kong

A new development on the use of adjudication is considered as vital to the building and construction industry. Adjudication is focusing on the building and construction industry in Hong Kong although the contractual parties can be overseas or in mainland PRC. Key backgrounds for the SOPL are highlighted below:

- The HKSAR Government and CIC had conducted a comprehensive and industry-wide survey on payment practices in local construction industry in 2011.
- The survey reviewed significant payment problems experienced by main contractors, sub-contractors, suppliers, consultants and sub-consultants etc.
- HKSARG proposed to adopt the experiences of overseas jurisdictions on use of SOPL to resolve some existing payment problems for Hong Kong, especially on late or default payment, pay-when-paid clauses and multiple sub-contracting systems unique in Hong Kong.

6.3.2 Scope of SOPL

- Statutory and Mandatory – contrast Construction Arbitration (based upon Parties Autonomy principles that rely on Arbitration Agreement) or Construction Mediation (voluntary in nature).
- Public sector – All future Government contracts and contracts for public sector
- Statutory/public bodies – Total of 31 specified statutory and/or public bodies and corporations.
- Private sector – All private developments intended as "new buildings" under Buildings Ordinance (Cap. 123) with contract value of $5,000,000 and supply contract or consultants' contract of value $500,000.

6.3.3 Key features and advantages of adjudication approach

(A) Payments

- Payment periods – Parties can agree payment periods between applications and payments but not exceeding 60 calendar days (interim payments) or 120 calendar days.
- A right to Adjudication – arises for non-payment and disputes about the value of works, services, materials or plant and/or financial claims about delay of time-related payments under the contract.
- A right to dispute resolution by Adjudication – a rapid procedure under which an adjudicator gives an independent decision on the dispute and amount of any payment due.

(B) <u>Prohibition of "pay-when-paid or pay-if-pay clauses"</u>

- <u>"Pay when paid" clauses</u> – shall not be effective or enforceable.
- <u>Prohibition of conditional payment clauses</u> – clauses having the effect of conditional payment or pay-when-paid effects shall not be effective or enforceable under the SOPL.

(C) <u>Suspension for non-payment</u>

- <u>Suspension or reduce rate of progress</u> – Unpaid parties have right to suspend or reduce rate of progress of work after non-payment of an adjudicator's decision or non-payment of amounts admitted as due.

(D) <u>Adjudication and enforcement</u>

- <u>Adjudication process</u> – speedy, robust and cost effective and may not follow court procedure.
- <u>Powers of adjudicator</u> – as defined by proposed SOPL of contractual adjudication under the contract.
- <u>Adjudicator's decision</u> – legally binding to parties for payments to ensure party's cashflows.
- <u>Short timeframe for adjudication process</u> – maximum period allowed from appointment of adjudicator to issue of adjudicator's decision will be 55 working days unless the parties both agree to a longer period. Straightforward cases will be decided quicker.
- <u>Right of parties to appeal</u> – Adjudicator's decision is not final but temporarily binding. Parties still have a right to refer the dispute to court or arbitration (if provided in the contract).
- <u>Legally enforceable</u> – Adjudicator's decision shall be binding and enforceable until the decision is challenged by parties and set aside by court /arbitral tribunal.

Due to the confidential nature of statutory and contractual adjudication, there are uncertainties or queries from the construction stakeholders whether these advantages as borrowed from overseas examples of SOPL are applicable to the context of Hong Kong. Similar to the arbitration and mediation approach, these proposed advantages will be investigated by questionnaires, interviews and case studies as discussed in Chapter 7 and 8.

7. Survey

7.1 Survey objectives

The objective is to obtain useful empirical data from different stakeholders in project management and ADR community for construction industry in Hong Kong regarding the merits of the dispute avoidance mechanism and ADR approaches investigated. The empirical data obtained will support the quantitative analysis of the research and the literature review. *As a humble contribution to the professional bodies, users and construction stakeholders etc., a simple user-guide would be prepared by the researcher that summarize the key finding and recommendations and send to the respondents of the survey (and the interviewee) for their kind reference on dispute avoidance and ADR approaches for construction industry in H.K., and thereby for the continuous improvement and learning in the society.*

7.2 Survey Methodology

Survey questionnaire was prepared and issued through an online survey platform that consists of questions based upon the research findings in literature review to address the research objectives. The survey was designed to include these areas: (i) background; (ii) conflict management and negotiation; (iii) arbitration approach; (iv) mediation approach; (v) statutory and adjudication approach; and (vi) recommendations. The survey was issued through colleagues from different institutions or business stakeholders such as colleagues in HKIA, HKIS, HKIE, HKICM, HKCA, APM (Greater China Branch), HKIAC, HKIArb, CIArb, IDRRMI, HKMC, various contractors, subcontractors, suppliers, consultants, specialists, academics, users in H.K. or overseas and ADR practitioners from project management's perspectives. The survey was followed up by emails and telephone etc. The results are used as quantitative analysis that would be supplemented by the qualitative analysis and literature review to give useful research findings. Among the total of 236 respondents, their backgrounds are summarized as follows:

a). Project manager or construction manager or project team members – 1st highest %.

b). Construction professionals – e.g. engineers, surveyors, architects, specialists – 2nd highest %.

c). Contractors, suppliers, subcontractors – 3rd highest %.

d). Dispute avoidance or dispute resolution practitioners – e.g. arbitrators, mediators, adjudicators. lawyers or ADR practitioners – 4th highest %.

[Note: participants may choose more than one background to reflect the construction industry characteristics.]

7.3 Empirical Survey Results

Table 4 – Survey on conflict management and negotiation skills (for dispute avoidance)

Questions	Greatest % of answer	Percentage of other answers	Observations
Q4 – Please choose skills useful for conflict management & negotiation in projects. [Can choose 1 to max. 4 skills]	Communication skills -79.2%.	Manage relationship/ conflict–78.4% Collaborative–75.8% Optimistic & positive – 52.5%.	Skills recommended by PMI are generally supportive by participants.
Q5 – Conflict management & negotiation skills are useful for stakeholders' engagement, communication & better management.	Agree – 45.8%.	Strongly agree – 37.3%. Neutral – 11.4%. Disagree & Strongly disagree – 4.2% & 1.3%	Skills recommended by PMI are generally supportive by participants.

Q6 – Integrated control on change management is useful to project success	Agree – 50.0%.	Strongly agree – 30.5% Neutral – 14.4% Disagree & Strongly disagree – 2.5% & 2.5%	Skills recommended by PMI are generally supportive by participants.
Q7 – Good change management is helpful to control project scope and project team carrying out duties	Agree – 44.5%	Strongly agree – 32.6%. Neutral – 18.6%. Disagree & Strongly disagree – 3% & 1.3%	Skills recommended by PMI are generally supportive by participants.
Q8 – Conflict management & negotiation skills are useful to obtain resources (e.g. time, money, staff).	Agree – 47.9%	Strongly agree – 31.4% Neutral – 11.9% Disagree & Strongly disagree – 6.4% & 2.5%	Skills recommended by PMI are generally supportive by participants.
Q9 – Conflict management & negotiation skills are useful for dispute avoidance.	Agree – 45.8%.	Strongly agree – 33.5%. Neutral – 14.8%. Disagree & Strongly disagree – 3.4% & 2.5%	Skills recommended by PMI are generally supportive by participants.

Table 5 – Survey on arbitration approach (for dispute resolution)

Questions	Greatest %	Percentage of other answers	Observations
Q10 – Parties can pre-agree to use arbitration with a private judge to resolve disputes	Agree – 42.4%	Strongly agree – 16.9% Neutral – 25.8%. Disagree & Strongly disagree – 12.3% & 2.5%	Arbitration advantage is supportive by participants. About 40% of participants choose neutral/disagree.
Q11 – Parties are free to choose arbitrator's background, qualifications, experiences or languages etc.	Agree – 43.2%.	Strongly agree – 24.6%. Neutral – 20.8%. Disagree & Strongly disagree – 8.1% & 3.4%	Arbitration advantage is supportive by participants. About 30% of participants choose neutral/disagree.
Q12 – Parties are free to choose language or venue for hearing etc.	Agree – 40.0%.	Strongly agree – 28.9% Neutral – 22.1% Strongly disagree & Agree – 6% & 3%.	Arbitration advantage is supportive by participants. About 30% of participants choose neutral/disagree.
Q13 – Arbitrator is neutral, impartial and/or independent	Strongly agree – 50%	Agree – 34.3%. Neutral – 10.6%. Disagree & Strongly disagree – 3.8% & 1.3%.	Arbitration advantage is generally supportive by participants.

Q14 – Arbitration procedures are flexible and not need to follow Court procedures/rules. Simple arbitration may be on documents-only.	Agree – 42.7%.	Strongly agree – 20.9%. Neutral – 24.8%. Disagree & Strongly disagree – 8.5% & 3%	Arbitration advantage is supportive by participants. About 36% of participants choose neutral/disagree.
Q15 – Arbitration is speedy and cost effective in comparison with open court litigation	Agree – 41.1%.	Strongly agree – 24.2%. Neutral – 22.0%. Disagree & Strongly disagree – 10.6% & 2.1%.	Arbitration advantage is supportive by participants. About 35% of participants choose neutral/disagree.
Q16 – Arbitration procedures are private and confidential that help to maintain business reputation & commercial relationships.	Agree – 42.7%	Strongly agree – 30.3%. Neutral – 19.7%. Disagree & Strongly disagree – 4.3% & 3.0%.	Arbitration advantage is supportive by participants. About 27% of participants choose neutral/disagree.
Q17 – Arbitration awards are final and binding and legally enforceable in HK, Mainland PRC and worldwide under New York Convention.	Agree – 40.7%	Strongly agree – 30.1% Neutral – 19.5%. Disagree & Strongly disagree – 8/5% & 1.3%	Arbitration advantage is supportive by participants. About 30% of participants choose neutral/disagree.
Q18 – Ranking of benefits (i.e. benefits in Q10 – Q16). [Note: Parties can choose from 1 benefit to a maximum of 8 benefits].	Choice 4 [Q13] – 78.8%.	Choice 2 [Q11] – 66.5%. Choice 3 [Q12] – 63.1% Choice 7 [Q15] – 62.3%. Choice 8 [Q16] – 62.3% Choice 6 [Q14] – 60.6% Choice 1 [Q10] – 51.7% Choice 5 [Q13] – 50.8%.	Choice 4 has highest %. Choice 1,2, 3, 5, 6, 7 & 8 all exceeds 50% support. Choice 1 & 5 are about 50% support.

Table 6 – Survey on mediation approach (for dispute resolution)

Questions	Greatest % of answer	Percentage of other answers	Observations
Q19 – Create a supportive & constructive environment to resolve disputes	Agree – 49.6%	Strongly agree – 28.4% Neutral – 15.7%. Disagree & Strongly disagree – 4.2% & 2.1%	Mediation advantage is generally supportive by participants.
Q20 – Mediation is flexible to allow parties to decide on the outcome or solutions of dispute as compared to litigation	Agree – 46.6%.	Strongly agree – 29.7% Neutral – 17.4%. Disagree & Strongly disagree – 4.2% & 2.1%	Mediation advantage is generally supportive by participants.
Q21 – Promote communication & discussion between parties for resolving disputes	Agree – 48.7%.	Strongly agree – 28.4% Neutral – 14.0% Strongly disagree & Disagree – 6.4% & 2.5%	Mediation advantage is generally supportive by participants.
Q22 – Mediation is time efficient and cost-effective to resolve disputes	Agree – 41.9%	Strongly agree – 36.9% Neutral – 13.1%. Disagree & Strongly disagree – 5.9% & 2.1%	Mediation advantage is generally supportive by participants.
Q23 – Mediation is private, voluntary and confidential and without prejudice in nature.	Agree – 42.8%.	Strongly agree – 32.2% Neutral – 18.2%. Disagree & Strongly disagree – 4.2% & 2.5%	Mediation advantage is supportive by participants. About 30% of participants choose neutral or disagree.
Q24 – Mediation can help parties to develop co-operative decision so as to maintain or even rebuild business/ commercial relationships	Agree – 49.6%.	Strongly agree – 28.0%. Neutral – 14.0%. Disagree & Strongly disagree – 5.9% & 2.5%	Mediation advantage is supportive by participants. About 25% of participants choose neutral or disagree.
Q25 – Ranking of benefits (i.e. benefits in Q19 – Q24). [Note: Parties can choose from 1 benefit to a maximum of 8 benefits].	Choice 3 [Q21] – 73.7%	Choice 2 [Q20] – 69.1%. Choice 4 [Q22] – 66.1% Choice 6 [Q24] – 65.3%. Choice 5 [Q23] – 64.4% Choice 1 [Q19] – 56.4	Choice 3 has highest %. Choice 1, 2, 4, 5 & 6 all exceeds 50% support.

Table 7 – Survey on statutory and contractual adjudication approach (for dispute resolution)

Questions	Greatest %	Percentage of other answers	Observations
Q26 – Awareness of statutory adjudication and/or contractual adjudication	Choice 3 – aware both statutory & contractual adjudication – 28.8%.	Choice 5 – wish to know more – 27.5% Choice 4 – not aware of adjudication – 22.5% Choice 1 – Aware statutory l adjudication – 11.9% Choice 2 – Aware contractual adjudication – 9.3%.	Choice 4 & 5 together exceeds 50% that represents overall awareness of either statutory or contractual adjudication in HK have areas for improvement.
Q27 – Pay when paid or pay-if-paid clauses shall not be effective or enforceable.	Agree – 39.4%.	Neutral – 30.1% Strongly agree – 22.9% Disagree & Strongly disagree – 4.7% & 3.0%.	Adjudication advantage is supportive by participants. About 38% of participants choose neutral/disagree.
Q28 – Parties can agree payment periods within statutory/contractual adjudication framework to ensure payment & cash-flows to stakeholders	Agree – 39.8%.	Strongly agree – 34.3% Neutral – 20.3% Strongly disagree & Disagree – 4.7% & 0.8%	Adjudication advantage is supportive by participants. About 26% of participants choose neutral/disagree.
Q29 – Right to dispute resolution by adjudication as a rapid procedure by an independent adjudicator's decision is useful to construction industry	Agree – 44.5%	Strongly agree – 28.8%. Neutral – 22.0%. Disagree & Strongly disagree – 3.8% & 0.8%	Adjudication advantage is supportive by participants. About 27% of participants choose neutral/disagree.
Q30 – Right to adjudication is useful for resolving (i) payment disputes; (ii) value of works, services, materials or plants (iii) time related disputes	Agree – 47.5%.	Strongly agree – 27.1%. Neutral – 20.3%. Disagree & Strongly disagree – 3.8% & 1.3%.	Adjudication advantage is supportive by participants. About 25% of participants choose neutral/disagree.
Q31 – If parties are unhappy with adjudicator's decision, they may appeal to court or arbitrator. Adjudication is widely used in other countries.	Agree – 41.1%.	Strongly agree – 28.8%. Neutral – 24.6%. Disagree & Strongly disagree – 4.2% & 1.3%.	Adjudication advantage is supportive by participants. About 30% of participants choose neutral/disagree.
Q32 – Unpaid parties have right to suspend or reduce rate of progress (for contractor, consultants, subcontractors, suppliers & subconsultants) in statutory & contractual adjudication.	Agree – 41.9%	Strongly agree – 26.3%. Neutral – 22.5%. Disagree & Strongly disagree –6.4% & 3.0%.	Adjudication advantage is supportive by participants. About 32% of participants choose neutral/disagree.

Table 8 – Survey on recommendations for dispute avoidance and dispute resolution

Questions	Greatest %	Percentage of other answers	Observations
Q33 – Training for conflict management and negotiation skills to construction stakeholders for dispute avoidance & better management	Agree – 44.1%	Strongly agree – 34.3% Neutral – 15.3% Disagree & Strongly disagree – 5.1% & 1.3%	Industry specific training is generally supportive by participants.
Q34 – Training for construction arbitration & mediation (from users' awareness and PM perspective) to construction stakeholders for dispute resolution & better project delivery	Agree – 44.1%.	Strongly agree – 35.2%. Neutral – 14.4%. Disagree & Strongly disagree -5.1% & 1.3%.	Industry specific training is generally supportive by participants.
Q35 – Training for statutory adjudication and contractual adjudication to construction stakeholders for dispute resolution & better project delivery	Agree – 44.1%.	Strongly agree – 33.9% Neutral – 16.1% Strongly disagree & Disagree -4.2% & 1.7%	Industry specific training is generally supportive by participants.
Q36 – Training for NEC 3, NEC 4 and/or other collaborative form of construction contracts to construction stakeholders for dispute resolution & better project delivery	Agree – 45.8%	Strongly agree – 33.1%. Neutral – 16.1%. Disagree & Strongly disagree – 3.8% & 1.3%	Industry specific training is generally supportive by participants.
Q37 – Training for Building Information Modelling (BIM) and/or other innovative technologies based upon collaboration & effective communication to construction stakeholders for dispute prevention and better project delivery	Agree – 46.6%.	Strongly agree – 27.1%. Neutral – 20.3%. Disagree & Strongly disagree – 4.2%. & 1.7%.	Industry specific training is generally supportive by participants.

8. Interviews and Case Studies

8.1 Interviews

Qualitative data is obtained to supplement the literature review and survey. It comprises of interviews and case studies. The interviews are classified into 3 groups: (i) Group 1: experts in project management for building & construction projects; (ii) Group 2: experts in ADR; and (iii) Group 3: experts in project management, technology & innovation or Building Information Modelling (BIM). They can choose more than one group.

Table 9 – Summary of interviews

Interviewee	Business or Professions	Group(s)
Mr. Stephen Ho	Assistant director of CIC and council member of HKIA	Group 3
Mr. Sunny Yeung	Architect, arbitrator and former director of developer	Group 1 & 2
Mr. Keith Wong	Real estate developer and contractor	Group 1
Prof. Philip Chan	Engineer, arbitrator, mediator, adjudicator and academics	Group 2
Prof. Francis Law	President of Hong Kong Mediation Centre & IDRRMI	Group 2
Anonymous	Engineer and client-side project management expert	Group 1
Mr. Robert Gerrard	NEC senior consultant and surveyor	Group 1 & 3
Mr. Raymond Kam	Quantity surveyor & contract expert	Group 1
Mr. Thomas Ho	Quantity surveyor & past president of HKIS	Group 1.
Mr. Felix Li	Architect and past president of HKIA	Group 1
Ms. Betty Lo	Engineer, surveyor and construction manager	Group 1
Mr. Roger So	Barrister and mediator	Group 2
Mr. Evenlyn Kwok	Quantity surveyor, PMP and ADR practitioner.	Group 1

A total of 13 experts had been interviewed and their expert opinions or comments being captured and analyzed. They had completed the questionnaire and were generally supportive to all the benefits identified in the literature review as summarized in Tables 4 to 7 above. Some of them gave additional views or comments during the interviews. These additional remarks are summarized in Table 10 to enrich the research findings.

Table 10 – Summary of specific views or remarks captured to enrich research findings

Interviewee	Additional views or remarks
Mr. Stephen Ho	Conflicts and disputes mainly come from insufficient use of more innovative or collaborative form of building and construction contracts. NEC contracts or Partnering contracts worth further development to enable behavioral changes to construction parties.
Mr. Sunny Yeung	Conflict management and negotiation skills are useful to dispute avoidance. In general, mediation has a lower successful rate and depends on parties' willingness. He prefers to use simplified form of arbitration and/or adjudication. Full arbitration may be costly and time-consuming process.
Mr. Keith Wong	Prefer statutory and/or contractual adjudication among the ADR approaches for fast and certainty in cashflows to contractors for speedy and cost-effective dispute resolution.
Prof. Philip Chan	Conflict management, negotiation skills and communication can integrate with ISO 9000 quality procedures. If dispute is unavoidable, dispute resolution advisor (DRA) and statutory/contractual adjudication can be used to prevent disputes to escalate. If disputes remain, mediation being low-cost and speedy can be tried. Arbitration with higher cost is the last resort. Adjudication, mediation & arbitration are confidential to preserve reputation and commercial relationships.
Prof. Francis Law	Conflict management and negotiation skills are useful to deal with their internal business or external environment for conflict avoidance and risk management. If dispute is unavoidable, mediation that is speedy, cost-effective are the 1st choice. Professional mediation can resolve 60 to 70% of disputes. Statutory and/or contractual adjudication is also speedy and cost-effective with the merit of legally binding and enforceable. Domestic and overseas experiences show that 95% of mediation settlement agreements or adjudicator's decisions will be followed by parties.
Anonymous	In his experience from client side, he supports using conflict management and negotiation to prevent/minimize disputes or conflicts. His company often come into negotiated final account with contracting parties using an independent subject expert advising on cost, time and quality etc. He recommends using international project management standards like APM or PMI.
Mr. Robert Gerrard	He supports using NEC contracts. It is useful to have simple and user-friendly contract clauses written in layman's terms. NEC contracts have built-in clauses and environment to foster collaboration and trust. Negotiation and mediation skills should be used in all NEC contracts to facilitate cooperation. Among the ADR approaches, mediation should always try first for every dispute. If mediation fails, he recommends using adjudication to give a speedy and binding decision. Arbitration should be avoided where possible.

Mr. Raymond Kam	Conflict management and sufficient communication are key success factors of all types of projects. It enables design and project changes to be managed. Government and public bodies may prefer adjudication and arbitration since they give certainty to binding decisions. Adjudication is suitable for small-to-medium size disputes with fast and cost-effective decisions. For large and complex disputes, arbitration has the merits of certainty but may be costly and time-consuming.
Mr. Thomas Ho	Conflict management and negotiation skills are essential to stakeholders' engagement and communication. If dispute is unavoidable, mediation that is speedy, cost-effective are always the 1st choice for small-to-medium size disputes. For complicated cases or when parties require binding decisions, adjudication and arbitration are the choices. Adjudication is speedy and cost-effective but has the disadvantage of being new to the industry as compared with arbitration.
Mr. Felix Li	Conflict management, negotiation skills and communication are essential in project management. For NEC projects, he considered that construction stakeholders in Hong Kong are relatively less prepared than U.K. on use of NEC or partnering contracts due to cultural and historical reasons. For adjudication, he has concerns on time-related disputes for complex situations.
Ms. Betty Lo	As a quantity surveyor and construction manager, she is in full support for conflict management, negotiation skills, use of NEC or other partnering contracts as well as BIM for dispute prevention.
Mr. Roger So	On the ADR approaches, he considers that mediation is speedy, cost-effective and flexible. Arbitration is close to litigation with final and binding decisions. Arbitration is expensive and a lengthy process but has the merit of confidentiality. Adjudication is something between mediation and arbitration. For NEC or collaborative form of contract, it depends on the organization's nature. A GAP analysis may be useful to consider whether NEC or BIM should be used.
Mr. Evenlyn Kwok	Conflict management and negotiation skills are useful for change management and dispute avoidance. They should be included in the training programme. Among the ADR approaches, adjudication is suitable when the work is in progress, mediation is preferred for settlement of dispute for the whole contract. Arbitration is the last resort if mediation fails. The industry stakeholders should be trained on adjudication before Contractual Adjudication is launched.

8.2 Case Studies

8.2.1 New Engineering Contract (NEC) – Happy Valley Underground Stormwater Storage Scheme by the Drainage Services Department (DSD) of the HKSAR Government

A. Background

Due to the major rainstorms in June 2008, April 2006 and August 2000, severe flooding incidents had occurred in the Happy Valley and its adjacent areas of Hong Kong. The urban context and dense population had imposed limitations to DSD to upgrade the underground drains. To avoid serious disruption to the public and minimize utilities diversion, a large underground stormwater storage tank was proposed to resolve the flooding problems.

B. New Engineering Contract (NEC) 3: ECC Option C

The project had adopted an innovative and collaborative contract arrangement using NEC 3 ECC Option C. The project commenced in September 2012 and completed in November 2017. Contract value was HK$1,068 million (i.e. about £107 million or USD $138.2 million). The contract scope involved building a 60,000m3 and 2.4-hectare underground storage tank, a pump house, 650 m in length twin-cell box culvert and associated improvement works together with green features and recreational features above ground for sustainability purpose. When the NEC project was completed, peak-flows during rainstorms could be controlled to increase the flood protection standard for protecting the 600,000 population in the Wan Chai District of Hong Kong.

The DSD and the NEC Project Manager had drafted an innovative clause to promote direct interaction with sub-contractors as stakeholders in addition to the Main Contractor to allow for partnering relationship. The project had more than 160 sub-contracts. There was a total cost saving of HK$ 60 million (i.e. about £6 million or USD$ 7.7 million). Besides, the total construction period was reduced, and this large and complex project could be completed 12 months in advance (i.e. commissioning of the stormwater storage tank as the main deliverable). The HKSAR Government was impressed by the achievements and introduced the general use of NEC 3 contracts for public works projects since 2016 where appropriate.

C. Conflicts management and negotiation skills

The NEC 3 project fosters a mutual trust and cooperation among contracting parties. The parties collaborate closely to resolve project, site or technical problems, minimize or avoid disputes, avoid delay and effectively control project cost. They adopt the "One Team, One Goal" approach in establishes common objectives and targets regarding cost, time, quality, safety and sustainability aspects. They also used "Knowing Me, Knowing You" as Stakeholders Engagement Workshops to develop team spirit. The merits and features are listed below:

Merits of the NEC project

· Team spirit and goal alignment.

- Pain-Gain sharing mechanism between Employer and Contractor to enhance cost effectiveness.
- Reduce risks of project delay and cost overrun.

<u>NEC Features</u>

- Mutual trust and cooperation
- Early problem resolution.
- Risk management.
- Timely coordination or reply.
- Subcontractor engagement & management.

8.2.2 Case Study on Construction Arbitration – *N v C* [2019] HKCFI 2292[6]

A. <u>Background</u>

Construction arbitration is confidential in nature and would be difficult to identify suitable case study. This recent construction arbitration case was reviewed by the Court of First Instance of Hong Kong SAR and so the decision by the Court is used as case study analysis.

The Plaintiff was the employer, and the Defendant was the main contractor for a residential development in Macau. The completion date was delayed for 360 calendar days. The architect granted an extension of time ("EOT") of 269 days with an entitlement of loss and expense and prolongation cost for 181 days EOT at MOP 100,000 per day. Delay related fluctuation costs of MOP 12 million was also granted. The architect had not allowed loss and expense for 88 days of the EOT grant. Disputes on the EOT, loss and expense etc. by the architect was entered into a construction arbitration and a sole arbitrator was appointed. The arbitrator made an award to the Defendant delay related loss and expense (prolongation costs) of MOP 25,500,000 and delay related loss and expenses (fluctuation costs) of MOP 12 million. The Plaintiff challenged and applied to the Court to seek for an order to set aside, or alternatively to remit the arbitration award.

B. <u>Application to set aside the arbitrator's award.</u>

The Plaintiff applied to the Court to set aside the arbitration award under s.4(1) and s.4(3) of Schedule 2 of HKAO (and section 81(1) of the HKAO) on the grounds that there was serious irregularity because:

(1) the Plaintiff was unable to present its case.

(2) the tribunal had exceeded its powers.

(3) the tribunal failed to conduct the proceedings in accordance with the procedures agreed by the parties, which was for the tribunal to decide on pleaded cases and agreed issues; and/or

(4) the tribunal failed to deal with all the issues which were put to it.

---

6 *N v C* [2019] HKCFI 2292; HCCT 3/2019.

The Plaintiff submitted that the Arbitrator knew the Defendant had no contractual entitlement to loss and expense under Clause 24 of the Contract for the EOT. However, the arbitrator had allowed the claim by the Defendant that the EOT which were awarded would give rise to entitlement of loss and expense (i.e. Agreement of Entitlement). Besides, the Plaintiff argued that the Agreement on Entitlement was never pleaded, nor identified as an issue with evidence and thereby the Plaintiff was deprived of the opportunity to present its case. The Plaintiff submitted that the arbitrator had exceeded its powers by doing so and failed to conduct the proceedings according to the parties' agreement ("Plaintiff's 1st arguments").

The Plaintiff submitted that the arbitrator had failed to deal with the issue regarding Defendant's failure to submit any application for loss and expense with the time of 2 months since there was a condition precedent under Clause 24 requiring the application for loss and expense to be submitted within 2 months. Hence, the claim should be time-barred ("Plaintiff's 2nd arguments").

C. The Decision

Hon Mimmie Chan J held that the Plaintiff's application to set aside the award under section 41(1) and (3) of the Schedule 2, or alternatively section 81 of the HKAO was dismissed.

The Court rejected the Plaintiff's 1st argument. Hon Mimmie Chan J found that the Plaintiff had confirmed there was an agreement to pay Defendant loss and expense at daily rate. The issue was whether there was a contention by the parties after the agreement was made to apply to EOT and the arbitrator had resolved the issue on the factual matrix. The Court concluded that the arbitration proceedings had given the Plaintiff full opportunity to present its case. In any event, the Court was not satisfied that the arbitrator's dealings were far from reasonable expectation and did not result in calling out for a serious irregularity or irregularity in due process.

The Court also rejected the Plaintiff's 2nd argument. Based upon the arbitrator's findings, the time limit under Clause 24 would not apply since the Defendant's claims were allowed based on the arbitrator's findings for the Agreement of Entitlement. The Court found that the arbitrator might not have given adequate reasons on this issue in the award or explained why the claim for time-bar was not in accordance with the existence, meaning and effects of the Agreement on Entitlement. The Court referred to *Secretary of State for the Home Department v Raytheon Systems Ltd.* [2014] EWHC 4375[7] and considered that "*....there was no need for the tribunal to provide any, or sufficient reasons for its decision is not the same as failing to deal with an issue...... A tribunal does not fail to deal with issues if it does not answer every question that qualifies for "an issue". If the tribunal decides all those issues put to it that were essential to be dealt with for the tribunal to come fairly to its decision on the dispute or disputes between the parties, it will have dealt with all the issue.*"

The Court reiterated that it was not proper to consider applications to set aside for serious irregularity regarding: (a) whether the arbitrator was right on the finding of facts and laws; (b)

---

7 *Secretary of State for the Home Department v Raytheon Systems Ltd.* [2014] EWHC 4375 (TCC), at para. 33(g).

whether the decision was supported by evidence; (c) whether there are sufficient reasons for the arbitrator's findings; and (d) the quality of the arbitrator's reasoning in his award. The Court would only consider whether the arbitrator committed a serious irregularity in arriving at its conclusions in the arbitration process.

D. Implications and Analysis

This important case decided clearly illustrates the Court is unwilling and reluctant to intervene or set aside any arbitration award unless there exists any clear serious irregularity. The case also shows that the disputed amount can be substantial involving unique building contract clauses like delay, EOT, loss and expenses, prolongation costs etc. based upon standard form of building contract as discussed in Section 5.3.3 of Chapter 5 above.

9.   Conclusions

The research finding involves a holistic analysis on the literature review, quantitative analysis in the survey and qualitative analysis through the interviews and case studies. The research findings are summarized as follows:

Table 11 – Benefits of conflict management and negotiation skills in dispute avoidance

Benefits	Quantitative	Qualitative observation
Skills useful for conflict management and negotiation in projects (based upon international standards from PMI)	Communication skills Manage relationship/conflict Collaborative Optimistic & positive	Supported by all experts in interviews. Conflict management, negotiation and communication skills can be integrated into ISO 9000 procedures.
Conflict management & negotiation skills are useful for stakeholder engagement and better management.	Confirmed & supported in survey [over 80%]	Supported by all experts in interviews. International standards of APM or PMI to be considered.
Integrated control on change management is useful to project success	Confirmed & supported in survey [over 80%]	Supported by all experts in interviews. International standards of APM or PMI
Good change management is helpful to control scope and project team carrying out duties	Confirmed & supported in survey [over 75%]	Supported by all experts in interviews. International standards of APM or PMI
Conflict management/negotiation skills are useful to get resources (e.g. time, money, staff).	Confirmed & supported in survey [over 75%]	Skills recommended by PMI are generally supportive by participants.
Conflict management & negotiation skills are useful for dispute avoidance.	Confirmed & supported in survey [over 75%]	Supported by all experts in interviews. Conflict management, negotiation can be integrated into ISO 9000 procedure

Table 12 – Benefits of arbitration approach in dispute resolution

Benefits	Quantitative	Qualitative observation
Parties can pre-agree to use arbitration with a private judge to resolve disputes	About 60% of responses supported this benefit.	Some experts consider that arbitration is more expensive than adjudication & mediation to be used as last resort.
Parties are free to choose arbitrator's background, qualification, experience or language etc.	About 68% of responses supported this benefit.	Some experts consider that arbitration is more expensive than adjudication & mediation to be used as last resort.
Parties are free to choose language or venue for hearing etc.	About 69% of responses supported this benefit.	Some experts consider that arbitration is more expensive than adjudication & mediation to be used as last resort.
Arbitrator is neutral, impartial and/ or independent	Confirmed & supported in survey [over 84%]	Some experts consider that arbitration is more expensive than adjudication & mediation to be used as last resort.
Arbitration procedures are flexible and not need to follow Court procedures/rules. Simple arbitration may be on documents-only.	About 60% of responses supported this benefit.	Some experts consider that arbitration is more expensive than adjudication & mediation to be used as last resort.
Arbitration is speedy and cost effective in comparison with open court litigation	About 64% of responses supported this benefit.	Some experts consider that arbitration for complex disputes may be costly and a lengthy process similar to litigation.
Arbitration procedures are private and confidential that help to maintain business reputation & commercial relationships.	About 65% of responses supported this benefit.	Some experts consider that this is the most important benefit for arbitration.
Arbitration awards are final and binding and legally enforceable in HK, Mainland PRC and worldwide under New York Convention.	About 70% of responses supported this benefit.	Some experts consider that this is the most important benefit for arbitration.

Table 13 – Benefits of mediation approach in dispute resolution

Benefits	Quantitative	Qualitative observation
Create a supportive & constructive environment to resolve disputes	Confirmed & supported in survey [78%]	Some experts consider that mediation is cost-efficient, speedy and confidential and should be tried as the 1st choice for ADR.
Mediation is flexible to allow parties to decide on the outcome or solutions of dispute as compared to litigation	About 76% of responses supported this benefit.	Some experts consider that mediation is cost-efficient, speedy and confidential and should be tried as the 1st choice for ADR.
Promote communication & discussion between parties for resolving disputes	Confirmed & supported in survey [77%]	Some experts consider that mediation is cost-efficient, speedy and confidential and should be tried as the 1st choice for ADR.
Mediation is time efficient and cost-effective to resolve disputes	Confirmed & supported in survey [79%]	Experts generally consider this benefit is most important. Mediation should be tried as the 1st choice for ADR especially for small-to-medium scale disputes.
Mediation is private, voluntary and confidential and without prejudice in nature.	About 75% of responses supported this benefit.	Some experts consider that mediation, adjudication, and arbitration are all confidential. Choice will depend on project size, project stage & stakeholder nature (e.g. public or private organization)
Mediation can help parties to develop co-operative decision so as to maintain or even rebuild business/commercial relationships	About 77% of responses supported this benefit.	Some experts consider that conflict management, negotiation skills and mediation can be considered holistically to foster a collaborative environment to resolve dispute/conflict.

Table 14 – Benefits of statutory & contractual adjudication in dispute resolution

Benefits	Quantitative	Qualitative observations
Awareness of statutory adjudication and/or contractual adjudication	Choice of "wish to know" or "not aware" together exceeds 50% that represents overall awareness of statutory or contractual adjudication have areas for improvement.	Experts shared the views to provide adjudication training to construction stakeholders (from awareness and project management perspectives)
Pay when paid or pay-if-paid clauses shall not be effective or enforceable.	About 70% of responses supported this benefit.	Some experts have concerns on use of adjudication for time related matters for complicated & large disputes
Parties can agree payment periods within statutory/contractual adjudication framework to ensure payment & cash-flows to stakeholders	About 74% of responses supported this benefit.	Some experts have concerns on use of adjudication for time related matters for complicated & large disputes
Right to dispute resolution by adjudication as a rapid procedure by an independent adjudicator's decision is useful to construction industry	About 74% of responses supported this benefit.	Some experts have concerns on use of adjudication for time related matters for complicated & large disputes
Right to adjudication is useful for resolving (i) payment disputes; (ii) value of works, services, materials or plants (iii) time related disputes	About 75% of responses supported this benefit.	Some experts have concerns on use of adjudication for time related matters for complicated & large disputes
If parties are unhappy with adjudicator's decision, they may appeal to court or arbitrator. Adjudication is widely used in other countries.	About 70% of responses supported this benefit.	Some experts have concerns on use of adjudication for time related matters for complicated & large disputes
Unpaid parties have right to suspend or reduce rate of progress. It may apply to contractor, consultants, subcontractors, suppliers & subconsultants for statutory adjudication & contractors/subcontractors for contractual adjudication.	About 68% of responses supported this benefit.	Experts suggest to provide adjudication training to construction stakeholders (from awareness and PM perspectives) DRA, adjudication may be integrated together as dispute prevention approach.

Table 15 – Integrating literature review, survey, interviews, case studies and research problems

The research findings are integrated in a holistic manner and illustrated as follows. The relationships between dispute avoidance via use of conflict management and negotiation skills at early stage and using main ADR approaches for disputes resolution are shown. Key recommendations are highlighted.

Dispute avoidance	Dispute resolution	Continuous improvement
· Conflict management & negotiation skills for dispute avoidance	· Investigate & unlock the advantages of ADR approaches and evaluate by quantitative & qualitative analysis.	· Identify limitations of the research and propose areas for further research and analysis.
· Conflict management & negotiation to improve project change management.	· Identify the unique characteristics and choice among the 3 ADR approaches – arbitration, mediation & adjudication in construction.	· Recommend areas for training on use of NEC contracts, partnering contracts or other collaborative form of contracts.
· International standards (e.g. PMI or APM) and industry standards in H.K. (e.g. CIC) to enhance stakeholder's change management, conflict management & communication.	· International standards (e.g. PMI or APM) and industry standards in H.K. (e.g. CIC) to enhance stakeholder's change management, conflict management & communication.	· Recommend areas for training on construction arbitration & mediation for industry stakeholders.
· Address Problem 1 – Conflicts resulting at core challenges to construction industry.	· Address Problem 1 – Disputes resulting at core challenges to construction industry.	· Recommend areas for training on statutory & contractual adjudication for the industry/
· Address Problem 2 – Conflicts unique in construction industry ecosystem.	· Address Problem 2 – Disputes unique in construction industry ecosystem.	· Recommend use of BIM and other technologies & innovation for early resolution of technical problems and conflicts.
· Address Problem 3 – Improper change management resulting in conflicts.	· Address Problem 3 – Improper change management resulting in disputes.	· Continuous improvement by supporting the industry development on conflict management and negotiation skills and the 3 major ADR approaches (i.e. arbitration, mediation & adjudication) to address the 4 problems initiating the study.
· Address Problem 4 – Support CIC's recommendation on dispute avoidance via unlocking the merits of conflict management.	· Address Problem 4 – Support CIC's recommendation on dispute resolution & unlocking the merits of conflict management.	
· Promotion on use of NEC contracts, partnering contracts or other collaborative form of contracts.	· Promotion on use of NEC contracts, partnering contracts or other collaborative form of contracts.	
· Promotion on use of BIM, technologies & innovation etc. for early resolution of technical problems and conflicts.	· Promotion on use of BIM, technologies & innovation etc. for early resolution of technical problems and conflicts.	

10. Limitations and Recommendations

10.1 Limitations

This Postdoctoral research paper has limitations in time and resources. Firstly, the research focuses mainly from the project management's perspective. It is recommended that a follow-up research may be developed in future that discuss and investigate mainly from ADR practitioners' perspectives (e.g. solicitors, barristers, arbitrators, mediators and adjudicators). Nevertheless, ADR's expert views and perspectives are considered in this research while the expert views of construction stakeholders, especially project managers are giving a higher importance.

Secondly, not all the dispute avoidance and resolution approaches are examined and analyzed. Only the more commonly used by both the public or private sector or those approaches the building and construction industry consider to be more essential are investigated. If resources are available in future, it is suggested that other types of ADR approaches such as Dispute Resolution Advisor (DRA) that only used in some large-size government projects, Independent Expert Certifier (IEC) and Expert Determination may be included in further research for the construction industry of Hong Kong.

10.2 Recommendations

10.2.1 Training – Conflict management & professional negotiation

There are generally no tailored-made conflict management and professional negotiation courses designed for the construction industry. The PMBOK (6th Edition) and the coming 7th edition both stress upon the importance of conflict management and negotiation skills for all types of project management. This is supported from the research findings of both the quantitative and qualitative analysis.

In the context of Hong Kong, there are conflict management and negotiation skills training programs in the markets for social workers or mediators etc. The professional institutions that may offer training are the Hong Kong Mediation Centre, the Hong Kong International Arbitration Centre and the Academy of International Dispute Resolution and Professional Negotiation. Based upon the research findings, the researcher recommended the construction industry major stakeholders (e.g. DevB, Hong Kong Contractors Association, HKIA, HKIS, HKIE, HKICM etc.) to explore the possibilities of developing tailored made conflict management and professional negotiation skills for the construction industry stakeholders to address the problems stated in core challenges in Section 2.1 of Chapter 2.

10.2.2 Training for NEC 3, NEC 4 and/or other collaborative form of building and construction contracts.

The research findings had supported the importance of collaborative approach in enhancing conflict management and negotiation skills for construction projects. The benefits include avoiding disputes or minimizes disputes, foster a cooperative and collaboration among all the stakeholders in the project, a fairer approach of profit and risk sharing between the employer and the contractor as well as enhancement in project programme and cost control. In some important and successful NEC projects (e.g. DSD Case Study as mentioned in Section 8.2.1), the main deliverable of the complex construction project can be delivered 12 months in advance and the overall cost is substantially reduced.

Based upon the research findings, the researcher recommends a wider and more systematic approach to launch both awareness and in-depth training on NEC 3, NEC 4 and other collaborative approach of building and construction stakeholders to cover different types of building and construction contracts, such as:

- Different types of main contracts between the employer and main contractor
- Different types of subcontracts between the main contractor and subcontractors down the tiers.
- Different types of supplier contract.
- Different types of consultancy services agreements.
- Different types of specialist construction contracts or specialist suppliers contracts.

10.2.3 Training for construction arbitration and mediation for industry stakeholders

The research findings had unlocked the merits of construction arbitration and mediation in the context of Hong Kong from project management perspectives. The merits or lessons learnt from individual arbitration case and mediation case for building and construction projects are difficult to share and communicate to the project management teams or construction teams due to the confidentiality principle and legal principles that are difficult to understand by ordinary construction stakeholders or users.

Based upon the qualitative and quantitative research findings, the researcher recommends the construction industry to develop 2 sets of training programs. The first set of training can focus on the general awareness including key highlights of the construction arbitration and mediation process; the merits or advantages to take note; and how to evaluate from clients or stakeholder's perspective a suitable choice of ADR approach. Factors to consider may include, cost effectiveness, duration, legal and professional expertise involved, certainty and finality of the approach, risk management, legal obligations and risks and overall impacts to the project completion or delivery. The second set of training is in-depth training to suitable construction industry professionals or stakeholders. They may be selected by their corporation based upon their experience, roles, job natures or seniority etc. The in-depth training can be to professional training to lawyers or ADR practitioners but shall be tailored-made for construction stakeholders. For instance, they may include:

- Principles and procedures for construction arbitration for construction professionals.
- Principles and procedures regarding construction arbitration and mediation for clients, contractors, consultants, subcontractors and suppliers for building and construction projects.
- Merits and risk management for construction arbitration.

- 40 hours general mediation training for construction professionals.
- 40 hours general mediation training for clients, contractors, consultants, subcontractors and suppliers for building and construction projects.
- Merits and risk management for construction mediation.

10.2.4 Training for statutory and contractual adjudication for construction industry

The research findings had unlocked the merits of adjudications. Due to the confidential nature of construction adjudication overseas and the statutory adjudication for Hong Kong is planning for submission to LegCo earliest in 2022 (with contractual adjudication to be launched in Q4 2021), the merits were more easily identified or aware by the legal and ADR practitioners instead of the construction industry stakeholders.

Based upon the qualitative and quantitative research findings, the researcher recommends the construction industry to develop 2 sets of training programs. The first set of training can focus on the general awareness including key highlights of statutory and contractual adjudication including its important impacts to construction stakeholders on payment, time-related disputes, cash-flows, right to adjudication and adjudication process etc. The merits as observed in Hong Kong and the international context can be explained and educated to the construction stakeholders to facilitate them to use statutory and contractual adjudication. The second set of training would be in-depth training to suitable construction stakeholders. They may be selected by their corporation based upon their experience, roles, job natures or seniority etc. The in-depth training on construction adjudication can be similar to those professional training to ADR practitioners but shall be tailored-made for construction stakeholders. For instance, they may include:

- Principles and procedures for statutory and contractual adjudication for construction professionals.
- Principles and procedures regarding statutory and contractual adjudication for clients, contractors, consultants, subcontractors and suppliers for building and construction projects.
- Merits and risk management for statutory and contractual adjudication.
- Overseas case analysis for statutory and contractual adjudication with demo of construction adjudication on public and private sector projects in Hong Kong.

10.2.5 Use of Building Information Modelling (BIM) and other technology innovation

It has been mandatory by Government since 2017 for public projects that BIM must be adopted on all capital works projects with a budget of HK$ 30 million or more[8]. DevB (2019) suggested that BIM has potentials to play significantly in enhancing the quality of information and decision-making capability in construction, time-and-cost monitoring, operation and maintenance planning and long-term asset management.[9]

---

8 Development Bureau (2017-2020), DEVB Technical Circular from No. 7/2017 to the latest DEVB TC(W) 12/2020.

9 Development Bureau (2018), "Construction 2.0 – Time to Change", published with assistance by KPMG.

By providing BIM and other technologies and innovation including data management and sharing, communication among construction project teams and stakeholders can be improved. Collaboration among employer, contractor, project team, consultants, subcontractors and suppliers can be improved. This would fundamentally change the "culture of the construction industry" towards a more collaborative, technology and data management driven method for construction. Technologies such as BIM, drone technology, SMART technologies, prefabrication, DfMA, Modular Integrated Construction (MIC), robotics etc. can be coordinated or integrated together using BIM. Since the culture of construction industry is changed, disputes can be avoided or prevented at the start through better collaboration as a team. Even if disputes appear or become unavoidable, the effects of the disputes may be minimized or quicker to resolve due to the BIM technologies and other innovation and technologies available. Some of these innovation and technologies are highlighted below[10]:

- Innovation – Modular Integrated Construction (MIC)/DFMA
- Innovation – Innovative procurement and contract forms.
- Innovation – Adoption of BIM technology.
- Innovation – Productivity indices to be used of KPI to be shared through BIM goals or BIM execution.
- Holistic project visibility – 3D BIM design; 4D and 5D BIM analysis on cost and schedules technique.
  Site surveying – BIM supported by Global Positioning System (GPS), LIDAR technique, smart cities.
- Robotics, Augmented Reality (AR)/Virtual reality (VR) – to be integrated with BIM through APIs.

References

A. Ordinances and Regulations

A.1 Hong Kong Ordinances and Regulations

1. Arbitration Ordinance (Cap. 609)
2. Buildings Ordinance (Cap. 123)
3. Buildings (Planning) Regulations, Regulation 41.
4. Mediation Ordinance (Cap. 620)
5. Model Law on the International Commercial Arbitration (1985), with amendments as adopted in 2006, of the United Nations Commission on International Trade Law ("UNCITRAL Model Law").
   [Note: Incorporated into the Hong Kong Arbitration Ordinance (Cap. 609)]

A.2 Overseas Statutes (and Acts)

6. Building and Construction Industry Security of Payment Act (Chapter 30B) (for Singapore)
7. Building and Construction Security of Payment Act 1999 (for New South Wales)
8. Construction Contracts Act 2002 (for New Zealand)

---

10 https://citac.cic.hk/en-hk/

9. Construction Contracts Amendment Act 2015 (for New Zealand)

10. Construction Contracts Act 2004 (or known as Security of Payment Legislation for Western Australia)

11. Construction Contract Regulation 2004 (or "Western Australia Regulations 2004")

12. Construction Industry Payment & Adjudication Act 2012 (Act 746)("CIPAA") (for Malaysia)

13. Housing Grants, Construction and Regeneration Act 1996 ("the UK 1996 Act")

B. Case Laws

B.1 Hong Kong Case Laws

14. *Chun Wo Building Construction Ltd. v Metta Resources Limited* (2016) 2 HKLRD.

15. *The Incorporation Owners of Triumph Court* (凱旋大廈業主立案法團) *and Law Ping Patsy* 羅平 [2018] CACV 51/2018

16. *N v C* [2019] HKCFI 2292; HCCT 3/2019.

B.2 Overseas Case Laws

17. *John Mowlem & Co. PLC v Eagle Star Insurance Co. Ltd.& Others* (No. 2) [1995] 62 BLR 126.

18. *Secretary of State for the Home Department v Raytheon Systems Ltd.* [2014] EWHC 4375 (TCC), at para. 33(g)

19. *Weldon Plant Ltd. v The Commission for the New Towns* [2000] BLR 496.

C. Publications, Articles, Books and Journals etc.

20. Datuk Professor Sundra Rajoo (2017), "Kuala Lumpur Regional Arbitration Centre (KLRCA) – A Practical Guide to Statutory Adjudication in Malaysia", published in 2017 by Kuala Lumper Regional Centre of Arbitration, Malaysia.

21. Deacons (2014), "A Simple Guide to Arbitration in Hong Kong", published in Hong Kong.

22. Deacons (2021), "Deacons Newsletter – Construction: Issue on 8 April 2021", articles prepared by Mr. K.K. Cheung, Mr. Justin Yuen, Mr. Joseph Chung and Mr. Stanley Lo, published in Hong Kong.

23 Development Bureau (2017-2020), DEVB Technical Circular from No. 7/2017.

24. Development Bureau (2017-2020), DEVB Technical Circular from No. 18/2018.

25. Development Bureau (2017-2020), DEVB Technical Circular from No. 9/2019.

26. Development Bureau (2017-2020), DEVB Technical Circular from No. 12/2020.

27. Development Bureau of the HKSAR Government (2019), "Construction 2.0 – Time to Change", published and produced with the assistance of KPMG in September 2018 in Hong Kong.

28. Development Bureau of the HKSAR Government (2016), "Practice Notes for New Engineering Contract (NEC) – Engineering and Construction Contracts (ECC) for Public Works Projects in Hong Kong", published in Hong Kong.

29. Development Bureau of the HKSAR Government (2015), "Proposed Security of Payment Legislation for the Construction Industry – Consultation Document", published in June 2015, printed by the Government Logistics Department.

30. Drainage Services Department of the HKSAR Government (2017), "Presentation power-points – NEC Implementation in the Happy Valley Underground Stormwater Storage Scheme", prepared by Senior Engineer Mr. Ellen Cheng, SE/DP1; Engineer Mr. C.L. Leung, E/D1; Engineer Mr. Kevin Cheung,

31. Gary Soo (2011), "Construction Arbitration in Hong Kong – A Changing Landscape", at pp 71-73.

32. Hon Mr. Justice Fung (2014), "Conference on Practical Approach to Resolving Disputes in the Construction Industry – Mediation," on 11 April 2014.

33. Hong Kong Judiciary (2011), "Summary of Mediation Reports filed in the Court of First Instance in 2011".

34. K.K. Cheung (2020), "Presentation slides by Deacons for HKIA-CDRC CPD Seminar – Legal and Practical Aspects of Time Related Claims and Unforeseen Risks (including COVID-19)".

35. Michael J. Fisher and Desmond G. Greenwood (2018); "Contract Law in Hong Kong", 3rd edition, HKU Press, printed in Hong Kong.

36. Professor Peter Malanczuk (2019), "Presentation power-points for the HKLTI Seminar: 'NEC Contracts in Hong Kong – The Legal Landscape and Beyond" on 25 June 2019".

37. Project Management Institute (2017), "PMBOK Guide: A Guide to the Project Management Body of Knowledge (6th Edition)", printed in the United States.

38. Research Office of Legislative Council Secretariat (2019), "Administration of Justice and Legal Services: Statistical Highlights: ISSH04/19-20 – Mediation Services in Hong Kong", published in Hong Kong on 29 October 2019.

39. The Government of the Hong Kong Special Administrative Region (1999), "General Conditions of Contract for Building Works."

40. The Government of the Hong Kong Special Administrative Region (1999), "General Conditions of Contract for Design and Build Contracts."

41. The Hong Kong Construction Industry Council (2010), "Guidelines for Dispute Resolution – September 2010", 1st edition, published in Hong Kong.

42. The Hong Kong Construction Industry Council (2015), "Reference Materials for Application of Dispute Resolution in Construction Contracts", revised edition, published in Hong Kong.

43. The Hong Kong Construction Industry Council (2016), "Reference Materials: NEC Case Book – Improvement of Fuk Man Road Nullah in Sai Kung", Version 1, published in Hong Kong in October 2016.

44. The Hong Kong Institute of Architects, the Hong Kong Institute of Surveyors and the Hong Kong Institute of Construction Managers (2005), "HKIA/HKIS/HKICM Standard Form of Building Contract (2005 Edition) (with Quantities)", printed in Hong Kong.

45. The Hong Kong Institute of Architects, the Hong Kong Institute of Surveyors and the Hong Kong Institute of Construction Managers (2005), "HKIA/HKIS/HKICM Agreement and Schedule of Conditions of Buildings Contract for use in the Hong Kong Special Administrative Region (With Quantities) (2005 Edition)", printed in Hong Kong.

46. The Hong Kong Institute of Architects, the Hong Kong Institute of Surveyors and the Hong Kong Institute of Construction Managers (2006), "HKIA/HKIS/HKICM Agreement and Schedule of Conditions of Buildings Contract for use in the Hong Kong Special Administrative Region (Without Quantities) (2006 Edition)", printed in Hong Kong.

47. The Hong Kong Institute of Architects, the Hong Kong Institute of Surveyors and the Hong Kong Institute of Construction Managers (2005), "HKIA/HKIS/HKICM Agreement and Schedule of Conditions of Nominated Sub-Contract for use in the Hong Kong Special Administrative Region (2005 Edition)", printed in Hong Kong.

48. The Hong Kong Institute of Architects, the Hong Kong Institute of Surveyors and the Hong Kong Institute of Construction Managers (2005), "HKIA/HKIS/HKICM Agreement and Schedule of Conditions of Nominated Supply Contract for use in the Hong Kong Special Administrative Region (2005 Edition)", printed in Hong Kong.

49. The Hong Kong Institute of Architects (2008), "The Code of Professional Conduct", 4th Revision, Rule 1.3 of the Principle 1.

50. The Law Society of Hong Kong (2019), "Hong Kong Lawyer – The Official Journal of the Law Society of Hong Kong", 18 November 2019 issue on N v C Arbitration Case Analysis.

51. The Office of the Government Economist, Financial Secretary of the HKSAR Government (2020),

52. Thomas Telford Ltd. of the Institute of Civil Engineers (ICE) (2020), "NEC 4 – Engineering and Construction Contract", 1st edition in June 2017, 4th edition with amendments in October 2020, printed in Great Britain by

Bell & Bain Limited, Glasgow, U.K.

[Note. Jointly recommended by The Government Construction Board, Cabinet Office, United Kingdom and The Development Bureau, HKSAR Government.]

53. Thomas Telford Ltd. of the Institute of Civil Engineers (ICE) (2020), "NEC 4 – Managing an Engineering and Construction Contract: Volume 4, User Guide", 1st edition in June 2017, 3rd with amendments in October 2020, printed in Great Britain by Bell & Bain Limited, Glasgow, U.K.

[Note: Jointly recommended by The Government Construction Board, Cabinet Office, United Kingdom and The Development Bureau, HKSAR Government.]

54. Thomas Telford Ltd. of the Institute of Civil Engineers (ICE) (2020), "NEC 4 – Preparing an Engineering and construction Contract: Volume 2, User Guide", 1st edition in June 2017, 3rd with amendments in October 2020, printed in Great Britain by Bell & Bain Limited, Glasgow, U.K.

[Note: Jointly recommended by The Government Construction Board, Cabinet Office, United Kingdom and The Development Bureau, HKSAR Government.]

55. Simon Lai and C.L. Leung (2018), "Presentation power-points by the Drainage Services Department of the HKSAR Government – Successful Application of NEC in Government (DSD) Projects: HKIS CPD Talk", presented and published on 31.7.2018.

56. Stanley Lo (2020), "Presentation slides by Deacons for HKIA-CDRC CPD Seminar – Legal and Practical Aspects of Time Related Claims and Unforeseen Risks (including COVID-19)".

57. Steven Cannon and Iain Black (2014), "Eversheds LLP – Statutory Adjudication", London 2014.

58. Sunny Yeung (2020), "Presentation slides by for HKIA-CDRC CPD Seminar – Legal and Practical Aspects of Time Related Claims and Unforeseen Risks (including COVID-19)."

59. Wilson Lam (2020), "Presentation slides by for HKIA-CDRC CPD Seminar – Legal and Practical Aspects of Time Related Claims and Unforeseen Risks (including COVID-19)."

D. Websites / Weblinks

60. https://constructionblog.autodesk.com/region/construction-trends/

61. https://www.pmi.org.hk/about-pmi-global/

62. https://www.pmi.org/certifications/project-management-pmp

63. http://www.hkmaal.org.hk/en/MediationinHongKong.php

64. https://mediation.judiciary.hk/en/figures_and_statistics.html#msfcjrrc

65. https://www.neccontract.com/NEC4-Products/NEC-Awards/Awards-2017/Awards-2017-Shortlist/Happy-Valley-Underground-Stormwater-Storage-Scheme

66. https://www.dsd.gov.hk/others/HVUSSS/en/innovation_nec.html

67. https://mediation.judiciary.hk/en/judgements.html

68. https://citac.cic.hk/en-hk/

Acknowledgement

The researcher would like to convey heartfelt thanks to the following professors for their guidance and advice:

Prof. Ir Philip K.I. Chan, Professor of my post-doctoral research

Prof. Eric Tao, California State University, Monterey Bay, USA.

Author's Biographical Notes

Dr. LAM WAI PAN, Wilson is an Adjunct Professor for ADR of the SABI University, Adjunct Professor of the Academy of International Dispute Resolution and Professional Negotiation and Guest Professor on Project Management in Peking.

For building and construction, he is a member of the International Postdoctoral Association, Member of HKIA, Corporate Member of RIBA, Authorized Person, Project Management Professional, BEAM Professional and NEC 4 Accredited Project Manager and Committee Member of APM (Greater Bay Area).

On information technology, he is a Prince 2 Practitioner, Certified Agile Practitioner and Scrum Master, Certified Blockchain Architect, Certified Big Data Consultant & Scientist, ITIL Expert, ITIL Digital & IT Leader, Certified PECB ISO/IEC27001 Lead Auditor, Certified PECB ISO 31000 Risk Manager, CIC Certified BIM Manager and HKIA BIM Professional & Assessor.

For Alternative dispute resolution, he is a fellow member of HKIArb and AIADR. He is the Council Member of HKIArb and ExCo member of HKMC and Arbitrators Appointment Advisory Board of HKIAC. He serves as Chairman of CDRC Committee in HKIA, Honorary Secretary and Alternate Director of the JMHO, Chairman of Construction & Adjudication Committee of HKIArb and the Editorial Committee of AIADR with experiences in many arbitration & mediation cases and listed in various panels. He was accredited as China Chief Legal Officer by the All-China Federation of Industry and Commerce in 2020. He was nominated as delegate of HKMC as "Observer" of the 54th Session of the United Nations UNCITRAL Commission for International Trade in 2021.

Ir Sr Cr Prof PHILIP CHAN PhD FEng

Postdoc

陳勤業 博導特聘教授、工程院士、博士後

Commerce & Industrial Engineering Experience 工程界經歷

現任法國北歐大學副校長；美國蒙特利灣加州州立大學特聘教授兼博士後導師，中國國家行政學院政府經濟研究中心專家委員會原副主任；勤業集團（前海股權交易中心新四板海外板掛牌代碼標準版：962820）主席，集團業務包括投資，教育，工程測量，仲裁調解，專家證人，智庫，品質環保等，上海聖瑞敕律所首席顧問，投身社會治理／工商業／教育四十年，也是香港註冊專業工程師 RPE 最多的一位共十一個界別。

美國蒙特利灣加州州立大學（CSUMB）和加拿大魁北克大學（UQAM）「博士後」專案大中華總裁兼博士後導師；曾任世界 500 強之瑞典雄霸兒童安全製品有限公司總裁及首席執行官（全球最大兒童汽車安全座安全部件製造商），世界最大手機充電器公司（諾基亞 Nokia Salcomp）全球品質及環保總經理，世界最大嬰兒車公司好孩子集團副總裁和世界三大電梯公司（包括中國迅達 Schindler 電梯有限公司——全球最大電扶梯製造商及通力 KONE 電梯有限公司——全球最大人行道製造商）總經理及總監等高級管理要職達二十多年。

Ir Sr.Prof. Chan is Vice President (HK & China) of SABI University, France and Distinguished Professor and Postdoctoral Advisor of California State University Monterey Bay (CUSMB), former Deputy Director of Expert Committee of Government Economics Research Center, National Academy of Governance (1/2017-3/2021), Chairman of Philip Chan & Associates Limited (QHEE962820), Principal Legal Advisor of Shanghai Singrights Law Firm, he obtained eleven Registered Professional Engineer (RPE) which is the most in Hong Kong.

Prof. Chan acquired forty years working experience as top executives from General Manager, Director, Vice President, President & CEO in world five largest multinational corporations including Car seat safety components (Swedish Holmbergs), Stroller manufacturer (Goodbaby Group, stock 1086), mobile phone charger (Nokia Salcomp), escalator (Schindler) and auto-walk (KONE).

Academic Achievement 學術成果

1) 法國北歐大學 副校長、特聘教授
   Vice President (HK & China) and Distinguished Professor, SABA University, France

2) 美國蒙特利灣加州州立大學 特聘教授、訪問學者
   Distinguished Professor and Visiting Scholar of California State University Monterey Bay (CUSMB)

3) 加州州立大學和加拿大魁北克大學 博士後導師
   Postdoctoral Advisor of CSUMB and UQAM

4) 中國國家行政學院政府經濟研究中心專家委員會 原副主任兼特聘高級研究員
   Former Deputy Director and Senior Special Researcher of Expert Committee of Government Economics Research Centre, National Academy of Governance (1/2017-3/2021)

5) 中國國家行政學院國際聯合培養博士後校友會 理事長
   International Postdoc Alumni of Government Economics Research Centre, National Academy of Governance

6) 魁北克大學中國社會經濟發展研究中心 客座教授
   Guest Professor, UQAM China Researcher Centre for Social Economics Development

7) 中國質量研究院 首任院長
   President of China Quality Institute

8) 法國北歐大學 法學榮譽博士
   Honorary Doctorate in Law, SABA University, France

9) 北京大學滙豐商學院／天津大學管理學院／北京師範大學 MBA 客座教授
   Guest Professor of Peking University HSBC Business College / Tianjin University / Beijing Normal University MBA

10) 北大博雅元培商學院 客座教授
    Guest Professor, The Bo Ya Organization of Peking University

11) 著作《持續改善──邁向成功企業的秘笈》獲香港大學饒宗頤學館收編供研究人員及讀者參閱

12) 部分已發表論文 Select Published Papers

	論文（著作）名稱、刊物（出版社）名稱、期號（出版日期）、字數
中國核心學術刊物	ISSN 1995-5316 WTO 下綠色壁壘條款的合法適用及局限， 《研訊學刊》19 期（2013 年 9 月），5993 字
國際學術刊物	部份學術刊物
國際學術會議論文	加拿大魁北克大學（UQAM）博士後論文在國際學術會議發表 UQAM's Postdoctoral papers published in international conference or journals 1) Asian Network For Quality Congress 2012 　ANQ2012 CONGRESS (31/7/2012 – 3/8/2012) 　Evaluation of The Practice of Quality Assurance in Greater China Construction 　Industry. 2) Annual Conference of Management Social Science 　ACMSS2013 (2-4/9/2013) 　DIMENSIONS OF CHIEF CUSTOMER LOYALTY OFFICER IN 　ELECTRONIC MANUFACTURING SERVICES IN GREATER CHINA 3) Annual Conference of Management Social Science 　ACMSS2013 (2-4/9/2013) 　The Success Factors of the Succession Process in Small and Medium-Sized Family 　Businesses in a One Child Context of China 4) Journal of Management and Research (April – June 2013) ISSN2278-0955 　Building Management in Hong Kong is a Trap for a Layman 5) Journal of Management and Research (April – June 2013) ISSN2278-0955 　Building Energy Saving Gensus Appraisal of the Drill of Building Energy Saving 　Management in Construction Industry in Greater China, Asia and Worldwide 6) The 4th Greater Pearl River Delta (GPRD) Conference 　on Building Operation and Maintenance – Sustainable 　Operation and Maintenance for Green Buildings 　12 December 2013 　Business Process Re-Engineering Innovation Management In Sustainable 　Operation And Maintenance For Green Building 　And Environmental Products Manufacturing 　Asian Network For Quality Congress 2014 Singapore 　Asian Network For Quality Congress 2014 Singapore 　ANQ2014 CONGRESS (5/82014 – 8/8/2014) 　ANQ2014 CONGRESS (5/82014 – 8/8/2014) 　Environmental Considerations in the Construction Industries in Hong Kong and 　China

7) Asian Network For Quality Congress 2014 Singapore

ANQ2014 CONGRESS (5/8/2014 – 8/8/2014)

Marketing Strategies of Jewelry Manufacture in Hong Kong: An analysis

8) Asian Network For Quality Congress 2014 Singapore

ANQ2014 CONGRESS (5/8/2014 – 8/8/2014)

Developing C2C Retail Entrepreneurship with Opportunities and Challenges: A Hong Kong example

9) Asian Network For Quality Congress 2014 Singapore

ANQ2014 CONGRESS (5/8/2014 – 8/8/2014)

Business Transformation: Case study of Hardware Company

10) Asian Network For Quality Congress 2014 Singapore

ANQ2014 CONGRESS (5/8/2014 – 8/8/2014)

Analysis of China Provate-Equity and Venture Capital Market

11) Asian Network For Quality Congress 2015 TaiwanANQ2015 CONGRESS (23-24/9/2015)

A research report on the necessity of popularizing

The use of CRS (Children Restraint System) in China

12) Asian Network For Quality Congress 2015 Taiwan

ANQ2015 CONGRESS (23-24/9/2015)

How KFC allocates the 4 Ps Marketing Strategies in China?

美國蒙特利灣加州州立大學 (CSUMB) 博士後論文在國際學術會議發表

CSUMB's Postdoctoral papers published in international conference or journals

13) Asian Network For Quality Congress 2016 Russia

ANQ2016 CONGRESS (21-22/9/2016)

China International Development Conceptual Strategy and Direction for the 21st Century – One Belt One Road Initiative

14) Asian Network For Quality Congress 2016 Russia

ANQ2016 CONGRESS (21-22/9/2016)

Poon's Formation of Seru-Flow Production on Manufacturing Medium Quantity

15) Asian Network For Quality Congress 2016 Russia

ANQ2016 CONGRESS (21-22/9/2016)

The Implementation Process and Key Success Factors Analysis of Lean Six Sigma

16) Asian Network For Quality Congress 2016 Russia

ANQ2016 CONGRESS (21-22/9/2016)

A Study of Innovation and e-Service Quality for developing e-Retailing Mass Entrepreneurship

17) Asian Network For Quality Congress 2018 Almaty

ANQ2018 CONGRESS (19-20/9/2018)

Factors affect to Customer Loyalty in Logistics Industry

	18) Asian Network For Quality Congress 2018 Almaty ANQ2018 CONGRESS (19-20/9/2018) Brief Discussion of Returnees' Employee Loyalty 19) Asian Network For Quality Congress 2019 Thailand ANQ2019 CONGRESS (21-25/10/2019) New Mechanically Automatic Inclinometer System 20) Asian Network For Quality Congress 2019 Thailand ANQ2019 CONGRESS (21-25/10/2019) A Study on discrepancy of interest rates in China's repo market 21) Asian Network For Quality Congress 2019 Thailand ANQ2019 CONGRESS (21-25/10/2019) A Study of PRC (Greater Bay Area) Guangdong-Hong Kong-Macao Construction Culture Difference 22) Asian Network For Quality Congress 2020 Korea ANQ2020 CONGRESS (21-25/10/2020) Enhancement of Corporate Governance in Small to Medium Size Non-Governmental Organizations (NGO) in Hong Kong from the Board's Perspective 23) Asian Network For Quality Congress 2020 Korea ANQ2020 CONGRESS (21-25/10/2020) Exploratory Study on Synergy of Sustainability Model between Time Bank and Innovative Elderly Healthcare Platform
學術專著	1) 持續改善（國學大師饒宗頤：選堂：書名提字）簡體版，書號：ISBN978-7-5130-3379-4，智慧財產權出版社，出版日期：2015 年 6 月，426 千字 2) 持續改善（國學大師饒宗頤：選堂：書名提字）繁體加強版，國際書號：978-988-8270-80-4，亞洲最大出版社（紅出版社），出版日期：2014 年 7 月，約 180 千字 3) 持續改善頂級企業實戰秘笈（國學大師饒宗頤：選堂：書名提字）繁體版，國際書號：978-988-8437-87-4，亞洲最大出版社（紅出版社），出版日期：2017 年 7 月，約 380 千字 4) 持續改善（國學大師饒宗頤：選堂：書名提字）英文版，書號：ISBN978-7-5130-0771-9/F，智慧財產權出版社，出版日期：2012 年 1 月，366 千字
	2014 年 12 月 8 日在香港理工大學獲香港工業專業評審局頒授首屆（工程院士）榮銜

Academic Qualifications (selected) 部份學歷

美國蒙特利灣加州州立大學 (CSUMB) 博士後

Postdoctoral Fellowship, California State University Monterey Bay (CUSMB)

中國國家行政學院政府經濟研究中心 經濟學博士後

Postdoctoral in Economics, Government Economics Research Centre, Chinese Academy of Governance

加拿大魁北克大學 管理與科技博士後

Postdoctoral Fellow in Management & Technology, University Of Quebec At Montreal (UQAM), Canada

中國天津大學 管理科學與工程博士後

Postdoctoral in Management Science & Engineering, Tianjin University, China

墨西哥阿茲特克大學 計算機科學與工程哲學博士 (PhD)

Doctor of Philosophy in Computer Science and Engineering (PhD), Azteca University, Mexico

法國北歐大學 工商管理博士

Doctor of Business Administration (DBA), SABI University, France

墨西哥阿茲特克大學 法學博士

Doctor of Law (LLD), Azteca University, Mexico

香港理工大學 品質管理碩士

Master of Science in Quality Management (MSc), Hong Kong Polytechnic University (HKPU)

澳門大學 工商管理碩士

Master of Business Administration (MBA), Macau University

澳州 Ballarat 大學 應用管理本科

Bachelor of Applied Management (BAM), University Of Ballarat, Australia

Professional Qualifications (selected) 部份專業資格

1) Legal, Mediation, Arbitration and Expert Witness（法律、調解、仲裁和專家證人領域）

Designation of Hong Kong Mediator appointed by Department of Justice HKSAR Government 香港 CEPA 調解員

DRAd – List of Disputes Resolution Advisors appointed by Architectural Services Department HKSAR Government 香港特區政府建築署糾紛調解顧問

Appeal Board Panel member of District Cooling Services Ordinance (Cap.624) appointed by Secretary for Environment of HKSAR Government 環境局局長已根據《區域供冷服務條例》第 24 條的規定，委任為上訴委員團成員

Disciplinary Tribunal Panel Member of Tower Working Platforms (Safety) Ordinance (Cap.470) appointed by Secretary for Development of HKSAR Government 發展局局長行使建築工地升降機及塔式工作平台（安全）條例（第 470 章）第 32 條所授予的權力委任為根據該條例成立的紀律審裁委員團成員

Arbitrator, Hong Kong International Arbitration Centre 香港國際仲裁中心仲裁員

Panel Arbitrator, Beijing Arbitration Commission 北京仲裁委員會仲裁員

Panel Arbitrator, Shanghai Arbitration Commission 上海仲裁委員會仲裁員

Panel Arbitrator, Nanjing Arbitration Commission 南京仲裁委員會仲裁員

Panel Arbitrator, Guangzhou Arbitration Commission 廣州仲裁委員會仲裁員

Panel Arbitrator, Zhanghai Arbitration Commission 珠海仲裁委員會仲裁員

Panel Arbitrator, Quanzhou Arbitration Commission 泉州仲裁委員會仲裁員

Panel Arbitrator & Mediator, Kuala Lumpur Regional Centre for Arbitration (KLRCA) 馬來西亞仲裁委員會仲裁員和認可調解員

Panel Arbitrator & Mediator, Indian Institute of Arbitration and Mediation (IIAM) 印度仲裁委員會仲裁員和認可調解員

FHKIArb 香港仲裁師學會資深會員

MCIArb 英國皇家特許仲裁學會會員

HKMC Accredited Mediator 香港和解中心認可調解員兼中港調解專家 HKMAAL Accredited Mediator 香港調解資歷評審局認可調解員

MAE (practicing) 英國認可執業專家證人

CPFAcct 註冊專業法證會計師

2) Engineering & Surveying（工程和測量）

FEng – Engineering Fellow 工程院士

FHKIE 香港工程師學會資深會員（飛機工程 Aircraft，屋宇裝備 Building Service，控制自動和儀器 Control，Automation & Instrument，土木工程 Civil，電機 Electrical，電子 Electronics，能源 Energy，資訊科技 Information，物流與運輸 Logistics & Transport，機械 Mechanical，製造與工業工程 Manufacturing & Industrial 等學部）

RPE 香港註冊專業工程師

FCIBSE 英國屋宇裝備工程師學會院士

CEng Chartered Engineer 英國皇家特許工程師（資訊科技，工程與科技，屋宇裝備）

China Registered Industrial Engineering Expert 中國工業工程專家

SCMES(Life) Registered China Senior Mechanical Engineer 中國機械工程學會高級會員（終身）與高級機械工程師專業技術資格證書

SCSEE 中國電機工程學會高級會員

FRICS 英國皇家特許資深測量師

MHKIS 香港測量師學會會員

PRS 香港註冊專業測量師

FCinstCES 英國皇家特許資深土木工程測量師

FCIOB 英國皇家特許資深建造師

MIFMA 國際物業管理協會會員

FHKICM 香港營造師學會資深會員

RCM 香港註冊專業營造師

3) Leadership & Corporate Governance（領導力和企業治理）

FCMI 英國皇家特許管理學會院士

CMgr. 英國皇家特許經理

F.CIM Chartered Manager 加拿大管理學會院士，加拿大特許經理

FHKIoD 香港董事學會資深會員

4) Marketing & Sales Leadership（市場及行銷領導力）

FCIM 英國皇家特許市務學會院士

Chartered Marketer 英國皇家特許市務師

FHKIM(Life) 香港市務學會資深院士（終身）

CPM (Asia Pacific) 亞太認許市務師

ProM Professional Marketer 香港專業市務師

China CMO 中華人民共和國市場總監業務資格證書

5) Financial Management（財務管理）

FCMA Certified Management Accountant, Australia 澳洲資深管理會計師

CPFAcct Certified Professional Forensic Accountant 加拿大註冊專業法證會計師

FITA 香港資訊財務師協會院士

6) **Logistics & Transport Management**（物流及運輸管理）

FCILT 英國皇家特許物流及運輸學會院士

7) **Quality & Environmental Leadership（品質與環保領導力）**

China Six Sigma Expert 中國六西格瑪專家

RQP(MBB & SSC) 註冊六西格瑪黑帶大師和宣導者

FCQI 英國皇家特許品質學會院士

CQP 英國皇家特許品質師

IRCA QMS 2000 Principal Auditor and Environmental Management System Lead Auditor 英國品質管制系統首席審核員和環保管理系統領導審核員

SCAQ 中國品質協會高級會員

CEnv Chartered Environmentalist 英國皇家特許環保師

FCIWEM 英國皇家特許資深水務環保師

8) **Information System Strategy**

FBCS 英國電腦學會院士

CITP Chartered IT Professional Fellowship 英國皇家特許資深資訊科技師

Awards & Honors 獲獎情況

2005 年，中國優秀質量人（首屆全國質量獎個人獎，劉源張工程院士，袁寶華，張瑞敏等獲此殊榮，在釣魚台國賓館接受全國政協常務副主席王忠禹頒獎）China Excellence Quality Person Award, 2005

2009 年，中華十大財智人物 Chinese Top Ten Wealth Wisdom Person Award 2009

2011 年，首屆都市盛世商界奇才 Prime Award for Outstanding Leader 2011

2018 年，華人楷模年度人物 Chinese Role Model Award 2018

2018 年，改革開放 40 周年——改革與創新商業領袖人物 China 40th Anniversary of Reform and Opening Up – Reform and Innovation Business Leader Award 2018

2021 年 5 月，加桑王子授予爵級大領勳章 Knight Grand Collar (KGCMA) of Sovereign Imperial and Royal House of Ghassan

THE AMERICAN BIOGRAPHICAL INSTITUTE (ABI USA)

American Medal of Honor,

Order of International Ambassador,

Research Board of Advisors,

Ambassador of Knowledge Medallion,

Universal Award of Accomplishment

INTERNATIONAL BIOGRAPHICAL CENTRE (IBC UK)

Honorary Doctorate of Letter,

Lifetime Achievement Award,

Honorary Director General,

Living Science Award,

Research Consultant

LISTED in Marquis Who's Who in Asia

"Continuous Improvement" towards world class Construction Industry secret

Philip K.I Chan[1,2,3*]

[1] Stanford University, Stanford, California, USA 94305
[2] University of California, Berkeley, Berkeley, CA, USA 94720-1234
[3] California Status University, Monterey Bay, 100 Campus Center, Seaside, CA, USA 93955-8001
* philip@pcconsulting88.com,

Abstract: It is always difficult to decide on an exact definition of 'quality' and this is particularly true of the construction industry. However, it is important to establish some form of consensus on quality criteria for construction industry. This paper makes a contribution to this field by through a survey which was conducted in Hong Kong and China to study how practitioners have implemented the concept of quality assurance in construction projects. The results suggest that there is a need for extensive training in order that a total quality approach is adopted. Construction industries strive for Continuous Improvement to become world class construction organization and adopt three area of improvement as corporate strategy including Quality, Innovation and Time To Market.

Key Words: Total Quality, Quality Assurance, Quality System, Survey Results, Quality, Innovation and Time To Market.

1. INTRODUCTION

1.1 What is a quality system

It is even more difficult to quantify quality element. As Ferry (1984) has acknowledged, "The definition of quality in building is much more difficult and it remains doubtful whether it will ever be possible to measure quality, although it may be possible to measure some of its attributes. It may even be possible to arrive at some sort of weighted index by assessing some factors and measuring others, but this would contain so many subjectivities both in the assessing and the weighting that its usefulness could be questioned".

2. MAJOR FACTORS DRIVING THE CHANGE

2.1 Change in socio-economic structure

The construction industry is an important element of the Hong Kong economy. It contributes bout 6% of Hong Kong's Gross Domestic Product based on economic activity. Of this, public sector construction contributes about 33% and private section 67%.

The general policy adopted by Pubic Authorities and the majority of private developers is "buying from the lowest bidder "i.e. the lowest tender gets the job. This drives contractors into competition by price rather than quality.

The traditional Chinese management style and business methods normally assemble hierarchical, informal and intimate structures. Decision-making power is centralized.

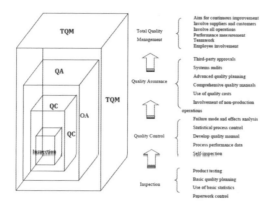

Figure 1.1 The Four Levels In the Evolution of Quality Management

There is less standardization of activities and fewer routine procedures and low level of specializations. The founder of this construction organization basically adopted this management style. Due to the rapid development in technology and the growth in the size and complexity of construction projects in the past decade, changes to the existing management structure become inevitable to enable the organization to remain competitive in this industry.

Hong Kong appears to be fortune enough to remain prosperous notwithstanding a continuous decline in the world economy. The impact on the local market is that more and more construction organizations from overseas such as Japan, Australia and European countries enter into the Hong Kong construction market as there is not much work in their home countries but there are still fair opportunities in Hong Kong. They enter the Hong Kong Market either on a joint venture agreement with a local organization or set up their own establishment based in Hong Kong. Some organizations also have support from their home countries. These international organizations bring with them their management venture, technology and expertise. It is not surprising to find that these establishments will provide a more sophisticated, systematic and better quality services to enable them to complete and survive in Hong Kong.

2.2 Public awareness and expectation

During the past decade Hong Kong has witnessed great changes with developing political awareness, more vocal environmental lobby, increasing sophistication and general raising of public expectations.

This movement is reflected not only in the volume of development but noticeably in many fine examples of buildings and urban design of international significance which have been built in recent years. These include the new generation public housing estates such as Ma On Shan, Junk Bay, commercial developments like Times Square, and the proposed development of the Wall City Garden.

Environmental protection issues arise from an awareness on the use of suitable and environmental friendly materials; reducing the use of timber as building materials, and increasing the use of energy conservation materials etc. This has a knock-on effect on management by reducing wastage and scrap, improving control of resources and minimizing pollution etc. All these call for a better quality standards and control.

A general rise in education, living standards and the growing awareness of the legal rights of the general public all give rise to higher expectations of the quality of the products. To provide just the basic standard will eventually put the organization under a disadvantageous competitive position.

2.3 Government encouragement

Government policy on getting value for money is always clear. The need to achieve value for money was stressed by the Governor is his opening address to the Legislative Council in 1988, when he emphasized the need to provide a framework of public sector management that would" encourage managers to get more from the resources they are consuming "

In March 1990, the Government launched a Quality Awareness Campaign, which emphasized the ISO 9000 International Standard Quality Management System. Besides enhancing the existing quality organizations, the campaign also introduced a quality management certification scheme to be administered by the HKQAA (Hong Kong Quality Assurance Agency) which is an independent agency that has been initially financed by the Government.

2.4 Housing Authority certification to ISO 9000

Originally, Housing Authority emphasis had been on meeting the construction program for public housings. However, a number of problems caused by sub-standard design, construction and maintenance practices ranging from unskilled workmanship to fraud developed over a couple of decades ago have raised both departmental and public concerns.

To alleviate problems with public housing, which became evident in the 1980s, the Housing Authority decided to adopt an ISO quality management system. It also requires its contractors, nominated sub-contractors and suppliers to implement ISO 9000 and to achieve registration with HKQAA by different target dates: main contractors by 31/03/93 concrete suppliers by 01/1992, nominated sub-contractors by '94 – '95, and consultants within 3 years.

Apart from the above, a quality assessment method, known as the "PASS" & "MASS" (The Hong Kong Polytechnic University 1992-1993). System has been developed since 08/1991 to record the performance of the contractor's installation and maintenance engaged in the Housing Authority's contracts. The aim of the system is to provide a fair means of comparing the performance of the contractors and aid in the decision to promote or downgrade a company and also to award preferential tendering opportunities. This may be moving from a traditional system of open tendering towards a selective tendering system.

3. CHANGES TO THE ORGANIZATION WITH THE INTRODUCTION OF QUALITY SYSTEM

The organization being impacted by the above factors, decided to introduce and implement a Quality System and to obtain ISO 9000 certification not later than the target date as set by the Hong Kong Housing Authority to maintain its competitive position.

The introduction of the quality system encountered obstacles and caused some changes to the existing organization. The major impact and barriers are discussed below.

4. PROBLEM STATEMENT

There is a common objective between quality assurance and elevator installation management-i. e error, defect or problem prevention.

The use of ISO9000 Quality system in construction industries, aims at achieving better project handling and control which is the essence of preventing problems and enhancing the quality of elevator installation. By implementation of a ISO9000 quality system, an elevator installation and maintenance team should be able to manage their task effectively to reach the pre-set quality standard in terms of time, cost and standard.

However, the QA concepts have slowly been making its present perception in the elevator industry in Hong Kong over the last two years. There has been a lack of study of the problem faced by practitioners in implementing the concept of Quality System for elevator projects. This lacuna means that the lessons and experience latent from the implementing QA in one particular installation project are not necessarily transferred to other project and shared for effective implementation. Apart from overcoming this gap, the aims of this research are to investigate what elevator contractors think about some quality management factors that are involved in their daily work, and the critical success factors for the elevator industry in Hong Kong.

5. AIMS AND OBJECTIVES OF THE STUDY

The aims of this paper are to:

a. Examine the concept of QA and the process of developing and implementing industry-wide QMS in accordance with recognized national and international standards for building projects,

b. Examine the attitude of developers, contractors, architects and engineers to wards the implementation of QA systems in the construction industry, and

c. Examine the need for QA program in building projects.

6. RESEARCH METHODOLOGY

For the purpose of this study, three hypotheses were formulated:

1. There does not appear to be a need to set up QA systems in construction projects as provisions have already been made within the technical and contractual framework of the construction industry.

2. The quality of construction projects will improve when the contractor is obliged to set up a QA system for the project contractually.

3. By virtue of the special nature of the construction industry, there is no necessity to adopt a 'total quality' approach in all construction projects.

To obtain an industry-wide response to the QA concept for the construction industry, a postal questionnaire survey was adopted as the methodology for the study. A questionnaire, with emphasis placed on soliciting empirical data for the above hypotheses, was drafted. This was pilot tested and sent to 241 contractors, 145 developers, 201 registered architects and 238 professional civi/structural engineers in mid-1998. A response rate of 18.7 percent (or 154 survey forms) was achieved from the 825 survey forms sent out. 19 returns were, however, rejected, because of their incompleteness. The responses that were acceptable came from the following groupings:

Developers: 29; Architects: 30; Engineers: 37; Contractors: 39; Total: 135

7. RESULT

A summary of main survey results is shown in Table 1. The findings of this postal questionnaire survey for the China and Hong Kong construction industry are shown in Table 1.

7.1 Concept of quality assurance

From the survey, it was found that 57 (42.2 percent) of the respondents agreed that QA is the same as quality control (QC). On the other hand, 78 (57.8 percent) of the respondents did not agree that QA is the same as QC.

Hence, a majority of the developers, architects, engineers and contractors surveyed not only disagreed that QA is the same as QC, they also felt that QA cannot be fully achieved by material testing alone: Nevertheless, the survey showed that some of the key players in the construction industry could not differentiate between QA and QC. Further more, the concept of QA does not seem to be understood readily in the construction industry.

7.2 QC circles

A QC circle is essentially made up of a voluntary group of people who come together to discuss quality-related problems and their possible solutions. Although QC circles appear to be a useful tool, which can be used, on construction sites, they are not a panacea for all quality problems.

From the survey, it was found that only 37 (27.4 percent) of the respondents (including the contractors) reported that their contractors have set up QC circles to identify, analyze and formulate solutions to quality problems. 60 (44.4 percent) of the respondents reported that their contractors do not implement any QC circles to resolve quality problems. Hence, although QC circles are well received in the manufacturing industry, they do not appear to be well received in the local construction industry. In the survey, one contractor reported the setting up of a QC circle to discuss how a piling record form could be designed to record all the essential piling information. It was agreed that, by the very act establishing QC circles, the persons involved in the circles would inculcate a quality culture and correspondingly increase their awareness of quality matters.

7.3 Establishment of a quality management unit

Table 1. Summary of main survey results

Description		Developers		Architects		Engineers		Contractors		Construction Industry	
		No.of firms	%	No.of firms	%	No.of firms	%	No.of firms	%	No.of firms	%
Do you agree that QA is the same as QC?	Agree	14	48.3	12	40.0	12	32.4	19	48.7	57	42.2
	Disagree	15	51.7	18	60.0	25	67.6	20	51.3	78	57.8
Did your contractor set up QC circles to resolve quality problem?	Yes	7	24.1	8	26.7	7	18.9	15	38.5	37	27.4
	No	10	34.5	14	46.7	13	35.1	23	58.9	60	44.4
	Not aware of	12	41.4	8	26.7	17	46.0	0	0	37	27.4
Do you agree that contractors should establish in-house Quality Management Units to enforce quality systems on site.	Agree	23	79.3	28	93.3	34	91.9	26	66.7	111	82.2
	Disagree	5	17.2	1	3.3	1	2.7	12	30.8	19	14.0
Do you agree that Quality Management Units should be organized within the consultants firms to ensure quality design and construction?	Agree	17	58.6	14	46.7	19	51.4	19	48.7	69	51.1
	Disagree	12	41.4	16	53.3	16	43.2	20	51.3	64	47.4
Do you agree that the clients should engage Quality Management Consultants to monitor the works undertaken by the consultants and the contracts?	Agree	9	31.0	7	23.3	16	43.2	15	38.5	47	34.8
	Disagree	20	68.9	23	76.7	19	51.4	24	61.5	86	63.7
Did your client require you to establish quality systems during construction?	Yes	5	17.2	4	13.3	7	18.9	8	20.5	24	17.8
	no	24	82.8	26	86.7	30	81.1	31	79.5	111	82.2
Do you engage a QA consultant to advise you on quality systems during construction?	Yes	2	6.9	1	3.3	1	2.7	2	5.1	6	4.4
	no	27	93.1	29	96.7	36	97.3	37	94.9	129	95.6
Did your contractor submit quality manuals before commencing work?	Contractually	5	17.2	4	13.3	7	18.9	8	20.5	24	17.8
	Voluntarily	2	6.9	1	3.3	1	2.7	6	15.4	10	7.4
	Not submitted	22	75.9	25	83.4	29	78.4	25	64.1	101	74.8
Did your contractor set up quality procedures during construction?	Yes	4	16.0	8	26.7	8	21.6	11	28.2	31	22.9
	No	25	84.0	22	73.3	29	78.4	28	71.8	104	77.1
Did your contractor use checklists for inspection of work before covering up?	Yes	9	31.0	10	33.3	11	29.7	23	58.9	53	39.3
	no	20	69.0	20	67.7	26	70.3	16	41.1	82	60.7

82 percent of the respondents agreed that the contractors should establish in-house quality management units to enforce quality systems on site. Only 14 percent of the respondents disagreed that contractors should set up quality management units to enforce quality construction on site.

8 percent of the respondents noted that it is part of the clients' requirements for contractors to establish in-house quality management units to enforce quality systems contractually. 30 percent of the respondents pointed out that, because the sub-contractors do not comply with the main contract

specifications, the main contractors should therefore establish quality systems to implement QA in their projects.

8 percent of the respondents noted that it is part of the clients' requirements for contractors to establish in-house quality management units to enforce quality systems contractually. 30 percent of the respondents pointed out that, because the sub-contractors do not comply with the main contract specifications, the main contractors should therefore establish quality systems to implement QA in their projects.

33 percent of the respondents observed that remedial works are required when the contractors' site supervisors put stress on speed of construction only, at the expense of quality. Hence, the main contractors must enforce their quality systems on site. 37 percent of the respondents felt that proper records of standards of workmanship would be available to the clients and the consultants if quality systems were established on site.

However, 32 percent of the responses from the developers, architects and engineers expressed the view that it is difficult for the consultants' site staff to enforce quality on site because of the negative attitudes of contractors towards quality construction. Hence, they felt that it is necessary for the contractors to establish their in-house quality management units to achieve quality construction.

29 percent of the respondents pointed out that the clients, designers and contractors realize that the consumers are now more aware of the quality of their products before marking any investment decisions. 65 percent of the respondents agreed that the contractors are beginning to realize that their reputation world be enhanced if the quality of their products were to satisfy the clients' requirements.

Like wise, 51percent of the respondents agreed that quality management units should be organized within the consultant firms to achieve quality construction. 47 percent of the respondents, on the other hand, disagreed that quality management units should be organized within consultant firms. In addition, 35 percent of the respondents agreed that the client should engage a quality management consultant to monitor the works undertaken by the consultants and the contractors.

From the above findings it was found that a majority of the organizations surveyed agree that the contractor should establish an in-house quality management units to enforce a quality system on site, and audit the contractor's quality system. It was found that a majority of the respondents do not agree that the client should engage a separate quality management consultant to check on works carried out by the contractors and the consultants.

7.4 Contractor's quality system

From the survey, only 24 (17.8 percent) out of the 135 respondents reported that it was their clients' requirements to ask contractors to contractually introduce quality systems during construction. Only 10 (7.4 percent) contractors have voluntarily set up quality systems on site to enforce quality construction. The remaining 101(74.8 percent) respondents reported that their contractors are not contractually required to set up quality systems during construction.

Six (4.4 percent) organizations reported that they have engaged QA consultants to advise them on quality systems during construction. The breakdown of these six organizations is as follows:

Survey group number of organizations which engaged QA consultants

Developers: 2; Architects: 1; Engineers: 1; Contractors: 2

From the above figures, it appears that these groupings are not, in the main, gearing up the construction industry for setting up QA systems during construction.

7.5 Factors affecting quality construction

10 factors that have an effect on the quality of construction were identified for the survey. The respondents were asked to rank these factors in their order of importance for achieving quality construction. The results of this ranking by all the respondents are shown as follows:

Ranking Factors

1. Poor workmanship by the contractors in completing the works results from low tender price.

2. The drawings and specifications do not specify clearly the intentions of the designers. Discrepancies are found between different consultants' drawings, which have resulted in poor co-ordination during construction.

3. The contractors pay more attention to completing the works on schedule and controlling the costs to within budget than to achieving quality in construction.

4. Poor co-ordination exists between the contractors and the sub-contractors, as well as the nominated sub-contractors.

5. The designers do not consider the 'build ability problems' in design. For example, the designers do not consider the use of special construction methods to achieve the tight tolerances caused by site constraints.

6. The contractors cannot plan and control the works. The contractors lack the skills to interpret the design and cannot provide the end products on site in accordance with the design and specifications.

7. The completion period fixed by the client and consultants is not realistic.

8. The design does not satisfy the relevant codes and standards. This has resulted in a large amount of remedial works for the contractors and delay in the completion of projects.

9. The contractors do not know how to establish a quality system to control the work.

10. The materials chosen by the consultants do not satisfy the standards or the Building Control Authority.

From the above rankings, it would appear that the quality of construction works is dependent to a large extent on the attitudes of the contractors and consultants. Hence, the quality of the products is adversely affected if the parties to the contract do not carry out their duties properly.

8. DISCUSSION

Research studies carried out by the Building Research Establishment showed that 50 percent, 40 percent and 10 percent of building defects have their origins in the design offices, poor workmanship and wrong choice of materials, respectively. Furthermore, it would cost four times as much to rectify an error at the construction stage than if the error could be identified and rectified at the design stage (BRE 1982). In addition, the cost of rectifying routine defects can amount to 7 percent of the project cost, and the cost of reworks is approximately five times the cost of getting it right first time.

The defects that are detected during construction contribute to unnecessary quality costs. These could be reduced if the parties involved in the project were to emphasis the achievement of quality construction in their works. This study reflects the quality problems, which occur in the construction industry. To help eliminate these problems and hence improve the image of the construction industry, QA must be practiced concurrently in the following areas:

Design services; Construction services & Manufacture of building products

A total approach to QA is also required for the quality concept to be implemented successfully in the construction industry. As Abdun-Nur (1970) noted, "The Quality System in construction is a system that deals with the procedures of obtaining the quality level of the construction needed for a project to perform the functions intended. And to do so within the various human, social, and environmental requirements and limitations. It encompasses determining the needs and will of the people, political considerations, human, social, and environmental factors, and how these influence design, specifications, contractual relationships, production, quality control, sampling, testing, charting, inspection, decision-making and feedback, and the interactions of all these facets of the system with one another" (Abdun-nur, 1970).

Hypothesis 1: There does not appear to be a need to set up QA systems in construction projects as provisions have already been made within the technical and contractual framework of the construction industry.

The evidence gathered from the survey showed that, although quality construction may have already been specified to some extent in specifications and other contract documentation, the latter may still be unable to achieve quality works first time every time. This leads to unnecessary reworks, delays and additional costs for rectifying defects. Hence, there is need to set up QA systems in building projects, even though provisions for quality construction may have already been made within the technical and contractual framework of the construction industry.

Hypothesis 2: The quality of construction projects will improve when the contractor is obliged to set up a QA system for the project contractually.

The survey results, however, showed that quality of construction projects will not necessarily improve even if the contractor is asked to set up a QA system contractually. This is because the attitude of the contractor is an important factor which governs the success or otherwise of a QA system. In addition, other factors, such as the quality of the design or clear understanding of the QA concept, are also critical for the successful implementation of QA system in the construction industry.

Hypothesis 3: By virtue of the special nature of the construction industry, there is no necessity to adopt a 'total quality' approach in all construction projects.

Although the construction industry is different in many difficult ways from the manufacturing industry where the QA concept originated, these difficulties actually reinforce and not undermine the applicability of the total quality approach in the former. The survey results showed that it is necessary to adopt a total quality approach in all construction projects in order to eliminate all factors, which have an adverse effect on the quality of construction works.

Apart from the above findings, there are, however, some limitations associated with the survey. Due to the inadequate sample size of the developers, this study could only attempt to interpret the survey results based on what was reported in the survey forms received. In addition, the person who filled in the completed survey form might not necessarily be in a responsible position in the organization. For example, the director of a contracting firm could have asked his site agent or any other inexperienced site engineer or foreman to complete the survey form. As a result, the views as expressed in the survey forms received may not truly reflect the views of the groupings, which the survey originally targeted. Furthermore, the person who the filled in the survey form might not fully comprehend the concept of QA and be able to differentiate it from QC. As a result, this perception would affect the way in which a respondent answered the questions in the survey. There is also a possibility that the architects, engineers, developers or contractors who were surveyed might be reporting on the same situation. The data may, therefore, overlap to some extent.

As QA is a relatively new concept in China and Hong Kong, there are at present no courses available to train works, supervisory staff, mangers and chief executive officers in the techniques of implementing quality systems in the construction industry. This is one area, which the construction industry in China and Hong Kong should address urgently. In the training programmers, some of the potential problems, as noted in this paper, which are likely to arise during the implementation of QA systems in the construction industry must be highlighted. This will help to building industry will function in the manner intended to achieve quality construction effectively.

References

Abdun-Nur, E. A. (1970) 'Control of quality a system' Journal of the Construction Division, ASCE, Vol. 96No. CO2, October

B.G.Dale and JJ.Plunkett (1990), Managing Quality, First published by Philip Allan, Simon & Schuster International Group.

Building Research Establishment, Quality in traditional housing: An investigation into faults and their avoidance (1982) (BRE).

Ferry, D. J. O., (1984), 'The role of the building professions in the achievement of quality', in Brandon, P. S. and Powell, J.A.(Eds), Quality and profitin building design, 92-98(Spon).

The Hong Kong Polytechnic University Quality and reliability Center, Quality Management Policies for Civil & Construction Contracts, Construction Industry Special Interest Group (1992-1993).

Authors' Biographical Notes:

Ir Prof. Dr. PHILIP CHAN Postdoctoral Fellowship, Stanford University, University of California, Berkeley and California State University Monterey Bay. Ir Prof Chan acquired academic and professional qualifications including H.ProfEng, HonDLitt, PhD, DBA, MSCJ, MSc, MBA, BAM, BSCE, BSEE, FEng, FHKIE RPE, CEng FCIBSE FIET FCIWEM FCInstCES, CEnv FSOE, CITP FBCS, FCIOB Chartered Builder, Chartered Surveyor, CMgr FCMI, Chartered Marketer FCIM, FCQI CQP, FCILT, FHKIM ProM, CPM(AP), CMO, CMA, FHKIoD, FHKQMA RQP(MBB), SCMES SCSEE, FITA, MCIArb, Accredited Witness Expert (MAE), Arbitrator (Hong Kong International Arbitration Centre, Beijing/Shanghai/Nanjing/Guangzhou/Zhuhai/Quangzhou Arbitration Commission, KCRLA, IIAM etc) and Accredited Mediator of Hong Kong Mediation Centre and HKMAAL. Ir Prof Chan have over 30 years professional experience in Five World Largest multi-national corporations including Autowalk, Escalator, Mobile Phone Charger, Stroller and Car Childseat's critical componments manufacturers as General Manager, Director, Vice President, President and CEO. Prof Chan is University of Quebec at Montreal (UQAM) and Chinese Academy of Governance (CAG) Postdoctoral Supervisor, Stanford/Berkeley/Monterey Postdoc Advisor and CSUMB iiED Advisor currently

Acknowledgement

The authors sincerely thank Prof. Eric Tao PhD, Director of iiED of California State University Monterey Bay for his kind assistance and valuable advice in this research. The authors also express their heartfelt thanks to Hong Kong Society for Quality (HKSQ) for their kind support and valuable advice in this research.

精益六西格瑪實施流程及
關鍵成功因素分析

陳勤業

天津大學管理學院博士後 300072

摘要

明確了精益六西格瑪的涵義，指出了實施精益六西格瑪的必要性和可行性；劃分了精益六西格瑪解決問題的種類；提出了關注系統、重視文化建設等 6 項實施精益六西格瑪的關鍵成功因素；精益六西格瑪實施中，對於簡單問題，可以直接用精益生產的方法和工具解決，對於複雜問題，給出了實施精益六西格瑪項目的 DMAIC II 流程和 14 個步驟。

關鍵詞：精益六西格瑪；精益生產；六西格瑪管理；生產模式

1. 精益六西格瑪（LEAN SIX SIGMA，簡寫為 LSS）的涵義

隨著人類需求的提高和社會生產力的發展，在企業生產與運作管理領域也進行著管理哲學和技術的進化。當今有兩種影響廣泛的管理方法，一種是來自於日本豐田生產系統的精益生產，另一種是來自於美國質量管理界的六西格瑪管理法。兩種方法為其實踐者帶來了巨大的效益，但隨著實踐的深入，也發現兩種方法還有需要改進的方面，二者可以相互促進。

精益六西格瑪是精益生產和六西格瑪的整合 [1-3]，它吸收了兩種生產模式的優點，彌補了單個生產模式的不足，已被許多企業實踐證明為有效的企業管理方法。它通過有效配置和優化資源，降低成本、縮短生產提前期和交貨期、提高質量和顧客滿意度來提高組織的競爭力。精益六西格瑪並不是精益生產和六西格瑪的簡單相加，而是二者的相互補充、有機結合。

按照所能解決問題的範圍，精益六西格瑪包括了精益生產和六西格瑪管理。根據精益六西格瑪解決具體問題的複雜程度和所用工具，我們把精益六西格瑪活動分為簡單精益改

善活動和精益六西格瑪項目活動，其中簡單精益改善活動全部採用精益生產的理論和方法，它解決的問題主要是簡單的問題。精益六西格瑪項目活動主要針對複雜問題，包括傳統意義上的精益項目、六西格瑪項目和狹義的精益六西格瑪項目，其中狹義精益六西格瑪項目需要把精益生產和六西格瑪的哲理、方法和工具結合起來。傳統六西格瑪項目主要解決與變異有關的複雜問題，例如控制一個過程的產品一次通過率；而精益六西格瑪項目解決的問題不僅包括傳統六西格瑪所要解決的問題，而且要解決那些與變異、效率等都有關的「綜合性」複雜問題，例如不但要控制一個過程的產品一次通過率，還要優化整個生產流程，簡化某些動作，縮短生產提前期，而且簡化這些動作和過程變異的控制有直接聯系。

通過實施精益六西格瑪管理，組織可以在以下方面獲得收益：減小業務流程的變異、提高過程的能力、提高過程或產品的穩健性；減少在制品數量、減少庫存、降低成本；縮短生產節拍、縮短生產準備時間、準確快速理解和響應顧客需求；改善設施布置、減小生產佔用空間、有效利用資源；提高顧客滿意度、提高市場佔有率 [4]。

2. 實施精益六西格瑪的必要性和可行性

實施精益六西格瑪是必要的。（1）六西格瑪優化的對像經常是局部的，缺乏系統整體的優化能力，所以它需要將自身需要解決的問題與整個系統聯系起來，然後優化流程；而精益生產理論的優點之一就是對系統流程的管理，它可以為六西格瑪的項目管理提供框架。系統中經常存在不能提高價值的過程或活動，無論員工如何努力，他們都無法超越系統流程的設計能力範圍之外，流程重新設計的目標就是盡量消除此類活動或過程，精益生產對此有一套完整有效的方法和工具。（2）精益生產依靠專家人才的特有知識，採用直接解決問題的方法，因此對於簡單問題，其解決問題的速度更快，但它缺乏知識的規範性，對於複雜的問題，它缺乏效率，無法保證其處於統計受控狀態；而六西格瑪管理更好地集成了各種工具，採用定量的方法分析、解決問題，解決問題有規範的 DMAIC 流程，為複雜問題提供了操作性很強的解決方法和工具 [5]。總之，精益生產告訴六西格瑪做什麼，六西格瑪管理告訴我們怎樣做，以保證過程處於受控狀態，對於複雜程度不同的問題，需要採用不同的方法去解決，二者是相互促進的，精益生產和六西格瑪管理相互促進關係見表1[6]，因此二者結合是必要的。

表 1　精益生產和六西格瑪管理相互促進關係表

精益生產對六西格瑪的促進作用	六西格瑪管理對精益生產的才促進作用
(1) 較少的庫存意味著較少的損壞 (2) 小的生產批量保證發貨之前的各工序及時地發現產生的質量問題。 (3) 當備用物料減少時，下游工序對上游工序產生的問題造成壓力（逼迫上游減少問題）。 (4) 全面設備維護可以保證漸少設備故障造成的變異。 (5) 為六西格瑪管理項目指出改進機會	(1) 提高質量可以減少生產批量；減少安全庫存；縮短生產提前期 (2) 可制造性設計減少準備時間，減小生產批量 (3) 再加工的減少可以縮短生產節拍 (4) 建立在質量之上的供應關係可以減少不增加價值的檢驗，並縮短節拍 (5) 加工過程中的檢驗減少，從而縮短生產節拍 (6) 保證過程處於受控狀態 (7) 為精益生產提供了具有操作性的工具和方法

將精益生產與六西格瑪管理進行集成形成精益六西格瑪是可行的。首先，精益生產追求通過不斷改善達到零庫存、零缺陷，而六西格瑪也是一種近乎完美的目標。所以兩者都是通過持續改進，追求完美理念的典範，這是兩者精髓上的同質性，正因為如此，兩者才能有結合的可能性。其次，精益生產和六西格瑪管理都與 TQM 有密切的聯系，它們的實施都與 PDCA 的模式大同小異，都是基於流程的管理，都以顧客價值為基本出發點，精益生產以價值流為研究對像，把流程與顧客需求緊密相連，而六西格瑪管理最大地吸收了現代質量管理理念，運用流程分析的方法提高顧客滿意度，這為兩種生產模式整合提供了基礎。第三，如前所述，精益的本質是消除浪費，六西格瑪的本質是控制變異，而變異是常常是引起浪費的一種原因，所以兩種模式關注的對像不是對立的，而是關注問題的著眼點不同。

3. 精益六西格瑪實施的關鍵成功因素（KSF）

實施精益六西格瑪的關鍵成功因素包括以下幾點：

（1）關注系統。精益六西格瑪的力量在於整個系統，而不是單個的項目，實施精益六西格瑪不能把注意力僅僅集中在項目上，而還要考慮運作系統整體的改進狀況，把短期財務績效與公司長期戰略平衡考慮。精益六西格瑪不是精益和六西格瑪簡單相加，而是要把精益和六西格瑪有機接合起來，處理整個系統的問題，對於系統中不同過程或同一過程的不同階段的問題，精益和六西格瑪相互補充，才能達到 1+1 > 2 的效果，例如當過程處於起始狀態，問題較為簡單，可以直接用精益生產的方法和工具解決，但隨著過程的發展，當問題處於複雜狀態時，就要用六西格瑪的方法解決。所以在實施中要關注於整個系統，用系統的思維方式、綜合考慮、恰當選用精益六西格瑪的方法或工具。現實一些企業實施精益六西格瑪時之所以沒有達到預期效果,就是因為他們雖然同時做了精益和六西格瑪,但是沒有把二者接合在一起,而是不同的部門分別使用不同的模式。

(2) 重視文化建設。不論是精益生產還是六西格瑪管理,文化對其成功都起到了重要的作用。同樣,實施精益六西格瑪也離不開文化建設。通過文化建設,使公司每一個員工形成一種做事的習慣,自覺地按精益六西格瑪的方式去做事情。精益六西格瑪的文化是持續改進、追求完美、全員參與的文化。只有追求完美,持續地對過程進行改進,才能不斷超越現狀,取得更大的績效;而現代的組織管理是一個非常複雜的系統,個人或一部分人的力量是有限的,只有靠全員參與,才能最大地發揮出集體的能力。

(3) 以流程管理為中心。精益生產和六西格瑪管理都是以流程為中心的管理方式,因此精益六西格瑪管理也必須以流程為中心,擺脫以組織功能為出發點的思考方式。只有以流程為中心才能真正發現在整個價值流中哪些是產生價值的,哪些是浪費,進行高效的管理。

(4) 領導的支持。精益六西格瑪需要處理整個系統的問題,同時要分析和解決的問題也更複雜,需要與不同的部門進行溝通,需要得到更多資源的支持,所以沒有領導的支持是不可能成功的。領導的支持應該是實實在在的支持,而不是僅僅有口頭上的承諾,所以這就要求領導也要參與到精益六西格瑪管理變革中去,只有參與其中,才能發現問題,有力地推動精益六西格瑪。

(5) 選擇正確的人員。精益六西格瑪實施涉及到整個公司範圍的所有人員,精益六西格瑪在意識、道德、技術水平等方面對人員素質都有較高的要求。在意識方面,要求員工具有不斷進取、追求完美的精神;在道德方面,要有高尚的職業道德觀,不要斤斤計較,要有團隊精神;在技術方面要求一般員工是多能工,領導者能熟練運用精益和六西格瑪的工具和技巧,接收過系統的管理知識的培訓,有良好的溝通能力。

(6) 正確使用方法和工具。在利用精益六西個瑪方法對系統分析之後,針對具體某一點的問題,可能僅僅用到的精益生產或者六西格瑪的方法或工具,也可能需要把兩個管理模式中的方法和工具結合起來使用。例如對於簡單問題,就應該用 Kaizen 的策略,用精益生產的方法和工具直接解決,如果還用六西格瑪的方法和工具,必然降低過程的速度;而對於複雜的問題,如果不用六西格瑪的方法和工具,就不能發現真正的原因,不能有效解決問題;還有一些複雜問題需要同時利用精益的和六西格瑪的方法和工具來解決,才能達到其目的。因此,精益六西格瑪管理要實現精益生產速度和六西格瑪的過程穩健性,必須確定問題的種類,針對具體問題選用恰當的處理方法和工具。

4. 精益六西格瑪項目實施流程

我們把精益六西格瑪活動可以分為簡單精益改善活動和精益六西格瑪項目活動。簡單精益改善活動主要是針對簡單問題，這類問題可以直接用精益的方法和工具解決，傳統的精益項目和六西格瑪項目按原來運用傳統的方法和工具處理，本文不再論述。

精益六西格瑪項目主要是針對於複雜問題，狹義的精益六西格瑪項目把精益生產的方法和工具與六西格瑪的方法和工具結合起來，實施流程採用新的「定義－測量－分析－改進－控制」流程，稱為 DMAIC II，它與傳統的 DMAIC 過程的區別是它在實施中加入了精益的哲理、方法和工具。

DMAIC II 各階段內容為：定義階段利用精益思想定義價值、提出流程框架，在此框架下，結合六西格瑪工具，定義改進項目；測量階段把精益生產時間分析技術與六西格瑪管理工具結合測量過程管理現狀；分析階段運用六西格瑪技術與精益流動原則結合，分析變異和浪費；改進階段以流動和拉動為原則，運用兩種模式中的所有可以利用的工具對流程增加、重排、刪除、簡化、合並，同時對具體流程穩定性和流程能力改進；最後是控制階段，除了完成六西格瑪管理控制內容外，還要對實施中產生的新問題進行總結，以便下一個循環對系統進行進一步完善。

精益六西格瑪項目的實施步驟為：

定義階段：（1）定義顧客需求，分析系統，尋找浪費或變異，確定改進機會；（2）分析組織戰略和組織的資源；（3）確定項目：包括項目的關鍵輸出、所用資源、項目範圍。

測量階段：（4）定義流程特性；（5）測量流程現狀（包括各流程或動作需要的時間）；（6）對測量系統分析；（7）評價過程能力。

分析階段：（8）分析流程，查找浪費根源或變異源；（9）確定流程及關鍵輸入因素。

改進階段：（10）確定輸入輸出變量之間的關係，提出優化方案；（11）制定改進計劃。

控制階段：（12）建立運作規範、實施流程控制；（13）驗證測量系統，驗證過程及其能力；（14）對實施結果進行總結，規範成功經驗，提出新問題。

5. 結論

精益六西格瑪是精益生產與六西格瑪管理的結合，其本質是消除浪費。精益生產和六西格瑪管理結合實施精益六西格瑪是必要的，精益生產為六西格瑪管理提供了改善的框架，保證了系統整體的優化目標，對於簡單問題，精益生產提供了快速有效的解決方法和工具；

六西格瑪管理保證過程處於統計受控狀態，對於複雜問題能夠發現變異的來源，從根本上解決問題；實施精益六西格瑪也是可行的，二者是互補的。精益六西格瑪吸取二者的優點，把二者有機結合，而不是精益生產和六西格瑪的簡單相加，精益六西格瑪活動包括簡單精益改善活動和精益六西格瑪項目活動。實施精益六西格瑪要注重其關鍵成功因素：關注系統、注重文化建設、以流程為中心、領導支持、選擇正確的人員、正確方法和工具。對於簡單問題，可以直接用精益的方法和工具解決，對於複雜問題的精益六西格瑪項目，可以用文中提出的 DMAIC Ⅱ 的方法和流程解決，在實施 DMAIC Ⅱ 時應正確選擇實施步驟和各個階段的工具。

參考文獻：

[1] 何楨 , 車建國 . 精益六西格瑪：新競爭優勢的來源 [J]. 天津大學學報 (社科版), 2005, 7(5): 321-325

[2] 周延虎 , 何楨 , 高雪峰 . 精益生產與六西格瑪管理的對比與整合 [J]. 工業工程 , 2006, 9(6): 1-4

[3] 何楨 , 張志紅 . 精益與六西格瑪的比較研究 [J]. 工業工程 , 2006, 9(1): 1-4

[4] Michael L.George. Lean Six Sigma: Combining Six Sigma Quality With Lean Speed [M]. New York: The McGraw-Hill Companies, Inc. 2002: 8-13

[5] Udit Sharma. Implementing Lean principles with the six sigma advantage. Journal of organizational excellence, Summer 2003: 43-52

[6] Edward D. Arnheiter and John Maleyeff. The integration of lean management and Six Sigma[J]. The TQM Magazine, 2005, 17(1): 5-18

作者簡介：

陳勤業教授小檔案：

陳勤業博士，博導，兼任天津大學管理學院客座教授和比立勤國立大學訪問教授，現任好孩子集團副總裁。政協委員。昆山市博士聯誼會副會長。畢業於香港理工大學機械工程專業，獲機械工程學士、工商管理及質量管理碩士，後獲工學及哲學博士，企業博士後。2005 年 10 月成為唯一的香港人當選首屆中國優秀質量人，獲英國劍橋國際名人傳記中心終生成就獎、科技成就獎和美國國際名人傳記中心世界成就獎。人民日報社中國經濟週刊專家顧問理事會常務理事。2006 年 3 月被香港特區政府委任為公共事務論壇委員會成員。香港聖約翰救傷隊東方救護支隊副會長。2006 中國十大管理實踐專家委員會專家委員。世界經理人雜誌專家及專欄作家。

在 2000 年入選首批中國工業工程專家，全國六西格瑪推進工作委員會專家委員，深圳市質量協會專家委員會副主任委員，中國質量協會理事，英國劍橋國際名人傳記中心名譽理事長，香港質量協會／香港六西格瑪學會名譽顧問，香港品質協會名譽顧問，註冊六西格瑪黑帶大師，英國註冊高級質量顧問師，國際註冊質量管理系統首席審核師，英國皇家特許資深工程師，英國皇家特許資深市務師，英國皇家特許資深經理，中國註冊市場總監，英國註冊資深財務總監，香港工程師學會資深院士（FHKIE），中國機械工程學會／中國電機工程學會／中國質量協會高級會員及許多英、美國、香港專業學會院士。（FCIM，FCMI，FCILT，FIQA，FIET，FITA，F.CIM，FHKQMA，FHKIM，MASME，MIEAust，PEng，Chartered Engineer，Chartered Marketer，Chartered Manager，CPM(HK & AP)，RPE etc.）。

陳博士從事質量管理及運作管理研究工作近三十年，在如何將全面質量，運作管理轉化為生產力方面頗有建樹。

作者聯系方式：

e-mail：philipchan88@yahoo.com

Doctor LAW Ping Keung Hely, Visiting Scholar

羅秉強博士、訪問學者

Education 教育經歷

2020 – Present

Visiting Scholar of Stanford_Berkeley_SCUMB online program

美國加州州立大學联合斯坦福大學、伯克利加州大學三大名校联合網上訪問學者項目

2017 – 2020

Azteca University Doctor of Philosophy in Civil Engineering Management

阿茲特克大學土木工程管理哲學博士

2014 – 2016

University of Sunderland Bachelor of Engineering (Hons)in Electronic and Electrical Engineering

新特蘭大學電子及電機工程榮譽學士

2011 – 2014

Heriot-Watt UniversityMaster of Science in Architectural Engineering (Building Services)

赫瑞瓦特大學屋宇裝備科學碩士

2000 – 2004

University of Central LancashireBachelor of Engineering (Hons) in Fire Engineering

蘭開夏大學消防工程榮譽學士

Professional Institution 專業學會

Think Tank Member, Dashun Foundation, Hong Kong
香港大舜基金智囊團成員

Member of International Postdoctor Association
國際博士後協會會員

Member of Institute of Fire Engineer
消防工程師學會會員

Member of Chartered Institute of Plumbing & Heating Engineering
給排水及採暖特許工程師學會會員

Member of Society of Operations Engineers
營運工程師學會會員

Member of Institute of Plant Engineer
廠房工程師學會會員

The Impact of Occupational Stress on Work Performance of Civil Engineering Project Managers

Law Ping Keung Hely *Ir Prof. Philip Chan

CHAPTER 1　INTRODUCTION

Research Background

Civil engineering is a professional discipline dealing with the design, construction, and maintenance of the physical built environment (Institution of Civil Engineers, n.d.). Construction of building, road, railway, and airport are examples of civil engineering project. The civil engineering project manager is a challenging profession. A civil engineering project manager is responsible for overseeing civil engineering projects from inception to completion (Leung, Chan, & Yu, 2009). Civil engineering projects are very unique human endeavors, having mixed goals and objectives for multiple stakeholders, including clients, suppliers, architects, surveyors, designers, contractors, and workers (CIOB, 1996). Hence, project managers should be equipped with diverse skills to cope with the demanding, complicated and dynamic situations that will occur during the process of civil engineering projects (Love, Edwards, &Irani, 2010; Munns&Bjeirmi, 1996).

Table 1 is a summary of the duties that a civil engineering project manager would be responsible for from pre-construction stage to completion stage. At the pre-construction stage, they are responsible for the evaluation of the project's feasibility, preparation for the construction program, gaining consent from clients about the design and specifications and quality assurance, and the organization of project team. At the construction stage, they are responsible for monitoring the progress, meeting with stakeholders, material supplies, and human resources management. At the post-construction stage, they still have duties in complying the final accounts, development of maintenance plan and facilities management plan, and reporting to the clients.

Table 1 Construction project managers' duties from pre-construction stage to completion stage.

Pre-construction Stage	Construction Stage	Completion Stage
• Ascertain the viability of the project in financial, technical, and time terms • Prepare master and construction programs for the design and construction phases • Obtain agreement from clients about the building design and specifications • Ensure that the quality, control, quantity assurance, and testing requirements are clearly defined and agreed with clients • Select and appoint consultants, contractors, sub-contractors • Develop the project team	• Review and monitor construction progress • Set up regular progress meetings with designers, contractors, sub-contractors, consultants, and suppliers • Coordinate a diverse range of material, plant, and labor inputs • Ensure site safety and security • Ensure the projects are environmentally friendly • Ensure work quality • Review, monitor, and measure the project budget and variation orders • Keep the project team motivated	• Ensure the final accounts comply with the contract conditions • Ensure all claims have been settled • Develop a maintenance program • Plan facilities management • Feedback to the clients

Source: Leung, Chan, & Yu, 2009.

A civil engineering project manager has a lot of duties during the whole construction project period. They have to equip with a diversified range of skills to meet the job demands. A civil engineering project manager should have skills such as scheduling, cost controlling, communicating, reporting, and negotiating, in order to plan, organize and supervise project tasks from the begin to the end of the project (Leung et al., 2009). It is not surprised that project managers are considered as the key persons for the success of civil engineering project (Chan, Scott, & Chan, 2004; Leung, Chan, &Dongyu, 2011; Munns&Bjeirmi, 1996). Hence, work performance of the project managers is very crucial in the civil engineering industry (Pheng&Chuan, 2006). The interest in the investigation into the factors affecting work performance of civil engineering project managers has been increasing (Dainty, Cheng, & Moore, 2005; Enshassi, Mohamed, &Abushaban, 2009).

An, Zhang, & Lee (2013) asserted that the success of a civil engineering project is dependent on the effective and efficient management of human resources. Zhang et al. (2013) explained that effective and efficient management of human resources in construction industry aims to improve employee job performance, specially the work performance of project managers. The performance of project managers is vital and indispensable in any civil engineering projects (Pheng&Chuan, 2006). Hence, the success of civil engineering project could be achieved through the maintenance and improvement of job performance of project managers (Chan et al., 2004). Project managers' job performance is a key indicator for measuring construction project success (Chan & Chan, 2004).

Pheng&Chuan (2004) believed that environmental factors affecting job performance of project managers in the construction industry. But stress has been more frequently cited as the major factor affecting job performance of project managers, as well as other professionals, in the construction industry (Bowen et al., 2014; Enshassi& Al Swaity, 2015; Leung, Chan, &Dongyu, 2011; Leung, Yu, & Chong, 2015).

Jobs in the civil engineering and construction industry are stressful (Loosemore&Waters, 2004; Naoum et al., 2018; Zhang et al., 2013). Due to the job nature of project manager, the stress level of this occupation would be higher. Civil engineering project managers occupy a crucial position for the project success (Munns&Bjeirmi, 1996) and have to cope with a lot of duties from the begin to the end (Leung et al., 2009). During the project management process, they have also toalign the diverse interests of different stakeholders and to complete the projects with certain resources within a limited time period (Dainty et al., 2003). It is inevitable that civil engineering project managers would face with the problems of workload and role conflict. In addition, project managers need to handle with a lot of uncertainties and instabilities in terms of material availabilities, weather conditions, site conditions, human resources, and financial situations. Therefore, project managers will encounter numerous occupational sources of stress and perceive high level of job stress. Chartered Institute of Building (CIOB) conducted a survey on the occupational stress in the construction industry and found that the majority of respondents had suffered from stress (CIOB, 2006). Studies have also been conducted to evaluate the occupational stress for construction project managers and other professionals in Hong Kong, Korea, and South Africa (Bowen, Edwards, & Lingard, 2012; Leung, Chan, Y &Dongyu, 2011; Leung, Liang, & Chan, 2016; Zhang et al., 2013).

It is noticed that the level of occupation stress for civil engineering project managers is high. Literature suggests that occupational stress may have negative effect on job performance (Jamal, 1985; Jex, 1998; Kazmi et al., 2008; Motowidlo et al., 1986; Nabirye et al., 2011). As discussed above, civil engineering project managers have many duties in the construction project. They play a critical role in the project success. The job performance of civil engineering project managers will influence the project performance. Therefore, it is necessary to understand how occupational stress affects their job performance. On the other hand, it seems very little literature on the effect of occupational stress on job performance in the context of civil engineering projects (Leung et al., 2011). The research issue of stress-performance relationship in construction industry has not well documented. Further research is required to explore how occupational stress affects project manager's job performance. Findings are expected to be helpful in improving their job performance (Zhang & Fan, 2013). This motivates the performance of the current study. The research problem is explained in the following section to provide a better understanding of the research's purpose and objectives.

Statement of Problem

Occupational stress occurs when employees have continually experienced some stressful conditions at work, which are also known as sources of occupational stress or occupational stressors.

Sources of occupational stress may include heavy workload, poor interpersonal relationship, and poor work environment. This research aims to identify the sources of occupational stress that significantly affect civil engineering project managers' work performance. It is understood that work performance of civil engineering project managers is critically important for project success. But their work performance would be undermined by occupational stress. It is because occupational stress has been found to have negative impact on employee work performance in many industries. Civil engineering project manager will encounter many sources of occupational stress that will increase their stress level. Hence, civil engineering project manager is a very stressful occupation. It is likely that their work performance could be undermined by the sources of occupational stress. Subsequently, project performance could also be negatively affected. The knowledge of relationship between sources of occupational and work performance of civil engineering project manager could be useful for the civil engineering industry. This research intends to address the following research problem:

What is/are the source(s) of occupational stress that significantly affect civil engineering project managers' work performance?

The research problem is justified in the following section.

Justification of the Problem

It is believed that occupational stress could deteriorate employees' work performance in various occupations (Kazmi, Amjad, & Khan, 2008; Nabirye et al., 2011; Slu, 2003; Wu, 2011). However, the empirical research on the relationship between occupational stress and work performance is still not sufficient (Bhatia & Goyal, 2018). Moreover, the impact of occupational sources of stress on work performance of civil engineering project managers has not yet been well documented and still need further research (Leung et al., 2016). This motivates the conduction of this research to evaluate the impact of some sources of occupational stress on work performance of civil engineering project managers, in order to fill the knowledge gap.

Sources of occupational stress is also known as occupational stressors. The set of significant stressors in civil engineering industry may be different from those in other industries. Work-related stressors in civil engineering industry may include work overload, role conflict, poor work environment, and poor interpersonal relationship. Some of them may have significant impact on job performance, but some may not. Practically, human resource management should acquire the knowledge about which occupational stressors affect job performance. The knowledge could help formulate and implement better managerial policies in order to reduce project manager's occupational stress. As a result, project manager's work performance is enhanced. It is expected that this research can identify the significant occupational stressors that undermine project manager's work performance. Then, the findings could contribute to practical managerial policies for improving project manager's work performance.

Research Aim and Objectives

This research aim is to identify the sources of occupational stress that significantly affect civil engineering project managers' work performance. Hence, it aims to address the following research problem:

What is/are the source(s) of occupational stress that significantly affect civil engineering project managers' work performance?

The research aim can be achieved by. firstly, exploring the sources of occupational stress in civil engineering industry by reviewing relevant literature, and, secondly, evaluating the effect of these sources of occupational stress on project manager's work performance. Research objectives are outlined as follows:

(1) To conduct a secondary research to explore the potential sources of occupational stress for civil engineering project managers;

(2) To conduct a questionnaire survey to assess civil engineering project managers' perception of sources of occupational stress and self-appraised job performance;

(3) To perform statistical analyses to evaluate the effect of sources of occupational stress on job performance;

(4) To conclude the significant factors undermining job performance of civil engineering project managers.

Plan of Research Methods

The research aim is to assess the sources of occupational stress for civil engineering managers and evaluate the effect of these occupational stressors on job performance. For accomplishing the research purpose, scientific and objective measurements of occupational stress and work performance are needed. Therefore, the assumptions of positivism are employed (Bell et al., 2018). The primary study is designed as a quantitative study. Quantitative data about occupational stress and work performance are gathered for statistical analysis (Burns & Burns, 2008).

Primary and quantitative data are obtained by employing the survey strategy. Hence, a sample of civil engineering project managers are selected. Data collected from the sample are analyzed to infer the findings for the target population (Bell et al., 2018; Ross et al., 2013). It is a research strategy that facilitates the collection of a relatively large sample of primary data with less time and resource (Fowler, 2013).Then, convenience sampling is used for choosing research subjects from the target population (Etikan et al., 2016). The sample size is about 200. It is believed to be adequate sample size in the investigation into occupational stress for civil engineering project managers in Hong Kong (Leung et al., 2011).

Questionnaire is employed as the data collection technique. It is a research instrument for collecting primary data by using close-end questions. Hence, the collected data can be easily translated into numbers for quantitative analysis (Hair et al., 2008). A questionnaire is developed by adapting from Enshassi& Al Swaity (2015) and Koopmans et al. (2014). Finally, primary data collected by the questionnaire are analyzed with statistical procedures by using SPSS 23.0 (Burns & Burns, 2008).

Structure of the Thesis

There are three chapters in the thesis.

Chapter 1: Introduction

This chapter introduces research problem and background in order to set the setting of this research.

Chapter 2: Data Analysis and Findings

This chapter presents results of statistical analysis on the collected primary data and gives interpretations on the results.

Chapter 3: Conclusion and Recommendations

This chapter draws conclusions based on research findings and provides recommendations for reducing occupational stress and improving work performance.

CHAPTER 2   DATA ANALYSIS AND FINDINGS

Data analysis to be performed

The following data analysis are to be performed. They are (1) descriptive demographic data analysis, (2) Cronbach's Alpha analysis, (3) mean, standard deviation, skewness and t-test analysis, (4) one-way ANOVA analysis, (5) correlation coefficient analysis and (6) multiple regression analysis.

Descriptive demographic data analysis

In the questionnaire, the following demographic information are asked. They are (1) gender, (2) age, (3) marital status and (4) years of working. The demographic data is described one-by-one below. There are totally 138 respondents in this survey. Table 5 shows all demographic information among the respondents.

From Table 5 below, there are more male respondents than female respondents. They are at the number of 79 and 59 or 57.2% and 42.8% respectively. It is quite normal to have more male respondents than female respondents in the engineering field.

Besides, a slight majority of respondents is aged between 40 and 49 years, amounting to 71 respondents or 51.4% of all respondents. The other three age groups account for a smaller percentage of respondents. They are 27 respondents or 19.6% for 18-29 years old, 28 respondents or 20.3% for 30-9 years and 12 respondents or a small 8.7% for those aged 50 years or above.

Furthermore, as for the distribution for marital status, most of the respondents are married.

They account for 84 respondents out of the 138 respondents or 60.9% of them. Another large group are those who are single, accounting for 45 respondents out of 138 respondents or 32.6%of them. Those who are married but single now only account for 9 respondents or 6.5% of them.

The final demographic information collected is years of working. They are evenly distributed between different years of working. They are 27 or 19.6%, 36 or 26.1%, 30 or 21.7%, 25 or 18.1% and 20 or 14.5% for those who work less than one year, 1-5 years, 6-10 years, 11-15 years and more than 15 years respectively.

Table 5 Demographic distribution

		n	%	Cumulative %
Gender	Male	79	57.2	57.2
	Female	59	42.8	100.0
	Total	138	100.0	
Age	18-29 years	27	19.6	19.6
	30-39 years	28	20.3	39.9
	40-49 years	71	51.4	91.3
	>50 years	12	8.7	100.0
	Total	138	100.0	
Marital status	Single	45	32.6	32.6
	Married	84	60.9	93.5
	Married but single now	9	6.5	100.0
	Total	138	100.0	
Years of working	Shorter than one year	27	19.6	19.6
	One to five years	36	26.1	45.7
	Six to ten years	30	21.7	67.4
	Eleven to fifteen years	25	18.1	85.5
	Longer than fifteen years	20	14.5	100.0
	Total	138	100.0	

Source: From author's data analysis

Cronbach Alpha analysis

The following table 6 shows the results of Cronbach's Alpha analysis. As a benchmark, Cronbach's Alpha is considered acceptable at a level above 0.700. A level of 0.800 is considered good level of Cronbach's Alpha while a level of higher or equal to 0.900 is considered excellent level of Cronbach's Alpha (Cohen, 1988). From table 6, all eleven variables in the research studies shows at least acceptable Cronbach's Alpha at a level of higher than 0.700. Therefore, the measurement scales are all internally consistent. Internal consistency means an acceptable correlations between different items on the same test.

Table 6 Cronbach's Alpha analysis

Variables	Cronbach's Alpha
Personality-home-work conflict	0.709
Relationship with others	0.800
Distrust	0.842
Work Overload	0.923
Role Conflict/Ambiguity	0.777
Poor Environment	0.789
Organizational Policies	0.832
Autonomy	N/A since only two items
Task performance	0.870
Contextual performance	0.902
Counterproductive behavior	0.757

Source: From author's data analysis

Mean and standard deviation analysis

Mean and standard deviation analysis measures the average score for each variable listed above in Table 6 and the dispersion of scores for each variable listed above in Table 6. The results of mean, standard deviation and skewness analysis is shown in Table 7.

Table 7 Mean, standard deviation and skewness analysis

Variables	Mean	Standard deviation (% of mean)	Mid-point score	Differences from mid-point	Sig. (2-tailed)	Skewness
Personality-home-work conflict	17.0942	3.08253 (18.03%)	18.0	-0.91	0.001	.916
Relationship with others	19.1522	4.63075 (24.18%)	18.0	1.15	0.004	-.192
Distrust	12.7899	3.79676 (29.69%)	14.5	-1.71	0.000	.649
Work Overload	33.7536	11.10127 (32.89%)	32.0	1.75	0.066	.145
Role Conflict/ Ambiguity	12.2029	3.37695 (27.67%)	14.5	-2.30	0.000	1.239
Poor Environment	17.0507	4.13693 (24.26%)	18.0	-0.95	0.008	.095
Organizational Policies	21.3043	5.00454 (23.49%)	21.5	-0.20	0.647	.132

Autonomy	7.4130	2.88142 (38.87%)	7.5	-0.09	0.723	.046
Task performance	22.4783	6.75669 (30.06%)	18.0	4.48	0.000	-.270
Contextual performance	34.8551	8.86356 (25.43%)	28.5	6.36	0.000	-.601
Counterproductive behavior	15.8913	4.19364 (26.39%)	18.0	-2.11	0.000	1.639

Source: From author's data analysis

From Table 7, in terms of the mean score, the eight independent variables such as personality-home-work conflict and autonomy are all having scores close to the mid-point score. The largest difference is found in role conflict/ambiguity which the score is much lower than the mid-point score of 14.5.

However, for task performance, contextual performance and counterproductive behavior, their differences from the mid-point score are 4.48, 6.36 and -2.11 respective. They signify that the 138 respondents believe that they have better job performance and do less counterproductive behavior than the mid-point score.

As for the other scores, the 138 respondents have a lower score in counterproductive behavior, role conflict/ambiguity, distrust and higher score in contextual performance and task performance. They are all good sign about the work stresses and work performance.

The column showing "sig. (2-tailed)" shows whether the mean scores are statistically significantly different from the mid-point score. Using $p<0.05$ as the benchmark, only work overload, organizational policies and autonomy are not statistically significant, meaning that all other variables have more than 95% of chances that the mean score is higher than the mid-point score except for work overload, organizational policies and autonomy.

The column "standard deviation (% of mean)" measures the amount of dispersion of a set of values in the variable (Mendenhall, Beaver & Beaver, 2012). The value of standard deviation shown in Table 7 represents one unit of standard deviation from the mean. For example, in personality-home-work conflict, the range of score from one standard deviation from the mean is 14.01 (17.09 minus 3.08) to 20.18 (17.09 plus 3.08), the range of score from one standard deviation from the mean is 22.65 to 44.85, etc. Assuming normal distribution, 68% of scores are within one standard deviation from the mean and 95% of scores are within two standard deviation from the mean. Therefore, in personality-home-work conflict, the range of score from two standard deviation from is 10.93 to 23.26 and 11.55 to 55.96 for work overload, etc.

The percentage of mean in one standard deviation measures the amount of dispersion of score. Among them, personality-home-work conflict has the smallest dispersion of score (i.e. 18.03%)

and autonomy has the largest dispersion of score (i.e. 38.67%). Workload has the second largest di0spersion of score from the mean (i.e. 32.89%). The larger the dispersion of score from the mean, the broader the range of scores and the greater the differences among the perspectives of respondents.

Skewness measures the asymmetry of the probability distribution of a real-valued random variable about its mean. A negative figure means the tail of the distribution curve is skewed to left while a positive figure means the tail of the distribution is skewed to the right. Therefore, a negative figure signifies that mean<median<mode while a positive signifies that mean>median>mode.

From the figures of skewness, most variables have a positive skewness, for example, counterproductive behavior (1.639), role conflict/ambiguity (1.239), personality-home-work conflict (0.916) and distrust (0.646). It means that there are more people score low in counterproductive behavior, role conflict/ambiguity, personality-home-work conflict and distrust. It is a good sign since the respondents generally have a good working environment and fewer stresses. For those in higher negative skewness, they are task performance, contextual performance. It is a good sign since more people score higher in work performance.

Overall, the scores in mean, standard deviation and skewness that are discussed as showing good signs are:

Mean:High scores of contextual performance and task performance/ low scores of role conflict/ambiguity, counterproductive behavior and distrust.

Standard deviation: High percentage of standard deviation in autonomy, work overload, task performance and distrust/ Low percentage of standard deviation in personality-home-work conflict, (poor) organizational policies and (poor) relationship with others.

Skewness: Positive skewness for counterproductive behavior, role conflict/ambiguity, personality-home-work conflict and distrust/ Negative skewness for task performance and contextual performance.

As a result, the score of mean and skewness shows positive signs for contextual performance and task performance, role conflict/ambiguity, counterproductive behavior and distrust among the 138 respondents. Besides, the score of standard deviation shows that there are wider dispersions for autonomy, work overload, task performance and distrust scores.

One-way ANOVA analysis

The one-way ANOVA analysis results for gender group and age group are worthy of discussion. In the following section, the mean score for different groups are shown first and, then with the one-way ANOVA results.

One-way ANOVA analysis for gender group

First, for the one-way ANOVA analysis in gender group, statistically significant differences ($p < 0.05$) (see sig. level of one-way ANOVA test) in scores of personality-home-work conflicts, (poor) relationship with others, distrust, work overload, poor environment, (poor) autonomy, task performance and contextual performance. While personality-home-work conflicts, (poor) relationship with others and distrustare generally not having differences between gender, the differences in other variables (i.e. work overload, poor environment, (poor) autonomy, task performance and contextual performance) can be explained by job nature and research studies about gender differences.

First, in terms of job nature, male workers in civil engineering projects generally work in more dangerous environment. It then results in (poor) working environment. Second, work overload results in male gender because there may be gender differences in job task allocation (De Pater, Van Vianen&Bechtoldt, 2010). Using a sample of 317 employees and 140 working at middle job levels and at senior job levels in a pharmaceutical company, male is more associated with challenging job experience such as work overload and being assigned of challenging job experience (De Pater, Van Vianen&Bechtoldt, 2010). Also, civil engineering is traditionally a job which is male-dominant. Therefore, it may explain why male gender has a higher score in work overload. Because of work overload, they may perceive themselves as deserving better job performance in both tasks and contextual performance (i.e. take on extra responsibility, take on challenging work tasks). However, the assignment of more challenging job and work overload may make employees in male gender feel that they do not have enough autonomy to finish their job.

Table 8 One-way ANOVA analysis for gender differences

Gender group	N	Mean	Standard deviation	Sig. level of one-way ANOVA test
Personality-home-work conflict				
Male	79	18.0759	3.26116	
Female	59	15.7797	2.25196	0.000
(Poor) Relationship with others				
Male	79	19.8608	3.68550	
Female	59	18.2034	5.54847	0.037
Distrust				
Male	79	13.3797	3.49837	
Female	59	12.0000	4.05990	0.034
Work Overload				
Male	79	38.3038	9.92453	
Female	59	27.6610	9.62685	0.000
Role Conflict/Ambiguity				
Male	79	11.9620	2.72897	
Female	59	12.5254	4.09102	0.334

Poor working environment				
Male	79	18.1772	3.96704	
Female	59	15.5424	3.90103	0.000
(Poor) Organizational Policies				
Male	79	21.4557	3.94773	
Female	59	21.1017	6.17474	0.683
(Low) autonomy				
Male	79	6.9747	2.43360	
Female	59	8.0000	3.32182	0.038
Task performance				
Male	79	23.8987	5.66500	
Female	59	20.5763	7.63206	0.004
Contextual performance				
Male	79	37.0886	7.60767	
Female	59	31.8644	9.58349	0.000
Counterproductive behavior				
Male	79	15.7089	3.79997	
Female	59	16.1356	4.69210	0.556

Source: From author's data analysis

One-way ANOVA analysis for different age groups

There are statistically significant mean differences across different age groups. Notably, they are personality-home-work conflicts, work overload, poor working environment, (poor) organizational policies, (low) autonomy and counterproductive behavior. They can be explained by the challenges of people faced at different life stages.

First, those aged 40-49 years old have higher personality-home-work conflict because they reach the age of having their own family, children and more responsibilities. They are thus more likely to feel highly preoccupied by work and home responsibilities. As for the younger generations aged 18-29 years old, they are more likely to experience work overload, feeling poor working environment, poor organizational policies, lower autonomy and doing more counterproductive behavior. It is because they are more concerned about their career development and act more aggressive to obtain more job responsibilities. Therefore, they experience work overload. They may also be more independent and therefore they may complain more about poor working environment, poor organizational policies and lower autonomy. Therefore, they may get disgruntled and do more counterproductive behavior than those older workers.

Table 9 One-way ANOVA analysis for different age groups

Age group	N	Mean	Standard deviation	Sig. level of one-way ANOVA test
Personality-home-work conflict				
18-29 years	27	18.2222	4.71767	
30-39 years	28	14.0000	4.06430	
40-49 years	71	21.7183	2.79378	
50 years or above	12	18.0833	3.28795	0.000
(Poor) Relationship with others				
18-29 years	27	18.2222	4.71767	
30-39 years	28	14.0000	4.06430	
40-49 years	71	21.7183	2.79378	
50 years or above	12	18.0833	3.28795	0.000
Distrust				
18-29 years	27	3.99929	.76966	
30-39 years	28	4.11395	.77746	
40-49 years	71	2.83723	.33672	
50 years or above	12	4.39697	1.26930	0.000
Work Overload				
18-29 years	27	37.7037	10.42119	
30-39 years	28	27.9643	7.62298	
40-49 years	71	34.2958	12.23624	
50 years or above	12	35.1667	7.42028	0.009
Role Conflict/Ambiguity				
18-29 years	27	13.2593	3.68565	
30-39 years	28	10.2500	3.40615	
40-49 years	71	12.5915	3.07328	
50 years or above	12	12.0833	2.81096	0.004
Poor working environment				
18-29 years	27	20.8519	4.20351	
30-39 years	28	13.2857	3.57830	
40-49 years	71	16.8873	2.88122	
50 years or above	12	18.2500	3.72034	0.000
(Poor) Organizational Policies				
18-29 years	27	24.9630	3.97571	
30-39 years	28	17.5357	5.38504	
40-49 years	71	21.7183	4.39865	
50 years or above	12	19.4167	2.50303	0.000

(Low) autonomy				
18-29 years	27	5.6296	3.20034	
30-39 years	28	5.2143	2.20029	
40-49 years	71	9.0704	2.05860	
50 years or above	12	6.7500	1.42223	0.000
Task performance				
18-29 years	27	20.9259	6.25070	
30-39 years	28	17.7500	5.09629	
40-49 years	71	25.5915	6.32360	
50 years or above	12	18.5833	4.01040	0.000
Contextual performance				
18-29 years	27	33.1481	9.76140	
30-39 years	28	36.2857	8.56287	
40-49 years	71	35.4789	8.78938	
50 years or above	12	31.6667	7.46304	0.306
Counterproductive behavior				
18-29 years	27	17.5926	5.37272	
30-39 years	28	13.9286	2.94302	
40-49 years	71	16.2394	3.95859	
50 years or above	12	14.5833	3.20393	0.000

Source: From author's data analysis

One-way ANOVA analysis for different marital status groups

From Table 10, there are statistically significant differences in (poor) relationship with others, work overload, role conflict/ambiguity, (poor) organizational polices, (low) autonomy, task performance and contextual performance.

First, in terms of (poor) relationship with others and (low autonomy), those who are married have higher score than those who are single and married but single now (i.e. divorced). In terms of work overload, those who are married have higher score than those who are single and married but single now (i.e. divorced). In terms of role ambiguity and (poor) organizational policies, those who are divorced have higher score than those who are single and married. In terms of both task performance and contextual performance, those who are married have higher score than those who are single and divorced.

Table 10 One-way ANOVA analysis for different marital status

Marital status	N	Mean	Standard deviation	Sig. level of one-way ANOVA test
Personality-home-work conflict				
Single	45	17.6000	3.86946	
Married	84	17.0238	2.37951	
Married but single now	9	15.2222	4.02423	0.101
(Poor) Relationship with others				
Single	45	16.6889	5.41808	
Married	84	20.7143	3.50068	
Married but single now	9	16.8889	3.62093	0.000
Distrust				
Single	45	12.2667	4.80246	
Married	84	12.9405	3.23889	
Married but single now	9	14.0000	2.78388	0.389
Work Overload				
Single	45	34.0000	11.36582	
Married	84	34.8333	10.47985	
Married but single now	9	22.4444	10.19940	0.006
Role Conflict/Ambiguity				
Single	45	12.6222	4.19030	
Married	84	11.6310	2.62876	
Married but single now	9	15.4444	3.28295	0.003
Poor working environment				
Single	45	18.0000	5.91608	
Married	84	16.5238	2.77022	
Married but single now	9	17.2222	3.59784	0.154
(Poor) Organizational Policies				
Single	45	21.9778	6.45082	
Married	84	20.5714	4.09903	
Married but single now	9	24.7778	1.98606	0.000
(Low) autonomy				
Single	45	5.6444	2.99360	
Married	84	8.2738	2.43148	
Married but single now	9	8.2222	2.33333	0.000
Task performance				
Single	45	19.0889	6.25526	
Married	84	25.2976	5.29100	
Married but single now	9	13.1111	5.41859	0.000

	N	Mean	Standard deviation	Sig. level of one-way ANOVA test
Contextual performance				
Single	45	35.0000	9.10295	
Married	84	36.3690	7.06728	
Married but single now	9	20.0000	9.94987	0.000
Counterproductive behavior				
Single	45	16.4444	5.79402	
Married	84	15.3452	2.97173	
Married but single now	9	18.2222	3.73423	0.082

Source: From author's data analysis

One-way ANOVA analysis for different years of working groups

There are statistically significant differences in personality-home-work conflict, (poor) relationship with others, distrust, work overload, (poor) working environment, (poor)organizational policies, (low) autonomy, task performance and counterproductive behavior.

For personality-home-work conflict, (poor) relationship with others, task performance and (low autonomy), those who worked for 1-5 years and 11-15 years have higher score than those of other years of working groups. For distrust and work overload, those who work for 11-15 years and 15 years or above have higher score than other years of working groups. For (poor) working environment, those who work for <1 year and 11-15 years have higher score than other groups. Finally, for (poor) organizational policies and counterproductive behavior, those who work for <1 year and 1-5 years have higher scores than other groups.

Table 11 One-way ANOVA analysis for different years of working groups

Years of working group	N	Mean	Standard deviation	Sig. level of one-way ANOVA test
Personality-home-work conflict				
<1 year	27	17.1481	2.67040	
1-5 years	36	17.9167	3.75214	
6-10 years	30	16.6667	2.10637	
11-15 years	25	17.8000	2.76887	
15 years or above	20	15.3000	3.27832	0.022
(Poor) Relationship with others				
<1 year	27	16.1481	2.86496	
1-5 years	36	21.9444	4.97390	
6-10 years	30	17.3000	5.35724	
11-15 years	25	20.5200	2.83019	
15 years or above	20	19.2500	2.89964	0.000

Distrust				
<1 year	27	12.1481	3.03446	
1-5 years	36	12.4167	4.17732	
6-10 years	30	11.1333	3.28773	
11-15 years	25	14.7200	2.89425	
15 years or above	20	14.4000	4.39378	0.001
Work Overload				
<1 year	27	33.4815	5.43047	
1-5 years	36	31.4444	13.08531	
6-10 years	30	31.9333	13.53395	
11-15 years	25	34.3200	9.69845	
15 years or above	20	40.3000	8.57229	0.049
Role Conflict/Ambiguity				
<1 year	27	11.5926	3.00332	
1-5 years	36	13.5278	3.43500	
6-10 years	30	12.0667	4.26642	
11-15 years	25	11.6000	2.90115	
15 years or above	20	11.6000	2.23371	0.093
Poor working environment				
<1 year	27	19.8148	4.46385	
1-5 years	36	16.6944	3.67866	
6-10 years	30	15.0000	4.14396	
11-15 years	25	17.4400	3.35510	
15 years or above	20	16.5500	3.54631	0.000
(Poor) Organizational Policies				
<1 year	27	23.6296	4.75676	
1-5 years	36	24.3611	4.40229	
6-10 years	30	18.9000	5.96166	
11-15 years	25	19.6400	2.30723	
15 years or above	20	18.3500	2.36810	0.000
(Low) autonomy				
<1 year	27	4.4074	1.92672	
1-5 years	36	9.2222	2.92878	
6-10 years	30	7.7000	2.84241	
11-15 years	25	8.8000	.91287	
15 years or above	20	6.0500	1.39454	0.000

Task performance				
<1 year	27	19.1852	3.35166	
1-5 years	36	25.1667	7.27226	
6-10 years	30	19.5000	7.56922	
11-15 years	25	24.8800	4.77249	
15 years or above	20	23.5500	6.94698	0.000
Contextual performance				
<1 year	27	31.8889	5.36609	
1-5 years	36	33.0278	9.50635	
6-10 years	30	37.2000	11.34536	
11-15 years	25	36.8000	7.17635	
15 years or above	20	36.2000	8.04330	0.078
Counterproductive behavior				
<1 year	27	16.2593	2.92985	
1-5 years	36	17.3889	4.69819	
6-10 years	30	16.0667	5.15239	
11-15 years	25	15.9600	1.36870	
15 years or above	20	12.3500	3.71731	0.000

Source: From author's data analysis

Correlation coefficient analysis

Correlation coefficient analysis measures how strong a relationship is between two variables. Therefore, as shown in Table 12, measurements of the relationship between two variables can be between every variable in the research model (i.e. PHWC, PR, DT, WO, RC, PE, POP, PA, TP, CP and CB). After that, the result is produced in Table 12. According to Cohen (1988), the range of effect size associated with different magnitude is suggested to be:

Small: r=.10-.30

Medium: r=.30-.50

Large: r=.50-1.00

Following the above guideline, those with effect size of medium or larger and positive correlation is highlighted in bold. Those with small effect size are highlighted in italic. Those with negative correlation coefficient are underlined. Those which are not statistically significant are neither highlighted nor underlined.

From Table 12, the first four variables i.e. PHWC, PR, DT and WO are highly related to each other at a medium to high degree. For example, PHWC is strongly related to PR and WO at a level of .569 (p<0.01) and .544 (p<0.01). PR is related to RC at level of .504 (p<0.01). For the other variables,

POP is strongly related to RC at a level of .588 (p<0.01). PA is strongly related to PR at a level of .714 (p<0.01).

As for task performance, it is moderately and strongly related to PHWC, PR, WO and PA at a level of .401, .707, .437 and .436 respectively (p<0.01). As for contextual performance, it is moderately and strongly related to PHWC, WO and TP at a level of .365, .522 and .624 respectively (p<0.01). Then, CP is negatively related to RC and POP at a level of -.314 and .307 (p<0.01). As for counterproductive behavior, it is positively moderately and strongly related to PR, DT, RC, PE, POP and PA at a level of .393, .472, .759, .533, .690 and .527 respectively (p<0.01) and negatively weakly related to CP at a level of -.202 (p<0.05).

As an overall analysis of the result of correlation coefficient analysis, most of the relationships are explainable. For example, distrust is associated with (poor) relationship with others (r=.332, p<0.01). Work overload is associated with personality-home-work conflict (r=.544, p<0.01), (poor) relationship with others (r=.439, p<0.01) and distrust (r=.417, p<0.01). Role conflicts/ambiguity is associated with (poor) relationship with others (r=.504, p<0.01) and distrust (r=.401, p<0.01).

(Poor) organizational policies is also associated with many variables, for example, role conflicts/ambiguity (r=.588, p<0.01),(poor) working environment (r=.512, p<0.01) and (poor) relationship with others (r=.380, p<0.01). It may be because (poor) organizational policies generate conflicts at work such as (poor) relationship with others and role conflicts/ambiguity because of task and resource allocation problems. Also, (poor) autonomy also generates (poor) relationship with others (r=.707, p<0.01) and role conflicts/ambiguity (r=.495, p<0.01). It may be because a person with little autonomy fight for more autonomy, resulting in conflicts with others. They may also have role conflicts/ambiguity because of the lack of autonomy. Lack of autonomy makes them difficult to know what to do and what their job are meant for. This generates role conflicts/ambiguity. In more succinct terms, it arises from incompatible demands placed upon a person relating to their job or position (Arnold et al, 2016).

As for task performance, contextual performance and counterproductive behavior, some relationships are normal, especially for those related to contextual performance and counterproductive behavior. However, it is not for the variables' relationship with task performance. First, for contextual performance, it is negatively related to role conflict/ambiguity and (poor) organizational policies. Second, for counterproductive behavior, it is related to personality-home-work conflicts (r=.296, p<0.01), (poor) relationship with others (r=.393, p<0.01), distrust (r=.472, p<0.01), role conflicts/ambiguity (r=.759, p<0.01) and (poor) environment (r=.527, p<0.01). These relationships are comprehensible and normal.

However, task performance is positively related to personality-home-work conflicts, (poor) relationship with others, distrust, work overload, (poor) organizational policies and (poor) autonomy. They are relatively incomprehensible. There is also no literature to understand and explain them.

Table 12 Correlation coefficient table for all variables

	PHWC	PR	DT	WO	RC	PE	POP	PA	TP	CP	CB
Personality-home-work conflicts (PHWC)	1.000										
(Poor) relationship with others (PR)	.569**	1.000									
Distrust (DT)	.398**	.332**	1.000								
Work Overload (WO)	.544**	.439**	.417**	1.000							
Role conflicts/ ambiguity (RC)	.356**	.504**	.401**	.273**	1.000						
(Poor) working environment (PE)	.229**	.212*	.480**	.181*	.222**	1.000					
(Poor) organizational policies (POP)	.312**	.380**	.257**	(0.029)	.588**	.512**	1.000				
(Poor) autonomy (PA)	0.160	.714**	.258**	0.026	.495**	0.048	.323**	1.000			
Task performance (TP)	.401**	.707**	.263**	.437**	(0.007)	0.154	.190*	.436**	1.000		
Contextual performance (CP)	.365**	0.166	0.114	.522**	-.314**	(0.150)	-.307**	(0.065)	.624**	1.000	
Counterproductive behaviour (CB)	.296**	.393**	.472**	0.091	.759**	.533**	.690**	.527**	0.079	-.202*	1.000

**. Correlation is significant at the 0.01 level (2-tailed).

*. Correlation is significant at the 0.05 level (2-tailed).

Source: From author's data analysis

Multiple regression analysis

Multiple regression analysis is the most important analysis for this research study. It is applied to accept or reject the eight hypotheses. Before beginning with the major multiple regression analysis for the eight hypotheses. An analysis for control variables is first presented for each of the TP, CP and CB.

Control variables analysis

The following table 13, 14 and 15 summarizes the result for the control variable for TP, CP and CB.

Table 13 Control variable analysis for TP

Model		Unstandardized Coefficients		Standardized Coefficients		
		B	Std. Error	Beta	t	Sig.
1	(Constant)	23.246	2.421		9.604	.000
	Gender	-4.070	1.153	-.299	-3.530	.001
	Age	2.207	1.088	.296	2.029	.044
	Marital status	.542	1.146	.046	.473	.637
	Years of working	-.497	.657	-.098	-.756	.451

Dependent Variable: TP

ANOVA: p=.001

Adjusted R2=.098

From Table 13, the overall impact for the control variables (i.e. demographic variables of gender, age, marital status and years of working) to TP, the dependent variable (DV) is only Adjusted R2=.098, signifying that the control variables (CVs) only explains few of the variances of TP. Therefore, the demographic variables (Dem. Vs) are likely to only be CV but not independent variables (IV) of TP.

Table 14 CV analysis for CP

Model		Unstandardized Coefficients		Standardized Coefficients		
		B	Std. Error	Beta	t	Sig.
1	(Constant)	42.631	3.101		13.747	.000
	Gender	-4.301	1.477	-.241	-2.911	.004
	Age	-.579	1.394	-.059	-.415	.679
	Marital status	-3.044	1.468	-.196	-2.073	.040
	Years of working	1.809	.842	.273	2.149	.033

Dependent Variable: CP

ANOVA: p=.000

Adjusted R2=.139

Table 15 CV analysis for CB

Model		Unstandardized Coefficients		Standardized Coefficients	t	Sig.
		B	Std. Error	Beta		
1	(Constant)	17.650	1.519		11.619	.000
	Gender	-.092	.724	-.011	-.127	.899
	Age	1.291	.683	.279	1.891	.061
	Marital status	-.362	.719	-.049	-.503	.616
	Years of working	-1.496	.412	-.476	-3.626	.000

Dependent Variable: CB

ANOVA: p=.005

Adjusted R2=.078

From Table 14 and 15, the overall impacts for the CV to CP and CB are only Adjusted R2=.139 and .078, signifying that the control variables (CVs) only explains few of the variances of CP and CB. Therefore, the Dem Vs are likely to only be CVs but not the independent variables (IV).

Overall, the Dem Vs are confirmed to not be IVs but CVs in this research model.

Multiple regression analysis for hypotheses testing

The following Table 16, 17 and 18 show the results for multiple regression analysis for TP, CP and CB.

Table 16 Multiple regression analysis for TP

Model		Unstandardized Coefficients		Standardized Coefficients	t	Sig.
		B	Std. Error	Beta		
1	(Constant)	4.387	1.836		2.390	.018
	PHWC	-.239	.132	-.109	-1.816	.072
	PR	1.053	.127	-.721	8.279	.000
	DT	.285	.099	-.160	2.885	.005
	WO	.220	.037	-.361	5.884	.000
	RC	-1.544	.123	-.772	-12.560	.000
	PE	-.261	.092	-.160	-2.846	.005
	POP	.550	.088	-.407	6.220	.000
	PA	.343	.179	-.146	1.915	.058

DV: TP

ANOVA: p=.000

Adjusted R2=.775

First, the model explains most of the variances of TP (adjusted R2=.775), signifying a high explanatory power of the model to TP. Second, since statistical significance are only achieved when p<0.05, only PR, DT, WO, RC, PE and POP are statistically significant. Therefore, the hypotheses are accepted or rejected as follows:

H1a: Personality-home-work conflict is negatively correlated with task performance (rejected) (p>0.05).

H2a: Relationship with others is negatively correlated with task performance (accepted)(β=-.721, p<0.01).

H3a: Distrust is negatively correlated with task performance (accepted) (β=-.160, p<0.01).

H4a: Work overload is negatively correlated with task performance (accepted)(β=-.361, p<0.01).

H5a: Role conflict and ambiguity is negatively correlated with task performance (accepted) (β=-.772, p<0.01).

H6a: Poor environment is negatively correlated with task performance (accepted) (β=-.160, p<0.01).

H7a: Organizational policies is negatively correlated with task performance(accepted) (β=-.407, p<0.01).

H8a: Autonomy is negatively correlated with task performance (rejected) (p>0.05).

Table 17 Multiple regression analysis for CP

Model		Unstandardized Coefficients		Standardized Coefficients		
		B	Std. Error	Beta	t	Sig.
1	(Constant)	27.425	3.214		8.534	.000
	PHWC	.869	.231	-.302	3.772	.000
	PR	-.213	.223	-.112	-.959	.339
	DT	.205	.173	.088	1.184	.238
	WO	.477	.065	-.597	7.296	.000
	RC	-1.788	.215	-.681	-8.312	.000
	PE	-.533	.161	-.249	-3.317	.001
	POP	.147	.155	.083	.953	.343
	PA	.770	.313	-.250	2.459	.015

DV: CP

ANOVA: p=.000

Adjusted R2=.600

First, the model explains most of the variances of CP (adjusted R2=.600), signifying a high explanatory power of the model to CP. Second, since statistical significance are only achieved when p<0.05, only PHWC, WO, RC, PE and PAare statistically significant. Therefore, the hypotheses are accepted or rejected as follows:

H1b: Personality-home-work conflict is negatively correlated with contextual performance(accepted) (β=-.302, p<0.01).

H2b: Relationship with others is negatively correlated with contextual performance (rejected) (p>0.05).

H3b: Distrust is negatively correlated with contextual performance (rejected) (p>0.05).

H4b: Work overload is negatively correlated with contextual performance(accepted)(β=-.597, p<0.01).

H5b: Role conflict and ambiguity is negatively correlated with contextual performance(accepted) (β=-.681, p<0.01).

H6b: Poor environment is negatively correlated with contextual performance(accepted)(β=-.249, p<0.01).

H7b: Organizational policies is negatively correlated with contextual performance(rejected) (p>0.05).

H8b: Autonomy is negatively correlated with contextual performance(accepted) (β=-.250, p<0.05).

Table 18 Multiple regression analysis for CB

Model		Unstandardized Coefficients		Standardized Coefficients		
		B	Std. Error	Beta	t	Sig.
1	(Constant)	-2.285	1.030		-2.219	.028
	PHWC	.214	.074	.157	2.891	.005
	PR	-.390	.071	.430	-5.467	.000
	DT	.021	.055	.019	.371	.711
	WO	-.010	.021	-.025	-.455	.650
	RC	.625	.069	.504	9.070	.000
	PE	.373	.052	.368	7.238	.000
	POP	.132	.050	.157	2.653	.009
	PA	.710	.100	.488	7.073	.000

DV: CB

ANOVA: p=.000

Adjusted R2=.816

First, the model explains most of the variances of CB (adjusted R2=.816), signifying a high explanatory power of the model to CP. Second, since statistical significance are only achieved when p<0.05, only PHWC, PR, RC, PE, POP and PA are statistically significant. Therefore, the hypotheses are accepted or rejected as follows:

H1c: Personality-home-work conflict is positively correlated with counterproductive behavior (accepted) (β=.157, p<0.01).

H2c: Relationship with others ispositively correlated with counterproductive behavior(accepted) (β=.430, p<0.01).

H3c: Distrust ispositively correlated withcounterproductivebehavior (rejected) (p>0.05).

H4c: Work overload is positively correlated with counterproductive behavior (rejected)(p>0.05).

H5c: Role conflict and ambiguity is positively correlated withcounterproductivebehavior (accepted) (β=.504, p<0.01).

H6c: Poor environment ispositively correlated with counterproductive behavior(accepted) (β=.368, p<0.01).

H7c: Organizational policies is positively correlated with counterproductive behavior(accepted) (β=.157, p<0.01)

H8c: Autonomy is positively correlated with counterproductive behavior(accepted)(β=.488, p<0.01).

Overall results

This research has 138 respondents who are civil engineering workers, of which 57.2% are male and 42.8% are female. The main results are which hypotheses are accepted or rejected. The hypotheses that are accepted are as follows:

H2a: Relationship with others is negatively correlated with task performance (accepted) (β=-.721, p<0.01).

H3a: Distrust is negatively correlated with task performance (accepted) (β=-.160, p<0.01).

H4a: Work overload is negatively correlated with task performance (accepted) (β=-.361, p<0.01).

H5a: Role conflict and ambiguity is negatively correlated with task performance (accepted) (β=-.772, p<0.01).

H6a: Poor environment is negatively correlated with task performance (accepted) (β=-.160, p<0.01).

H7a: Organizational policies is negatively correlated with task performance (accepted) (β=-.407, p<0.01).

H1b: Personality-home-work conflict is negatively correlated with contextual performance (accepted) (β=-.302, p<0.01).

H4b: Work overload is negatively correlated with contextual performance (accepted) (β=-.597, p<0.01).

H5b: Role conflict and ambiguity is negatively correlated with contextual performance (accepted) (β=-.681, p<0.01).

H6b: Poor environment is negatively correlated with contextual performance (accepted) (β=-.249, p<0.01).

H8b: Autonomy is negatively correlated with contextual performance (accepted) (β=-.250, p<0.05).

H1c: Personality-home-work conflict is positively correlated with counterproductive behavior (accepted) (β=.157, p<0.01).

H2c: Relationship with others is positively correlated with counterproductive behavior(accepted) (β=.430, p<0.01).

H5c: Role conflict and ambiguity is positively correlated with counterproductive behavior (accepted) (β=.504, p<0.01).

H6c: Poor environment is positively correlated with counterproductive behavior(accepted) (β=.368, p<0.01).

H7c: Organizational policies is positively correlated with counterproductive behavior(accepted) (β=.157, p<0.01)

H8c: Autonomy is positively correlated with counterproductive behavior(accepted) (β=.488, p<0.01).

Other important findings from other analysis are:

Male workers in civil engineering projects generally work in more dangerous environment, have higher workload and have better TP and CP.

The younger generation workers aged 18-29 years old are more likely to experience work overload, feeling poor working environment, poor organizational policies, lower autonomy and doing more counterproductive behavior.

The score of mean and skewness shows positive signs for contextual performance and task performance, role conflict/ambiguity, counterproductive behavior and distrust among the 138 respondents. Besides, the score of standard deviation shows that there are wider dispersions for autonomy, work overload, task performance and distrust scores.

Chapter conclusion

This chapter illustrates and summarizes the results and findings for this research study. The next chapter shall discuss the research findings and offer the (1)theoretical implications, practical implications, limitations of this study and future research directions.

CHAPTER 3   CONCLUSION AND RECOMMENDATIONS

Overview

This chapter is divided into a discussion of theoretical implications, practical implications, limitations of studies and recommendations for future studies.

Theoretical implications

This section is divided into (1) overview of key findings, (2) theoretical implications, (3) practical implications, and (4) limitations of studies and recommendations for future studies.

Overview of findings

The findings have been reasonable so far with most of the hypotheses receiving adequate support. The major hypotheses that received strong supportand worthy of discussion are:

H2a: Relationship with others is negatively correlated with task performance (accepted) (β=-.721, p<0.01).

H4a: Work overload is negatively correlated with task performance (accepted) (β=-.361, p<0.01).

H5a: Role conflict and ambiguity is negatively correlated with task performance (accepted) (β=-.772, p<0.01).

H6a: Poor environment is negatively correlated with task performance (accepted) (β=-.160, p<0.01).

H7a: Organizational policies is negatively correlated with task performance (accepted) (β=-.407, p<0.01).

H8b: Autonomy is negatively correlated with contextual performance (accepted) (β=-.250, p<0.05).

H4b: Work overload is negatively correlated with contextual performance (accepted) (β=-.597, p<0.01).

H5b: Role conflict and ambiguity is negatively correlated with contextual performance (accepted) (β=-.681, p<0.01).

H2c: Relationship with others is positively correlated with counterproductive behavior(accepted) (β=.430, p<0.01).

H5c: Role conflict and ambiguity is positively correlated with counterproductive behavior (accepted) (β=.504, p<0.01).

H6c: Poor environment is positively correlated with counterproductive behavior(accepted) (β=.368, p<0.01).

H7c: Organizational policies is positively correlated with counterproductive behavior(accepted) (β=.157, p<0.01)

H8c: Autonomy is positively correlated with counterproductive behavior(accepted) (β=.488, p<0.01).

Discussion of findings

First, (poor)relationship with others has a strong negative correlation with task performance (β=-.721, p<0.01).Also, (poor) relationship with others has a strong negative correlation with counterproductive behavior (β=.430, p<0.01). As mentioned in the literature review, workplace often involves many politics. Apart from negative things such as power struggle and competition, good relationship with others may lead to better job performance. It is because good relationship with others lead to ease of cooperating with colleagues and is key to teamwork. Better teamwork and relationship lead to better communication, fairer division of works and every team member are then empowered to achieve quality (Natale et al., 1995). Park &Deitz and Leung et al. (2008a) all suggested significant effect of relationship quality, specificity, and teamwork on improving job performance. It implies that poor relationship quality has a negative correlation with task performance.

In fact, the work of civil engineering needs many collaborations and interactions with clients, contractors, sub-contractors, consultants, and suppliers. Therefore, it requires good relationship not only with internal parties such as colleagues, but also external parties such as clients, consultants, sub-contractors. Therefore, it is reasonable to expect that poor relationship with others has a strong and negative relationship with task performance. It may also lead to counterproductive behavior such as complaining about unimportant matters, made troubles at work, focusing on the negative aspects of a work situation, instead of positive impacts, speak about negative aspects of the organizations with both colleagues and external parties. Therefore, this result is reasonable.

Second, as mentioned in Chapter 2 literature review, role conflict is a situation when there are incompatible demands placed on a person relating to their job or position. In layman's term, it means the need to work on things that are unnecessary and meaningless things to achieving organizational goal (Kahn et al, 1964; Rizzo et al., 1970). Role ambiguity is situation when the role does not have enough clarity, certainty and/or predictability one might expect in a job. In layman's term, it means that the work instruction is not clear enough for employees with difficulties to clarify (Rizzo et al., 1970; Parasuraman et al., 1992). Work overload is a situation when an employee faces multiple roles to perform at the same time and lacks the resources in performing them (Arnold, et al., 2016; Caplan & Jones, 1975). Research studies such as Anton (2009), Fried et al (2008), Tarrant & Sabo (2010), Leung et al. (2005b), Leung et al. (2008b), Leung et al. (2007) and Leung & Chan (2012) suggested that work overload, role conflict and role ambiguity were important stressors to work. Brown et al. (2005), Gilboa et al. (2008), Fried et al. (2008), Jamal (2011) and Jaramillo et al. (2006) found that work overload, role conflict and role ambiguity were significantly negatively related to job performance.

The job nature of civil engineering managers includes, for example, preparation of master and construction programmes for design and construction phases, ensuring that the quality, control, quality assurance, and testing requirements are unambiguously defined and agreed with clients, reviewing and monitoring construction progress, coordination of a diverse range of material, plant and labor inputs, ensuring site safety, etcetera (Leung, Chan & Yu, 2009). These works always have clear program and procedures to follow. Time is tense and accuracy at work are required. Therefore, role conflict (the need to work on unnecessary and meaningless things to achieving organizational goal), role ambiguity (work demands are ambiguous) are detrimental and then stressful to work performance. It therefore has great negative relationship with task performance(β=-.772, p<0.01) and contextual performance (β=-.681, p<0.01). The reason may be because role ambiguity and role conflict generategreat amount of stress causes great disturbance to workers at the core substantive or technical tasks central to one's job (Borman&Motowidlo, 1993). Similarly, it causes disturbances to contextual performance (CP)(i.e. behaviour that supports the organisational, social and psychological environment in which the core tasks must function (Borman&Motowidlo, 1993). Besides, it is normal to expect that work overload has a negative relationship with task performance and contextual performance. However, it is not significantly related to counterproductive behaviour. It may be because those with work overload are highly focused on their tasks and have no time and resources to do things that counterproductive.

Third, (poor) organizational policies should have a larger effect size to TP negatively. As mentioned in the literature review, Katou&Budhwar (2009) found that HRM policies components such as resourcing and development, compensation and incentives and, involvement and job design were significantly positively related to employee skills (Effect size: 0.36), employee attitudes (effect size: 0.94), and employee behaviors (effect size: 0.97). All these are significantly related to organizational performance (Employee skills to organizational performance: 0.90, employee attitudes to organizational performance: 0.48, employee behavior: 0.44). However, the effect size f2 of the relationship between POP and TP was extremely small at .002. As for POP's relationship with CP and CB, the effect size f2 was small at .120 and .098 on CP and CB.

It is reasonable to have a significant relationship for task performance by (poor) organizational policies. It is because organizational policies such as HRM policies has serious impact on attracting, retaining, and motivating employees. HRM policies include resourcing and development, compensation and incentives, and involvement and job design. All these policies are important to attract, retain and motivate employees (DeCenzo, Robbins &Verhulst, 2016). For example, different employment training programme and employee development programme (such as job rotation, committee assignment), clear career path given to employees, and establishing good remuneration and recognition to employees are often instrumental to their task performance (DeCenzo, Robbins &Verhulst, 2016). Therefore, it is easy to understand the significant negative relationship between (poor) organizational policies and task performance.

Besides, it is reasonable to expect that organizational policies does not have significant relationship to contextual performance (CP) (i.e. which means behaviour that supports the organisational, social and psychological environment in which the core tasks must function)

(Borman&Motowidlo, 1993). It may be because task completion is the main value from an employee. Besides It is reasonable to have a significant relationship from organizational policies to CB. It is normal to expect that those companies with poor organizational policies should have more CB since the organizational policies if an organization treat an employee unfairly, with low generosity on financial incentives and allowances, and lack of support for their work. Therefore, it is easy to understand why poor organizational policies have a smaller effect size on CB than CP.

In addition, it is reasonable to expect that (poor) physical environment is positively correlated with counterproductive behavior. (Poor) physical environment (PE) leads to physical discomfort, functional discomfort and then turns to psychological discomfort to a worker (Vischer, 2007). Leung et al. (2005b), Leung et al. (2008b), and Leung et al. (2009) suggested that poor working environment contributed to stress to workplace. Since stress leads to reduction in work performance (Abramis, 1994; Jamal, 1984; Jamal, 1985; Jamal, 2007; Leung et al, 2011; Motowidlo et al., 1986; Siu, 2003; Sullivan &Bhagat, 1992; Wu, 2011) in studies in USA, Hong Kong, Taiwan, Malaysia, and Pakistan, there is good evidence that poor physical environment shall lead to reduction in task performance and increase in counterproductive behavior. As civil engineering project managers, it is a requirement to work in bad and dangerous environment such as construction sites.Although they may get used to it, it is of human nature that it reduces task performance and leads to counterproductive behavior (such as complaining, focusing on the negative aspects of tasks and speak negatively about the jobs to both internal and external parties).

(Low) autonomy is, as expectation, negatively related to job performance. Autonomy is defined as the degree to which the job provides substantial freedom, independence, and discretion to employee scheduling the work and determining the procedures to be used in carrying it his/ her duties (Hackman & Oldham, 1975).Hackman & Oldham (1976) suggested that autonomy is positively related to positive individual and work outcomes such as intrinsic motivation among employees and high level of job satisfaction.Gagne& Deci (2005) asserted that intrinsic motivation is only achieved by the impact on need satisfaction or the perception of autonomy, competence, and relatedness. Also, the need for autonomy is one of the most fundamental human needs and necessary to fulfil intrinsic motivation to emerge or be sustained. Therefore, because employees are intrinsically motivated, employees are more interested and persistent in their work. It improves their work quality. Since their work should have a higher quality than those who are not intrinsically motivated, their job performance thus improved (Dysvik&Kuvaas, 2011).Therefore, it is reasonable to expect that autonomy is negatively related to contextual performance and positively related to counterproductive behavior.

Finally, distrust is negatively correlated with task performance (β=-.160, p<0.01). Distrust means a confident and negative expectations regarding another person's conduct (Lewicki et al., 1998). In layman's terms and in organizational context, it means that a colleague is convinced that another colleague has ominous intentions. The colleague is then inclined to maintain distance from another colleague's behavior.Research studies from Dirks &Ferrin (2001) and Dirks &Ferrin (2002) found that trust is positively related to job performance if followers trust their leaders. However, they also asserted that if followers do not trust their leaders, they shall spend more time in guessing the true intention of their leaders' words and action. They are then distracted from their work and resulted in reduction in job performance. This implies a negative relationship between distrust and job performance. The degree of such relationship ranges from -.26 to -.43 (Colquitt et al., 2007; Verburg et al., 2018). Therefore, this finding is reasonable.

Overall, apart from the above findings, only the following findings are away from expectation:

(Low) autonomy is not significantly related totask performance,

Distrust, (poor) relationship with others and (poor) organizational policies are not significantly related to contextual performance, and

Distrust is not significantly related to counterproductive behavior.

Apart from the above findings, all other findings are reasonable. Besides, the adjusted R2 are .775, .600, and .816 for the task performance, contextual performance, and counterproductive behavior model. It means that the model has great explanatory power to the three factors of job performance. Although other factors such as intelligence, personality, and social skills/political skills may influence job performance, this model did not account for them. There may be further research studies that can consider these factors.

Practical implications

Based on the findings, work overload, role conflict/ambiguity, (poor) organizational policies, (low) autonomy and poor physical environments are the major culprits for lower job performance and more counterproductive behavior. Therefore, the recommendations for practice shall focus on these areas.

First, for work overload, there should first be improvement in the planning and forecasting of workload so that employees face more realistic deadlines. As civil engineering managers, there are many duties in pre-construction stage, construction stage, and completion stage. Some tasks involve many cooperation and interaction with other parties both internally and externally. These tasks include ensuring quality, control, quantity assurance, and testing requirements are clearly defined and agreed with clients, coordinate a diverse range of material, plant, and labor inputs, and review and monitor construction progress (Leung, Chan & Yu, 2009). The working hours of these tasks are difficult to estimate. Therefore, work teams should always rely on experience of the normal duration in determining the time needed to complete these tasks. Therefore, there need to be improvement in planning and forecasting of workload and be realistic.

Second, to ensure reasonable workload, managers should always have good analysis of subordinates' knowledge, skills, and abilities to ensure that they are equipped to do the job and offer opportunities for them to develop further. They should also adjust staffing levels to reflect peaks and troughs in workload. The skills and abilities of subordinates are important because managers depend on them to be successful. If there are inadequate number of staff and the subordinates are not skillful and able, it will waste many times of managers to coach those staff and coordinate the works. Therefore, these steps have to be ensured and conducted well in every projects.

Third, to ensure reasonable workload, managers should also utilize information technology to reduce cognitive load on staff. Examples are systems that help staff monitor the progress of tasks that are carried out for several days or weeks, or when they are dealing with many simultaneous tasks. Therefore, project management software should be used to assist in monitoring progress and allocate workload among staff evenly.

As for role conflict/role ambiguity, managers can do the following:

Review, update and publicize job descriptions,

Allow teams and subordinate having more latitude and control on how roles and responsibilities are allocated,

Strive to minimize the chance for competing or conflicting demands in roles when designing and revising job descriptions, and

Have good allocation of tasks in a team to ensure that conflicting demands do not have conflicts for individual employees (Arnold, 等 , 2016).

As higher-level staff who supervise civil engineering managers, they should ensure the above to reduce role conflict. As for role ambiguity, higher-level staff should ensure the tasks are clearly defined to avoid such things.

As for (poor) organizational policies, human resources management policies should make a good effort in training, developing and motivating employees. They include providing reasonable and good remunerations, providing relevant trainings in both technical skills and soft skills, providing recognition and awards for outstanding staff, and committee assignments, etcetera (Arnold, 等 , 2016).

As for (low) autonomy, more control should be provided to employees/managers for their tasks. Higher-level staff should increase employee control over how work is allocated, give staff control and freedom to identify and rectify common problems quickly and without unnecessary approval from senior managers. In some ways, higher-level staff should review and, in appropriate time, change the guidance the staff are given about completing tasks and having more control and discretion about how tasks are completed (Arnold, 等 , 2016).

Finally, as for (poor) physical environment, physical environment must be ensured to be less dangerous, more sanitary, and spacious enough for working. Safety equipment and environmental design should be configured to ensure these aspects.

Since these culprits of lower job performance are more prevalent among younger generations, high-level staff should have more care about young civil engineering managers about their workload, working environment and autonomy.

Limitations and recommendations for future studies

The first limitation is a lower sample size in ensuring the representativeness and generalizability of findings. While power analysis only ensures that there are adequate samples to increase the probability of detecting variance among independent variables and dependent variable, sample size calculation ensures that there are enough samples to generalize to a wide circumstance. The formula is as follows:

Sample size=$((z^2 \times p(1-p))/e^2) /(1+((z^2 \times p(1-p))/(e^2 N)))$

Where z=z score, e=confidence interval, N=population size, p=0.5

Assuming that there are 7,300 civil engineering professionals in Hong Kong (ICE Hong Kong, 2020), the sample size required to have a confidence interval of 5%, z-score of 1.96 is 365. Therefore, the sample size is still inadequate to provide generalizability to the population of civil engineers.

The second limitation is the absence of mediation and moderation analysis. Mediation analysis means an understanding of a known relationship by investigating the mechanism by which one or several independent variables in influencing another variable through a mediator (Darlington & Hayes, 2016). Moderation analysis is to investigate how a categorical variable affects the direction and/or strength of the relation between independent variables and dependent variable (Darlington & Hayes, 2016). In practice, mediators can be added in between the relationship of the eight stressors and job performance such as job satisfaction, organizational commitment, organizational citizenship behavior(Arnold, 等 , 2016). Moderator that are applicable include age, years of experience, levels of position in organizations, etcetera (Arnold, 等 , 2016).

The third limitation is that the study is not a mixed method case study. Apart from recognizing the statistical relationship between the eight stressors and job performance, it is difficult to address the reasons behind the relationship and strength of relationship. Therefore, instead of similar conducting statistical analysis, mixed methods study which include qualitative methods such as semi-structured interviews to civil engineering managers may be adopted to investigate the remaining questions of the study (Saunders et al., 2012).

The final limitation is that organizational environment, behavioral factors and stressors may not be the only category of factors that influence job performance. Other factors such as intelligence, personality and social skills may also act as independent variables or moderators for the relationship (Arnold, 等 , 2016).

Therefore, it is recommended that future studies should:

Increase the sample size,

Have mediation and moderation analysis using factors such as job satisfaction, organizational commitment, organizational citizenship behavior, and age, years of experience, levels of position in organizations,

Use mixed method studies with both quantitative and qualitative methods, and

Consider the impact of intelligence, personality, and social skills on job performance.

BIBLIOGRAPHY

Abidin, R. R., &Abidin, R. R. (1990). Parenting Stress Index (PSI) (Vol. 100). Charlottesville, VA: Pediatric Psychology Press.

Abramis, D. J. (1994). Relationship of job stressors to job performance: Linear or an inverted-U?. Psychological reports, 75(1), 547-558.

AbuAlRub, R. F. (2004). Job stress, job performance, and social support among hospital nurses. Journal of nursing scholarship, 36(1), 73-78.

Bakker, A. B., Demerouti, E., &Euwema, M. C. (2005). Job resources buffer the impact of job demands on burnout. Journal of occupational health psychology, 10(2), 170-180.

Bell, E., Bryman, A., & Harley, B. (2018). Business research methods. Oxford university press.

Betoret, F. D. (2006). Stressors, self-efficacy, coping resources, and burnout among secondary school teachers in Spain. Educational psychology, 26(4), 519-539.

Chan, A. P., & Chan, A. P. (2004). Key performance indicators for measuring construction success. Benchmarking: an international journal, 11(2), 203-221.

Chan, A. P., Scott, D., & Chan, A. P. (2004). Factors affecting the success of a construction project. Journal of construction engineering and management, 130(1), 153-155.

Chartered Institute of Building (CIOB). (1996). The code of practice for project management for construction and development. Harlow, U.K: Longman.

Dainty, A. R., Cheng, M. I., & Moore, D. R. (2003). Redefining performance measures for construction project managers: an empirical evaluation. Construction Management & Economics, 21(2), 209-218.

Dainty, A. R., Cheng, M. I., & Moore, D. R. (2005). Competency-based model for predicting construction project managers' performance. Journal of Management in Engineering, 21(1), 2-9.

Darlington, R., & Hayes, A. (2016). Regression Analysis and Linear Models: Concepts, Applications, and Implementation. New York: The Guilford Press.

Edwards, J. R., Caplan, R. D., & Van Harrison, R. (1998). Person-environment fit theory. Theories of organizational stress, 28, 67-78.

Enshassi, A., & Al Swaity, E. (2015). Key Stressors Leading to Construction Professionals' Stress in the Gaza Strip, Palestine. Journal of Construction in Developing Countries, 20(2), 53-79.

Enshassi, A., Mohamed, S., &Abushaban, S. (2009). Factors affecting the performance of construction projects in the Gaza strip. Journal of Civil engineering and Management, 15(3), 269-280.

Folkman, S., & Lazarus, R. S. (1984). Stress, appraisal, and coping (pp. 150-153). New York: Springer Publishing Company.

Fowler Jr, F. J. (2013). Survey research methods. Sage publications.

Gagné, M., & Deci, E. L. (2005). Self-determination theory and work motivation. Journal of Organizational behavior, 26(4), 331-362.

Garrosa, E., Moreno-Jiménez, B., Rodríguez-Muñoz, A., & Rodríguez-Carvajal, R. (2011). Role stress and personal resources in nursing: A cross-sectional study of burnout and engagement. International journal of nursing studies, 48(4), 479-489.

Gilboa, S., Shirom, A., Fried, Y., & Cooper, C. (2008). A meta-analysis of work demand stressors and job performance: examining main and moderating effects. Personnel Psychology, 61(2), 227-271.

Hackman, J. R., & Oldham, G. R. (1975). Development of the job diagnostic survey. Journal of Applied psychology, 60(2), 159-170.

Hackman, J.R., & Oldham, G. (1976). Motivation through the design of work: Test of a theory. Organizational behavior and human performance, 16(2), 250-279.

Hair, J. F., Celsi, M., Ortinau, D. J., & Bush, R. P. (2008). Essentials of marketing research. New York, NY: McGraw-Hill/Higher Education.

Institution of Civil Engineers (n.d.). What is Civil Engineering? Available at: https://www.ice.org.uk/what-is-civil-engineering [Accessed: December 18, 2018].

Jackson, S. E., & Schuler, R. S. (1985). A meta-analysis and conceptual critique of research on role ambiguity and role conflict in work settings. Organizational behavior and human decision processes, 36(1), 16-78.

Jamal, M. (1984). Job stress and job performance controversy: An empirical assessment. Organizational behavior and human performance, 33(1), 1-21.

Jamal, M. (1985). Relationship of job stress to job performance: A study of managers and blue-collar workers. Human Relations, 38(5), 409-424.

Kahn, R. L., Wolfe, D. M., Quinn, R. P., Snoek, J. D., & Rosenthal, R. A. (1964). Organizational stress: Studies in role conflict and ambiguity. New York: Wiley.

Lazarus, R. S. (1993). From psychological stress to the emotions: A history of changing outlooks. Annual review of psychology, 44(1), 1-22.

Lazarus, R. S. (1995). Psychological stress in the workplace. Occupational stress: A handbook, 1, 3-14.

Lazarus, R. S. (2000). Toward better research on stress and coping. The American psychologist, 55(6), 665-673.

Lazarus, R. S., & Folkman, S. (1984). Stress, appraisal, and coping. New York: Springer.

Margolis, B. L., Kroes, W. H., & Quinn, R. P. (1974). Job stress: An unlisted occupational hazard. Journal of Occupational and Environmental Medicine, 16(10), 659-661.

Marzuki, N. A. &Ishak, A. K. (2011). Towards healthy organization in correctional setting: Correctional officers' wellness, occupational stress and personality. International Journal of Social Science and Humanity Studies, 3, 355-365.

McGrath, A., Reid, N., &Boore, J. (2003). Occupational stress in nursing. International journal of nursing studies, 40(5), 555-565.

Nabirye, R. C., Brown, K. C., Pryor, E. R., & Maples, E. H. (2011). Occupational stress, job satisfaction and job performance among hospital nurses in Kampala, Uganda. Journal of nursing management, 19(6), 760-768.

Naoum, S. G., Herrero, C., Egbu, C., & Fong, D. (2018). Integrated model for the stressors, stress, stress-coping behaviour of construction project managers in the UK. International Journal of Managing Projects in Business, 11(3), 761-782.

Natale, S. M., Libertella, A. F., & Rothschild, B. (1995). Team performance management. Team Performance Management: An International Journal, 1(2), 6-13.

Orth-Gomer, K., Wamala, S. P., Horsten, M., Schenck-Gustafsson, K., Schneiderman, N., &Mittleman, M. A.

(2000). Marital stress worsens prognosis in women with coronary heart disease: The Stockholm Female Coronary Risk Study. Jama, 284(23), 3008-3014.

Osipow, S. H., Doty, R. E., & Spokane, A. R. (1985). Occupational stress, strain, and coping across the life span. Journal of Vocational Behavior, 27(1), 98-108.

Osipow, S. H., & Spokane, A. R. (1998). Occupational stress inventory-revised. Odessa, FL: Psychological, 1-15.

Parasuraman, S., Greenhaus, J. H., &Granrose, C. S. (1992). Role stressors, social support, and well-being among two-career couples. Journal of Organizational Behaviour, 13(4), 339-356.

Park, J. E., &Deitz, G. D. (2006). The effect of working relationship quality on salesperson performance and job satisfaction: Adaptive selling behavior in Korean automobile sales representatives. Journal of Business Research, 59(2), 204-213.

Payne, N. (2001). Occupational stressors and coping as determinants of burnout in female hospice nurses. Journal of advanced nursing, 33(3), 396-405.

Richardson, K. M., & Rothstein, H. R. (2008). Effects of occupational stress management intervention programs: a meta-analysis. Journal of occupational health psychology, 13(1), 69-93.

Rizzo, J. R., House, R. J., &Lirtzman, S. I. (1970). Role conflict and ambiguity in complex organizations. Administrative science quarterly, 150-163.

Roohafza, H., Feizi, A., Afshar, H., Mazaheri, M., Behnamfar, O., Hassanzadeh-Keshteli, A., &Adibi, P. (2016). Path analysis of relationship among personality, perceived stress, coping, social support, and psychological outcomes. World journal of psychiatry, 6(2), 248-256.

Sanne, B., Mykletun, A., Dahl, A. A., Moen, B. E., & Tell, G. S. (2005). Testing the job demand–control–support model with anxiety and depression as outcomes: the Hordaland health study. Occupational medicine, 55(6), 463-473.

Saragih, S. (2011). The effects of job autonomy on work outcomes: Self efficacy as an intervening variable. International Research Journal of Business Studies, 4(3), 203-215.

Saunders, M., Lewis, P., & Thornhill, A. (2012). Research methods for business students. New York: Pearson education.

Taris, T. W., Schreurs, P. J., & Van Iersel-Van Silfhout, I. J. (2001). Job stress, job strain, and psychological withdrawal among Dutch university staff: Towards a dualprocess model for the effects of occupational stress. Work & Stress, 15(4), 283-296.

Tarrant, T., & Sabo, C. E. (2010). Role conflict, role ambiguity, and job satisfaction in nurse executives. Nursing administration quarterly, 34(1), 72-82.

Teasdale, E. L. (2006). Workplace stress. Psychiatry, 5(7), 251-254.

Vagg, P. R., & Spielberger, C. D. (1998). Occupational stress: Measuring job pressure and organizational support in the workplace. Journal of occupational health psychology, 3(4), 294-305.

Van der Doef, M., & Maes, S. (1999). The job demand-control (-support) model and psychological well-being: a review of 20 years of empirical research. Work & stress, 13(2), 87-114.

Wahba, M. A., & Bridwell, L. G. (1976). Maslow reconsidered: A review of research on the need hierarchy theory. Organizational behavior and human performance, 15(2), 212-240.

Weiten, W. (2008). Psychology: Themes and Variations. Belmont, CA: Thomson.

Wu, Y. C. (2011). Job stress and job performance among employees in the Taiwanese finance sector: The role of emotional intelligence.

Prof. LAM, Huen Sum (PhD, PhD Supervisor, Fellow)

林絢琛博士、教授、博導、院士

Li Ka Shing Faculty of Medicine,
the University of Hong Kong, Hong Kong

Phone(Mobile): +852 6542 3388
+19896539363
Email: michael@united-edu.org
Wechat: michaellam888888888

Prof. Michael Lam, with his PhD graduated in Li Ka Shing Faculty of Medicine, the University of Hong Kong, is a very experienced director, accreditor and quality controller of local and collaborative academic programmes and vocational related certificates, a healthcare researcher, a community public health practitioner with solid experience in investment and management of healthcare and education related corporate, including listed companies.

He has published more than 50 education related and health related quality papers in SCI and SSCI journals and attained several engineering patents. He has been appointed by Hong Kong Council for Accreditation of Academic and Vocational Qualifications and Universities to accredit and re-accredit the local and non-local academic programmes. With substantive experience in programme quality control and management, he has been appointed by Universities and institutes in Hong Kong, United Kingdom and Asian countries as professor and advisory board member.

With a dual track experience in theory and practice, he has been appointed by the Hong Kong Government to vet and to approve the fund for certain health related constructions and health promotion areas. He has also been appointed by Hong Kong government and professional associations to be the judge the management of government and private facilities. Knowing his strength in academic programme accreditation and industry practice, he is invited by the National Higher Vocational College Principal Joint Meeting as a keynote speaker for 3000 principals of vocational education on how to make good synthesis of education and industrial practice, hereafter, promoting them to one belt one road countries.

Apart from the attainment in Health and Education area, Prof. Michael Lam acted as the Chairman of United Education Investment Group Limited, Chinese Overseas Student Federation Charity Foundation and Executive Deputy Chairman of the Association of Chinese Entrepreneurs in Scotland, has established more than 5 friendship cities between China and Scotland.

A. Professional Recognition and Certificates

Recognition	Organization	Year
Project Manager Professional	Project Manager Institute	2020
Fellow of Chartered Management Institute	Charter Management Institute	2019
Member of Institute of Public Accountant	Institute of Public Accountant	2018
Member of Institute of Financial Accountant	Institute of Financial Accountant	2018
Principal Fellow of Chartered Institute of China Family Office	Chartered Institute of China Family Office	2017
Fellow and Accreditor of Medical and Health Board	International Industry and Professional Accreditation Association	2016
Certified Doctor of Natural Medicine	Health Practitioner of Canadian Examination Board	2013
Certified Community Sport Planner	Hong Kong Sport Education Association	2012
Registered Teacher	Education and Manpower Bureau	2008

B. Academic Working Experience

Item	Institution and Job Description	Full Time/ Part Time/ Appointed Consultancy (Year)
1.	Judge of Excellent Teaching Award, City University of Macau, Macau, China	2019
2.	Faculty Advisory Board Member, Hong Kong College of Technology (University License)	06/2017 (Appointment)
3.	Consultant of Hong Kong College of Technology (University License) QF: Level 4 (Higher Diploma, Professional Diploma) and 5 (Bachelor Degree)	03/2016 (Appointment)
4.	Advisory Board Member, Open Institute, City University of Macau http://www.cityu.edu.mo/oi/advisory-committee	02/2016 (Appointment)
5.	Appointed Specialist of Hong Kong Council for Accreditation of Academic and Vocational Qualifications (HKCAAVQ) http://www.hkcaavq.edu.hk/en/specialists/specialists-register/result	02/2016 (Appointment)
6.	Adjunct Professor, Jinan University, Shandong Province, China	12/2015 (Appointment)
7.	External Panel Member of Hong Kong College of Technology QF: Level 4 (Higher Diploma, Professional Diploma)	12/2015 (Appointment)

8.	Accreditation Panel Member of The University of Hong Kong QF: Level 4 (Higher Diploma, Professional Diploma)	12/2015 (Appointment)
9.	External Reviewer of Hong Kong Baptist University and Hong Kong Recreation Management Association QF Level 3	07/2014 (Appointment)
10.	External Reviewer of Professional Diploma in Leisure, Recreation and Sports Administration of Hong Kong Baptist University (Hong Kong SAR Qualification Framework Level 4)	11/2013-Present (Appointment)

C. Voluntary and Commercial Working Experience

Government Appointment

	Name of Organization	Position Held	Date
1.	Zibo Government and Zibo Education Department, Shandong Province	Power of Attorney (appointed representative)	2019
2.	City of East Lothian Council, Scotland	Power of Attorney (appointed representative)	2019
3.	Friendship city between China and Scotland UK	Representative or coordinator	2017-2020
4.	Work Improvement Team Award for Sport Center, Leisure and Cultural Department, Hong Kong Special Administrative Region	Judge	2018
5.	Hong Kong Council for Accreditation of Academic and Vocational Qualifications (Appointed by HKCAAVQ Hong Kong)	Specialist of Education, Recreation and Leisure Management, Facility Management, Sport, Physical Education, Sport Science and Health	2016
6.	Sir David Trench Fund, HKSAR http://www.hab.gov.hk/en/related_departments_organizations/asb37.htm (Appointed by HK Government)	Trustee/Board Member (Vetting and Approving Funding for Sports and Recreation Sector in Hong Kong-for facility construction, reconstruction, renovation and maintenance)	2013-2019
7.	Fujian Province (Appointed by PRC Government)	City Municipal Councilors (Education Sector)	2012-Present
8.	Fujian Province, Putian City (Appointed by PRC Government)	Vice Chairman of Federations of Youth Associations	2012-Present
9.	North District Council, Hong Kong (Appointed by HK NDC)	Steering Board Member of Healthy City Project (Health Promotion)	29/06/2009-30/03/2010

International Association and Professional Association (Award)

	Name of Organization	Position Held	Date
1.	Rosebud Primary School	Chairman of Supervisory Board	2019-Present
2.	Dodgeball Association of Hong Kong	Vice President	2019-Present
3.	Best Residential Clubhouse Management Award, Hong Kong Property Management Association & Hong Kong Recreation Management Association	Judge	2018
4.	Asian Sport Psychology Professional Alliance	Principal Adviser (Accreditation)	2017-Present
5.	Hong Kong Association of Floorball Association (Member of International Floorball Association)	Founding Executive Vice President	2016-Present
6.	International Industry and Professional Accreditation Association	Secretary General	2015-Present
7.	• Chinese Overseas Student Federation Charity Foundation (Charity Exempt from tax under section 88 of the Inland Revenue Ordinance) • Taoxingzhi Education Foundation and Chinese Overseas Student Federation Charity Foundation Joint Charity Fund (A joint Chinese Government fund under Education Bureau and Hong Kong Government Approval)	Chairman Co-Chairman	03/2015-Present 2018-Present
8.	Hong Kong Recreation Management Association and Hong Kong Association of Property Management Company	Judge of Hong Kong Residential Clubhouse Management Award 2011-2015	2011-Present
9.	Hong Kong Recreation Management Association	• Executive Member • Chairman of Membership Committee • Team Member of Timothy Fok Scholastic Award • Editorial Board Member of Journal of Hong Kong Recreation Review	2009-Present
10.	Hong Kong Institute of Adventure Counselling	Consultant/ Advisor	2009-Present
11.	HK Backpackers	Vice President	2011-Present

12.	• Hong Kong Sports Education Association	Chairman	2007-2010
	• Joint Branch Associations of 6 districts and 2 sub association (Kwai Tsing District, North Districts, Shatin District, Sai Kung District, Tsuen Wan District, and Wan Chai District, Hong Kong Integrated Football Association, Hong Kong Active Orienteering Association) • Certified Community Sport Planner Committee	Chairman Chairman	2010-Present 2013-Present
13.	Hong Kong Special Olympics (Athlete Leadership Program Board)	Board Member	10/2009-Present
14.	International Special Olympics Committee	Mentor of the International Global Messenger	2006-2009

Commercial and Consultancy

	Name of Organization	Position Held	Date
1.	United Education Investment Group Limited	Chairman	2003-Present
2.	United Marco Polo Club	Chairman	2003-Present
3.	Sunway International Holding Limited (Main Board Listed Company #58) Investment fund for Health and Education	Independent Non-Executive Director	2017-2019
4.	Good Fellow Healthcare Holding Limited (Gen Board Listed Company #8143) Hospital Management	Independent Non-Executive Director and shareholder (family)	2017-Present
5.	International Greenland Group (Land Developing)	Non-Executive Director and owner (family)	2016-Present
6.	Pulse Group Ltd (Property and Fitness Management) • Manage 7 real estate (over 10000 apartments and 7 residential clubhouses) • Manage 5 fitness clubs for international insurance companies (AIA and Manulife) • Manage 8 government facilities	Non-Executive Director and owner (family)	2018-Present
7.	Chartered Institute of China Family Office	Chancellor	2018-Present

The Effectiveness of Health Literacy Oriented Programs on Physical Activity Behaviour in Middle-Aged and Older Adults with Type 2 Diabetes: A Systematic Review

1.1 INTRODUCTION

The concept of HL emerged in the early 1990s (Cutilli, 2007), but it began to receive greater attention after the USDHHS ranked HL as one of the top four public health priorities (Galson, 2008). Some researchers have suggested that limited HL tends to result in poor self-care behaviours (including PA behaviour) and eventually leads to poor health outcomes among middle-aged and older adults (Gazmararian et al., 2003; Osborn et al., 2011; Paasche-Orlow & Wolf, 2007). PA behaviour was successfully modified by HL-oriented programs for middle-aged or older adults with hypertension (Joyner-Grantham et al., 2009; Osborn et al., 2011; Ussher et al., 2010) and coronary heart diseases (Ussher et al., 2010). It may be assumed that HL-oriented programs can also improve PA behaviour of T2DM patients, but no reviews of empirical studies have proved this assumption.

1.2 AIMS OF THIS CHAPTER

This chapter review existing literature on the effectiveness of HL-oriented programs in influencing PA behaviour particularly in the context of diabetic self-care behaviours of middle-aged and older adults with T2DM, so as to summarize and critically evaluate the format, the study design, intervention content, limitations and implications of previous HL-oriented programs. This review should permit better planning of the HL-oriented program to be applied in the present study (see Chapter 5) (objective 1 of the thesis).

1.3 METHODS

1.3.1　Design

In order to compare outcomes among HL-oriented programs addressing PA behaviour of the diabetic self-care behaviours, only interventional articles were screened and reported in accordance with the guidance of Preferred Reporting Items for Systematic Reviews and Meta-Analyses (PRISMA). PRISMA, which is commonly used in conceptual or practical systematically reviews in health care, can scientifically integrate the results of related studies (Moher, Liberati, Tetzlaff, Altman, & Group, 2010). The searching process took place in August 2012 and was updated in April 2013.

1.3.2　Search Methods

Articles were identified from eight electronic databases between January 1990 and April 2013: Medline, PubMed, CINAHL, PsycINFO, Proquest, Sport DISCUS, ISI Web of Science and ERIC. The key search terms were generated from three systematic reviews on HL, PA behaviour, middle-aged and older adults with T2DM (Conn et al., 2003; Cutilli, 2007; Hamel, Robbins, & Wilbur, 2011). The key search terms selected were health literacy, literacy, numeracy, reading ability, rapid estimate of adult literacy in medicine, test of functional health literacy in adults, wide range achievement test, diabetes numeracy test, type 2 diabetes mellitus, glycemic control, T2DM, TIIDM, older adult, older people, senior, elderly, elder, geriatric, aged, middle-aged, physical activity, exercise, physical exercise, Tai Chi, walking, brisk walking, strength training, aerobic training, aerobic exercise, running, aquatic exercise, self-management, self-care, compliance and the scale of summary of diabetes self-care activities. Hand searching was performed on the reference lists of all articles identified. In order to ensure a comprehensive search of international articles, the assistance of librarians from the University of Hong Kong was obtained. Table 1 lists the inclusion and exclusion criteria for selecting articles for review.

Table 1

Inclusion and Exclusion Criteria for Screening of Health Literacy Oriented Programs on Physical Activity Behaviour
Inclusion criteria
1. Articles with middle-aged (age 45-64) or older (age > 65) adults
2. Articles written or translated in English
3. Articles with at least 10 participants
4. Articles related to health literacy and physical activity
5. Participants included in the articles were patients with type 2 diabetes mellitus
6. Year of publication between 1990 and 2013
7. Full Text
Exclusion criteria
1. Publication based on authors' opinion only.
2. Articles did not include physical activity among self-care behaviours.

1.3.3 Search Outcomes

According to PRISMA's 2010 flow diagram of the screening process (Moher et al., 2010), 366 references were located, 22 were duplicated, 25 non-full text articles were excluded, and 313 articles did not meet the inclusion criteria and therefore were excluded. Six articles were included in the qualitative synthesis. Figure 2 illustrates the screening process.

1.3.4 Quality Appraisal

The articles were appraised according to the developed interventional review guidelines developed by Schulz, Chalmers, Hayes and Altman (1995) adapted by Jadad et al. (1996) and subsequently modified by DeWalt and Hink (2009). As described in Table 2, all six interventional studies were evaluated on 11 items of quality and strength of evidence in reviewed studies, with scoring from 0 to 2 (0 = poor, 1 = fair, 2 = good) after removing items not applicable to this review. The average score, with all factors equally weighted, provided the final quality and strength grades (less than 1 = poor; 1 to 1.5 = fair; 1.5 or more = good) (Berkman et al., 2004). No article was excluded at this quality appraisal stage.

The final included articles were formatted, assessed and analyzed by an assessment instrument used in previous systematic reviews, which is shown in Table 3 (Arden-Close, Gidron, & Moss-Morris, 2008; Montgomery & McCrone, 2010).

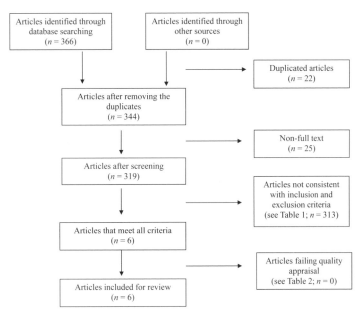

Figure 2. Study flow from identification to final selection of articles on health literacy oriented programs including physical activity behaviour.

Table 2

Summary of the Quality and Strength of Evidence in Reviewed Studies of Health Literacy Oriented Programs Including Physical Activity Behaviour						
Studies						
Description	Rosal et al. (2011)	Cavanaugh et al. (2009)	DeWalt et al. (2009)	Gerber et al. (2005)	Kim et al. (2004)	Rosal et al. (2005)
1) adequacy of study population	2	2	2	2	2	0
2) adequacy of randomization procedure	2	2	0	2	0	1
3) describes the withdrawals and dropouts	1	2	2	2	1	1
4) intention to treat	0	0	0	0	0	0
5) outcome assessors blind to the intervention	1	1	0	0	0	0
6) comparability of subjects across comparison groups	2	2	1	2	1	2
7) validity and reliability of the literacy measurement	1	2	2	2	2	1
8) maintenance of comparable groups	2	2	1	2	1	2
9) appropriateness of the outcome measurement	2	2	2	2	2	2
10) appropriateness of statistical analysis	1	2	1	1	2	2
11) adequacy of control of confounding	2	2	1	1	2	1
Total Score	16	19	12	16	13	12
Average Score	1.45	1.73	1.09	1.45	1.18	1.09
Quality Rate	Fair	Good	Fair	Fair	Fair	Fair

1.3.5   Data Abstraction and Synthesis

The two-reviewer approach was adopted to review the titles, abstracts and content of the selected articles. A data extraction form was developed and circulated for achieving agreement between the two reviewers. As the extracted articles contained vast variations in their study aims and methods, a systematic review rather than a meta-analysis was carried out (Whittemore, 2005).

1.4 RESULTS

1.4.1   Descriptive Information

Among the six reviewed interventional articles, four were randomized controlled trial (RCT) studies (Cavanaugh et al., 2009; Gerber et al., 2005; Rosal et al., 2011; Rosal et al., 2005) and two were one-group pre-post comparisons. (DeWalt et al., 2009; S. Kim et al., 2004).

A total of 980 middle-aged and older adults with T2DM (ranging in age from 45 to 67.2) were involved in the reviewed studies. All studies were conducted in the US. Five of them were published in the mid to late 2000s, while the last one appeared in 2011.

The quality appraisal result showed that only one study was rated as good (Cavanaugh et al., 2009). Five studies met the fair standard (DeWalt et al., 2009; Gerber et al., 2005; S. Kim et al., 2004; Rosal et al., 2005; Rosal et al., 2011). Table 2 presents the outcomes of the quality appraisal of the reviewed articles.

1.4.2   The Effectiveness of Health Literacy Oriented Programs on Physical Activity Behaviour

Intervention to improve HL seemed to have a somewhat suggestive effect on PA behaviour in middle-aged and older adults with T2DM. In Rosal et al. (2005) observed highly significant differences in PA behaviour and walking habits in the intervention group after the HL-oriented programs. Cavanaugh et al. (2009) affirmed the effectiveness of intervention with numeracy, a component of HL, on self-care behaviours including PA behaviour in the intervention group. Kim et al. (2004) also confirmed that HL-oriented programs were effective in modifying the PA behaviour, especially in T2DM patients with higher HL levels.

The changes in PA behaviour accelerated with the duration of HL-oriented programs. Rosal et al. (2005) showed that the differences in the amount of calories burned through PA behaviour between the intervention group and the control group was wider at 6-month than at 3-month.

Moreover, PA behaviour was nearly the most popular choice among subjects, when compared to medication and diet in HL-oriented programs in two reviewed articles (Cavanaugh et al., 2009; DeWalt et al., 2009).

However, Rosal et al. (2011) did not find HL-oriented programs to have an effective impact on PA behaviour in an RCT study. Although there was a greater trend toward walking habits in the intervention group, the difference was not significant. The total time expended in exercise and walking did not increase significantly after the HL-oriented programs.

HL-oriented programs were found to significantly enhance the self-efficacy of self-care behaviour (including PA behaviour) and the knowledge of self-care behaviour (including PA behaviour) in two reviewed articles. Four perceived barriers to PA behaviour namely, lack of prompting by clinical staff, time constraints, availability of technology, and personal factors were identified in one RCT study (Gerber et al., 2005).

However, no studies showed a mediating effect of self-efficacy, knowledge of PA or perceived barriers on the relationship between HL-oriented programs and PA behaviour (Cavanaugh et al., 2009; S. Kim et al., 2004; Rosal et al., 2011)

Table 3

Analysis of the Six Reviewed Articles of Health Literacy Oriented Programs on Physical Activity Behaviour

Study (year); Country	Design	N; Age	Aims	Intervention	Measures (HL & PA)	Statistical tests	Result
Rosal et al. (2011); USA	RCT	N=211 (Intervention group N=106; Control group N=105) Age: ≥ 45	To test whether a theory-based HL and culturally tailored self-management intervention improves glycemic control and SMB among low-income, low-literacy Latinos with type 2 diabetes	Intervention group: 12-session weekly, theory-based and culturally tailored, HL-enhanced education program including picture-based food guide, soap opera, step counter guide, color-coded charts of glucose level warnings, meetings with health specialists, and 8 monthly follow-ups Control group: usual care in community health centres Measurement: baseline; 4 months; 12 months	Self-reported PA Self-efficacy of diet and PA change	t-test Mixed-effect logistic regression model Linear regression model	The intervention group had a greater but non-significant trend of walking habits after the intervention when compared to the control group. No significant change was found regarding total time of PA, time of walking and time of sitting after the intervention Intervention significantly changed the self-efficacy of diet and physical activity (p = 0.001), diabetes kilocalories burned (p < 0.001), and percentage of fat (p = 0.003)
Cavanaugh et al. (2009); USA	RCT	N=179 (Intervention group N=85; Control group N=94) Median age: 52	To examine the effectiveness of literacy and numeracy intervention on diabetes care	Intervention group: received usual care including 1-to-6 and face-to-face visits conducted by diabetes nurse practitioner, certified diabetes educator and registered dietician; 1-to-2 didactic training session about HL, numeracy and effective communication; and DLNET Control group: usual care in local enhanced diabetes care program	DNT REALM SDSCA PDSMS	Descriptive Wilcoxon's rank-sum Multivariable model Regression analysis	The most commonly used sections of DLNET were glucose testing (88%), PA (83%), general nutrition (77%) and foot care (63%) Intervention group significantly improved the self-efficacy of diabetes self-management score (p < 0.05) and self-care behaviours (including PA) (p < 0.01)
DeWalt et al. (2009); USA	Quasi-experimental study	N=229 Mean age: 56	To evaluate the usefulness of a diabetes self-management guide in helping patients to set and achieve their behavioural goals	12- to 16-week goal-setting intervention that consisted of an in-person introduction and counseling session based on the American College of Physicians Foundation Living with Diabetes Guide	Goal-setting domains S-TOFHLA	t-test Chi-square (both tests compared within subjects)	Exercise was the most popular goal with more than 40% participants' votes, among other goals such as diet, insulin control, and medication No difference was observed on goal achievement (including PA) across various HL levels and ability of language

	Study design	Sample/Age	Purpose	Intervention	Instruments	Statistical analysis	Results
Gerber et al. (2005); USA	RCT	N=244. Lower HL intervention group: N=68, mean age=57.7. Higher HL intervention group: N=54, age=49.4. Lower HL control group: N=67, mean age=60.4. Higher HL control group: N=55, mean age=51.8	Evaluate a clinic-based, low-literacy computer multimedia intervention for diabetes patients and evaluate the barriers and facilitators to implementing the computer-based education	Intervention group: 19 low-literacy computer multimedia interventional lessons apart from receiving standard care. Control group: standard care	S-TOFHLA	t-test, Mann-Whitney U test, Fisher's exact tests	Four barriers to PA were identified: lack of prompting by clinical staff, time constraints, availability of technology, and personal factors. Positive PA behavioural changes were self-reported by patients in intervention group
Rosal et al. (2005); USA	RCT (pilot study)	N=25 (intervention group, N=15, Mean age=62.7; control group, N=10, Mean age=62.4)	Examine the effectiveness of a pilot program on SMB of illiterate Spanish-speaking patients with diabetes	Intervention group: 10-week cognitive behaviour intervention on SMB (material was tailored for the low-literacy patients). Control group: a commonly used booklet describing the benefits and recommendations of SMB	Community Healthy Activities Model Program for Seniors PA questionnaire, IMSES	General linear model	An increasing trend of PA in the intervention group as compared to control group was found; at three months, the difference was 640kcal/week, $p = 0.08$; at six months, the difference was 789kcal/week, $p = 0.06$
Kim et al. (2004); USA	Intervention study (no control group)	N=92, Age: mean age=58.2 (adequate HL); mean age=67.2 (limited HL)	Examine the association between HL and SMB and determine the effectiveness of diabetes literacy education in improving self-management behaviour	Individual meeting with diabetes educator and three weekly, three-hour group classes on diabetes education	SDSCA, S-TOFHLA, DKQ	ANCOVA	Knowledge of SMB (including PA) improved ($p < 0.05$). Patients with higher literacy level exercised more than those with low literacy after the intervention ($p < 0.05$)

Note. DKQ=Diabetes Knowledge Questionnaire; DLNET=Diabetes Literacy and Numeracy Education Toolkit; DNT=Diabetes Numeracy Test; HL=health literacy; PA= physical activity; PDSMS=Perceived Diabetes Self-Management Scale; RCT= Randomized Controlled Trials; REALM=Rapid Estimate of Adult Literacy in Medicine; SMB=self-management behaviour; SDSCA=Summary of Diabetes Self-Care Activities scale; S-TOFHLA= Short Test of Functional Health Literacy in Adults

1.5 DISCUSSION

Based on these six studies, it would be premature to draw conclusions on the effectiveness of HL-oriented programs in modifying PA behaviour among middle-aged and older T2DM patients.

The evidence supporting the impact of HL-oriented programs on PA behaviour is untenable. This review consisted of four RCTs and two one-group, pre-test/post-test interventional studies, of which only one study design was rated as good in a quality appraisal. The supportive findings revealed by DeWalt et al. (2009) and Kim et al. (2004) involved only within-group comparisons, making the examination of these HL-oriented programs difficult since their effective influence on PA behaviour could be implicated by confounders.

Moreover, there were conflicting findings on the effect of HL-oriented programs on PA behaviour. Rosal et al. (2011, N = 211) rejected their own positive findings about the impact of HL-oriented programs on PA behaviour in a previous pilot RCT study (N = 25) conducted in 2005. The great difference in the sample size of the two studies could have influenced the investigations. Since the earlier study recruited only illiterate T2DM patients, sample bias could have confounded the positive evidence of HL-oriented program impact on PA behaviour (Rosal et al., 2005).

Although positive effects of HL-oriented programs on PA behaviour were found in two other RCT studies (Cavanaugh et al., 2009; Gerber et al., 2005), and also in the one-group study by Kim et al. (2004), PA was not distinctly and objectively investigated as a dependent variable. Most of the studies considered PA behaviour as one of a group of diabetes self-care behaviours (along with diet, self-monitoring of blood glucose, foot care, medication adherence and so forth) with subjective self-reported measurements. There were no articles on HL-oriented programs designed specifically to modify PA behaviour of T2DM patients. The effects of combined self-care behavioural modification programs (addressing multiple behaviours such as diet, foot care and medications) were questioned by a recent meta-analysis on non HL-oriented community-based PA behavioural modification programs for adults with T2DM. Exclusively PA-focused behavioural modifications were found to be more effective in changing the duration and intensity of PA behaviour, as objectively measured by METs (Plotnikoff et al., 2013). The value of a PA-only behavioural modification program was supported by an earlier meta-analysis measuring the metabolic effects of behavioural modification programs to modify the PA behaviour of adults with T2DM (Conn et al., 2007).

Furthermore, the frequency and duration of PA behaviour were collected as outcome measures in the reviewed articles while the intensity of PA, the most important factor affecting the blood glucose control (Colberg et al., 2010; Shiraki, Sagawa, & Yousef, 2001; Sigal et al., 2004), was not measured. In addition, objective measurements using pedometers and accelerometers to monitor the number of steps taken by patients could provide a reliable outcome measure of PA behaviour; hence these objective measurements would enable researchers to elucidate the relationship between HL-oriented programs and PA behaviour in middle-aged and older T2DM patients.

In addition, according to the IOM (2004), HL is manifested differently in various countries because of their differing cultures and languages, and thus it is questionable to generalize from a series of US studies of HL-oriented programs on PA behaviour to what might occur elsewhere.

None of the six articles reviewed took the importance of statistical power, effect size and sample calculation into account before the initiation of HL-oriented programs; in fact, some authors of the studies reviewed considered inadequate statistical power to be a limitation of their programs.

It is possible that this literature review failed to identify other useful articles because the screening process focused only on full-text articles published in English and available in the above-mentioned eight academic databases. Three of the studies selected (Cavanaugh et al., 2009; S. Kim et al., 2004; Rosal et al., 2011) contained a small portion of patients with type 1 diabetes mellitus (T1DM) (n = 60 out of 1,040), but this is not a significant limitation because the 60 patients with T1DM were successfully excluded in the present review since their results were given separately in the three studies. Finally, the extent to which the results were affected by the use of different HL instruments could not be estimated.

1.6 CONCLUSION

The effectiveness of HL-oriented programs in modifying PA behaviour among middle-aged and older T2DM patients remains unconfirmed. The causal pathway between HL and health outcomes in the HL model (Paasche-Orlow & Wolf, 2007; Osborn et., 2011) indicated that self-care behaviours, including PA behaviour, could be modified only through theoretically grounded and systematically planned HL-oriented programs, through mediating factors such as knowledge, the components of health beliefs (e.g. self-efficacy and perceived barriers), and motivation. Based on the articles reviewed, factors such as knowledge of PA, self-efficacy of PA and perceived barriers to PA could affect the impact of HL-oriented programs on PA behaviour; however, the mediating effects were not properly investigated. Only Rosal et al. (2011) intended to align mediating factors (knowledge, attitude and self-efficacy) with the related constructs of social cognitive theory to design HL-oriented programs tailor-made for Latinos in the U.S. More sophisticated, theory-based, HL-oriented programs should be planned for different countries, races, cultures and languages (Zarcadoolas et al., 2005; IOM, 2004).

1.7 IMPLICATIONS FOR FUTURE STUDIES

Future HL-oriented programs on PA behaviour on middle-aged and older patients with T2DM shall focus on (a) measuring PA behaviour alone, separate from other diabetic self-care behaviours; (b) using valid, reliable, objective, culturally appropriate and age-specific measurements of PA behaviour; (c) the effectiveness of culturally and language-appropriate HL-oriented programs in different countries; (d) the duration of the program and follow-up period; (e) more culture- and language-specific measurement of HL; and (f) the calculation of effect size and sample size before the initiation of programs.

COGNITIVE AND PSYCHOLOGICAL DETERMINANTS OF THE HEALTH LITERACY ON MIDDLE-AGED AND OLDER ADULTS: A SYSTEMATIC REVIEW

2.1 INTRODUCTION

In order to maximize the effects of HL-oriented programs on health outcomes, some researchers tried to summarize and to explain the potential mechanism between HL and health outcomes. They concluded that limited HL is the first step on the causal pathway, which affected the knowledge, psychological determinants of health care behaviours and eventually lead to poor health outcome. Cognitive determinants (verbal working memory, processing speed and cognitive function) were interpreted as the abilities affecting HL (Osborn et al., 2011; Paasche-Orlow & Wolf, 2007; Pignone & DeWalt, 2006). Over the past twenty years, the research on HL has been developed in a narrow scope, as most of the HL studies focused on demographics (age, ethnicity and education), health knowledge, health outcomes (including self-care behaviours and health outcome) (Cutilli, 2007; Osborn et al., 2011; Paasche-Orlow et al., 2005). Different negative health outcomes, such as poor physical and mental health (Wolf, Gazmararian, & Baker, 2005); poor glycemic control (Schillinger et al., 2002) and higher mortality (Tan et al., 2004; Baker et al., 2007) have been found closely related to limited HL in ageing population. The diabetic self-care knowledge has been widely supported to be the mediating factor affecting the relationship between HL and self-care behaviours or health outcomes (DeWalt, Boone, & Pignone, 2007; Osborn et al., 2010). Some specific conceptual models of health knowledge on the relationship between HL and self-care behaviours or HL and health outcomes have been established (DeWalt et al., 2007; Osborn et al., 2011; Pignone & DeWalt, 2006), however, health knowledge alone is rarely adequate to cause a change on self-care behaviours or health outcome, not all programs based on these models can consistently change the self-care behaviours and health outcomes (Day, 2000; Friedman, Corwin, Dominick, & Rose, 2009; Jaarsma et al., 1999; Jimenez et al., 2005). Psychological and cognitive determinants on causal pathways between HL and health outcomes were other important components advocated by Paasche-Orlow and Wolf (2007), however, inadequate attentions were drawn to elucidate the importance of psycho-cognitive determinant on the causal pathway, especially on middle-aged and older adults (Baker, 1999; DeWalt et al., 2007; DeWalt & Pignonc, 2005; S. H. Kim & Yu, 2010).

This chapter was based on a peer-reviewed paper published in Journal on Nursing (Appendix 1).

2.2 AIMS OF THIS CHAPTER

The aim of this chapter is to systematically review and evaluate the psychological and cognitive determinants of the causal pathways between HL and health outcome on middle-aged and older adults (objective 1 of the thesis).

The purposes of this chapter are a) to consolidate and explain the least investigated psychological and cognitive mechanism on the causal pathways between HL and self-care behaviours or health outcomes and hence, properly incorporate the required psychological determinants into the content of this HL-oriented program and; b) to reveal the importance of HL related studies on middle-aged and older population due to the existence of cognitive functional degradation caused by biological ageing.

2.3 METHODS

The studies were reviewed and reported in accordance with the guidance of PRISMA (Moher et al., 2010). The systematic review included the studies published from January 1990 to December 2010 in research and evidence based database. The searching process started from January to February 2011 and updated in March 2011.

In this systematic review, the following key questions were examined:

1. Is HL of middle-aged and older adults related to psychological determinants?

2. Is HL of middle-aged and older adults related to cognitive determinants?

2.3.1 Searching Criteria

The searching terms used in this review includes health literacy, literacy, psychological, cognitive, moderator, mediator, predictor, determinant, factor, association, relationship, casual pathway, correlate, correlation, middle-aged, older adult, older people, senior, elderly, elder, geriatric and aged. To be included, studies had to (a) be written in English; (b) be published between 1990 and 2010; (c) be a study more than 10 participants; (d) be with middle-aged and older adults aged 50 or over; (e) be full text available; and (f) be included the psychological or cognitive determinants of HL. We excluded the studies solely investigated the relationship between demographics, health status, medication, mortality, health outcome, knowledge and HL, as these studies were affluent and their relationship with HL had been widely explored (Cutilli, 2007).

2.3.2 Literature Search and Review

Databases including CINAHL, Medline, PubMed, Proquest, PsycINFO and ISI were searched.

Following the searching procedures described, CINAHL yielded 17 results and 2 of them were selected after the screening. Medline yielded 33 results and 3 of them were selected. PubMed yielded 88 results and 3 of them were selected. Proquest yielded 53 results and 7 of them were selected. PsycINFO yielded 39 results and 4 of them were selected. ISI yielded 18 results and none of them were selected. Only unique studies were investigated and evaluated after the final review. After adjusting the two studies which included the same older adults (approximate 2700 older adults), the 19 studies yielded a total of 12,444 middle-aged and older adults.

2.3.3 Quality and Strength of Evidence

We graded the studies according to the instrument developed to test the relationship between HL and health outcomes (Berkman et al., 2004). DeWalt and Hink (2009) adopted this grading criterion in another systematic review investigating HL and child health outcomes. Although the instrument adopted had been applied in several systematic reviews, the results were considered as a reference only as the result was not formally validated. All studies were graded according to (a) the adequacy of study population; (b) the comparability of subjects across comparison groups; (c) the validity and reliability of the literacy measurement; (d) the maintenance of comparable groups; (e) the appropriateness of the outcome measurement; (f) the appropriateness of statistical analysis and (g) the adequacy of control of confounding (Berkman et al., 2004; DeWalt & Hink, 2009).

A score was given to each above items and the items which were not applicable to this study designs were excluded (0=poor, 1=fair; 2=good). The average score of these items on equal-weighting base would be the final quality and strength grades (<1=poor; 1 and <1.5 =fair; 1.5 and >1.5 =good) (Berkman, et al., 2004).

2.4 RESULTS

2.4.1 Study Characteristics of Key Questions

Table 4 shows the overview of included studies. All articles reviewed were cross sectional and longitudinal studies. It ranged in sample size from 32 to 3260 middle-aged and older adults. HL was often measured by Test of Functional Health Literacy in Adults (TOFHLA) (Parker, Baker, Williams, & Nurss, 1995) and Rapid Estimate of Adult Literacy in Medicine (REALM) (T. C. Davis et al., 1993). Most studies embraced descriptive information, like age, gender, ethnicity, socio-economic status, knowledge level and insurance status, which may affect the health outcomes.

2.4.2 Key Question 1: Is Health Literacy of Middle-aged and Older Adults Related to Psychological Determinants?

Health Literacy and Self-Efficacy

Six out of eight studies supported the existence of positive relationship between HL and self-efficacy of middle-aged and older adults (DeWalt et al., 2007; Ishikawa, Nomura, Sato, & Yano, 2008;

S. H. Kim & Yu, 2010; Osborn et al., 2010; Peterson et al., 2007; Sarkar, Fisher, & Schillinger, 2006; Torres & Marks, 2009; von Wagner, Knight, Steptoe, & Wardle, 2007; von Wagner, Semmler, Good, & Wardle, 2009). Kim and Yu (2010) showed that the general self-efficacy of Korean was correlated with HL and general self-efficacy was found to be a mediator between the HL and physical or mental health. Ishikawa et al. (2008) and Osborn et al. (2010) also confirmed this relationship on diabetes patients. Sarkar et al. (2006) showed that the self-efficacy was associated with self-care behaviours across all HL levels. Moreover, the new colorectal cancer patient in UK (von Wagner et al., 2009) and the community clinic patients (Torres & Marks, 2009) also supported the positive behaviour between self-efficacy and HL. von Wagner et al. (2009) showed that HL was a predictor of self-efficacy for participating in the colorectal cancer screening after controlling the demographics and information processing variables. Inconsistently, DeWalt et al. (2007) did not find any relationship between HL and self-efficacy on diabetes patients. However, unlike the other studies reviewed, the patients were recruited from the same internal medicine practice, hence the generalizability of the sample to the other clinics or regions were inadequate. Peterson et al. (2007) could not confirm the association between HL and self-efficacy for completing the Fecal Occult Blood Test (FOBT) or colonoscopy on patients with colorectal cancer after adjusting the age, sex, race, insurance status and literacy status, however, the small sample size limited its power to detect the differences.

Health Literacy and Attitude or Belief

Three studies measured the relationship between HL and attitude or belief of middle-aged and older adults (T. C. Davis et al., 1996; Dolan et al., 2004; Guerra, Dominguez, & Shea, 2005). Guerra et al. (2005) revealed that among four attitudes (being embarrassing, harmful, painful and troublesome) towards mammography screening, only being embarrassing was not associated with participants with lower HL. Dolan et al. (2004) supported this relationship by showing that male veterans with lower HL 1) would not use a FOBT kit even though they were recommended by their physician, 2) was less concern and worried that FOBT was messy, 3) was less concern and worried that FOBT is inconvenient. Guerra et al. (2005) disavowed this relationship as they found that Hispanic women's HL was not associated with the attitude or the belief about mammography examination. However, the small sample size limited its power to detect the differences.

Health Literacy and Perceptions

Two studies measured the relationship between HL and perceptions of middle-aged and older adults (Dolan et al., 2004; Peterson et al., 2007). Perceived barriers were found on lower literate patients with colorectal cancer to complete FOBT and colonoscopy (Peterson et al., 2007). Similarly, male Veterans with lower HL had a higher perceived barrier on FOBT (Dolan et al., 2004). However, perceived risk to get a worse colorectal cancer in the next 10 years regardless the level of HL was not found (Peterson et al., 2007).

Health Literacy and Motivation, Desire and Intention

Four studies measured the relationship between HL and motivation, desire or intention of middle-aged and older adults (DeWalt et al., 2007; Lillie et al., 2007; Miller, Gibson, & Applegate, 2010; Torres & Marks, 2009) were reviewed. Stage I and II primary breast cancer patients with higher HL had a higher intention and preference for active participation in decision making (Lillie et al., 2007). Torres and Marks (2009) support such behaviour intention concerning the hormone therapy on community clinic patients. For desire, inconsistent conclusions were drawn. DeWalt et al. (2007) showed that low HL patients with T2DM had a lower desire to participate in medical decision; however, Lillie et al. (2007) showed that patients' desire for additional information of breast cancer was not associated with HL. Miller et al. (2010) revealed that limited HL has no association with patients' motivation on nutritional concerns. Nutrition motivation could associate with HL only through the mediator of attention.

2.4.3 Key Question 2: Is Health Literacy of Middle-aged and Older Adults Related to Cognitive Determinants?

Health Literacy and Cognitive Function

Six studies measured the relationship between HL and cognitive function of middle-aged and older adults (Baker, Gazmararian, Sudano, & Patterson, 2000; Baker et al., 2002; Baker et al., 2007; Federman, Sano, Wolf, Siu, & Halm, 2009; Levinthal, & Wu, 2008; Morrow et al., 2006). One study on patients with heart failure showed that HL was associated with cognitive function (processing speed and verbal working memory) (Morrow et al., 2006). Five studies on community dwelling older adults also supported such association. After adjusting the age, sex, ethnicity, English language proficiency, household income, self-reported general health and any deficiency in activities of daily living, older adults with lower HL had a lower cognitive function (abnormal immediate recall, abnormal delayed recall, abnormal verbal fluency and abnormal score of Mini Mental State Exam) (Federman et al., 2009). Baker et al. (2000) and Levinthal and Wu (2008) supported this relationship since they found that lower HL resulted in lower cognitive function of middle-aged and older adults after adjusting sex race and education. More interestingly, Levinthal and Wu (2008) who entered all of the cognitive and sensory variables in the multivariate regression analysis of HL, found that the cognitive and the sensory variables accounted for an additional 24% (cognitive more than sensory) of the total variance in Short Test of Functional Health Literacy in Adults (STOFHLA) and cognitive function was acted as a mediator of age, race, and education on HL (beta values decreased 75%, 40%, and 48%, respectively), this implied that cognitive function was very important in constructing the HL of older adults. Similarly, an experiment conducted by (Baker et al., 2007) showed that cognitive function was directly merged into the multivariate model of HL and mortality when the hazard ratio of TOFHLA dropped to 1.27 and TOFHLA was then lost its surrogate measure for cognitive ability. It inferred the close relationship between HL and cognitive functional tests (Baker et al., 2002) proved that functional HL was linearly associated with cognitive function across the entire range of TOFHLA scores.

Table 4

Overview of Included Studies of Psychological and Cognitive Determinants in Health Literacy Related Studies

Author/Types of Study	Instrument	Age/ Number of Participants/ Diseases	Outcome Assessed	Literacy Relationship	Quality
Psychological Determinants					
Davis et al. (1996) cross sectional	REALM	Mean age=56 N=445 Women visited outpatient clinics	Attitude to Mammography Screening: 1. Being Embarrassing, 2. Being Harmful 3. Being Painful 4. Being Troublesome	-LHL no association on embarrassment -LHL associated with harm (p<0.01) -LHL associated with pain (p<0.05) -LHL associated with troublesome (p<0.05)	G
DeWalt et al. (2007) cross sectional	REALM	Mean age >58 N=268 Type 2 diabetic patient	-Self-efficacy of Diabetes Self-management -Trust -Desire to Participate in Medical Decision -Perception of Doctor's Facilitation of Patient Involvement	-HL no association on SE of diabetes self-management -HL no association of Trust -HL no association of facilitation of patient involvement -LHL lower desire to participate in medical decision (p<.001)	G
Dolan et al. (2004) cross sectional	REALM	Mean age=67.3 N=377 male veterans received general medical care	Beliefs or Attitude: 1. Perceived Susceptibility 2. Perceived Severity 3. Perceived Benefits 4. Cues to Action 5. Perceived Barriers	No difference between HHL and LHL on 1. perceived susceptibility, 2. perceived severity, 3. perceived benefits Difference between HHL and LHL on 4. Cues to action -LHL would not use a FOBT kit even if recommended by their physician (p<.05) 5. perceived barriers -LHL less concern and worried that FOBT is messy (p<.01) and inconvenient (p<.05) -LHL for not getting flexible sigmoidoscopy within past 5 years (p<.01)	G
Guerra et al. (2005) cross sectional	S-TOFHLA	Mean age = 59 N=97 Hispanic ethnicity with no prior history of breast cancer	Attitude or Belief about Mammography Examination	HL no association on attitude or belief about mammography examination	F
Ishikawa et al. (2008) cross sectional	New diabetic HL instrument in Japan	Mean age = 65 N=138 Type 2 diabetic patients	Self-efficacy of Diabetes Self Care	LHL lower SE (p<.001)	G

Study	Tool	Sample	Variables	Findings	Grade
Kim et al. (2010) cross sectional	K-TOFHLA	Age > or = 60 N=103 Community dwelling senior	-General Self Efficacy -Physical and Mental Health	HL is a predictor of GSE (p<.01)	F
Lillie et al. (2007) cross sectional	REALM	Mean age = 59 N=163 stage I or II primary breast cancer patients	- Desire for Additional Health Information - Preference for Active Participation in Decision Making	-HHL higher preference for active participation in decision making (p<.001). -HL no association to desire for additional health information	G
Miller et al. (2010) cross sectional	4 nutrition texts at grade levels between 8-10 Flesch-Kincaid criteria.	Mean age= 71.1 N=32 Community dwelling senior	The Moderation of the Followings on Nutritional Comprehension: 1. Motivation 2. Knowledge of Nutrition 3. Attention Span	-LHL no direct association of motivation -LHL associated with attention (p<.05) -LHL associated with knowledge (p<.05) -Knowledge associated with motivation (p<.001) -Attention associated with motivation (p<.05)	F
Osborn et al. (2010) cross sectional	REALM	Mean age = 54 N=383 Diabetic patients	Decisional Diabetes Self Efficacy	LHL lower SE (p<.01)	G
Peterson et al. (2007) cross sectional	REALM	Mean age > or = 60 N=99 Colorectal cancer patients	Cancer Screening: Perceived Barrier Perceived Benefit Perceived Risk Reported Self Efficacy	-LHL higher barrier on completing FOBT (p<.001) and colonoscopy (p<.01). -HL no association of perceived benefits -HL no association of perceived risk -HL no association of reported self-efficacy	F
Sarkar et al. (2006) cross sectional	S-TOFHLA	Mean age=58.1 N=408 Type 2 diabetic patient	Self-efficacy of Diabetes Self Care	-SE associated with self-management behaviours across HL levels (p<.01) -HL is a moderator of SE and self-management (association is implied)	G
Torres et al. (2009) cross sectional	S-TOFHLA	N=106 Mean age= 52.58 (SD=5.35; 45-65) Community clinic patients	-Decisional Self Efficacy -Behaviour Intention Concerning Hormone Therapy -Knowledge about Hormone Therapy	-LL lower knowledge (p<.01) -LL lower behaviour intention (p<.01) -LL lower SE (p<.01) -Knowledge no associated with behaviour intention -SE explained 66% of behaviour intention -HL explained 9% of behaviour intention	F
Wagner et al. (2009) cross sectional	UK-TOFHLA	Age 50-69 (Average =54.2) N=96 New colorectal cancer patients in UK	Self-efficacy of Colorectal Cancer Screening	-LHL lower SE (p<.001) -HL is a predictor of SE for participating in the Colorectal Cancer Screening	F

Cognitive Determinants

Baker et al. 2000 Cross sectional	S-TOFHLA	Age > or = 65 N=2774 Community dwelling senior	-Cognitive Function	-LHL lower cognitive function (p<.001)	G
Baker et al. (2002) cross sectional	S-TOFHLA	Aged > or = 65 N= 2787 Community dwelling senior (at least one chronic disease)	Cognitive Function	FHL linearly associated to cognitive function across the entire range of STOFHLA scores (R2 = 0.39, p <.001).	G
Baker et al. (2007) cohort study 1997-2003	S-TOFHLA	Age > 65 N=3260 Community dwelling senior	-Cognitive Function -Mortality	-HL is a predictor of mortality (hazard ratio was 1.50 [95% CI 1.24-1.81]) -Cognitive function is a predictor of mortality (hazard ratio was 1.74 [95% CI 1.30-2.34]) -LHL lower cognitive function (p<.05)	G
Federman et al. (2009) cross sectional	S-TOFHLA	Age >75 N=414 Community dwelling senior	Cognitive Function: -Memory -Verbal Fluency -Cognitive Function	-LHL abnormal immediate recall (p<0.001) -LHL abnormal delayed recall (p<0.01) -LHL abnormal verbal fluency (p<0.01) -LHL abnormal MMSE(p<0.001)	G
Levinthal et al. (2008) cross sectional	S-TOFHLA	Age >60 N=180 Community dwelling senior	Cognitive Ability: -Letter and Pattern Comparison -Listening Span	-LHL lower letter and pattern comparison (p<.001) -LHL lower listening span(p<.001) -LHL lower speech discrimination (p<.001)	G
Morrow et al. (2006) cross sectional	S-TOFHLA	Mean age =62.9 N=314 patient with heart failure (67% were female)	Cognitive Function: -Verbal Working Memory -Processing speed	-LHL lower processing speed (p<.001) -LHL lower verbal working memory (p<.001)	G

Note. HL- Health Literacy; FHL- Functional Health Literacy; HHL- High Health Literacy; LHL- Low Health Literacy; SE-Self Efficacy; GSE- General Self Efficacy; K-TOFHLA- Korean Test of Functional Health Literacy in Adults; S-TOFHLA- Short Test of Functional; Health Literacy in Adults; REALM, Rapid Estimate of Adult Literacy in Medicine; FOBT-Fecal Occult Blood Test; MMSE- Mini Mental State Exam; CI- Confidence Interval; UK-TOFHLA-British Version of Functional Health Literacy in Adult.

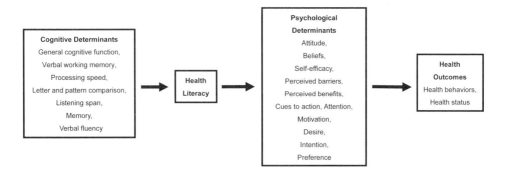

Figure 3: Causal pathway between health literacy and health outcome concerning the cognitive and psychological determinants

2.5 DISCUSSION AND CONCLUSION

Evidence assimilated within this review showed the trend to support the relationship between HL and psychological and cognitive determinants on middle-aged and older adults. The majority of the studies demonstrated the positive relationship between HL and psychological determinants (self-efficacy; attitude or belief; perceptions; motivation and intention) and cognitive determinants (verbal working memory, processing speed, letter and pattern comparison, listening span, memory, verbal fluency, global cognitive function) across various diseases and community dwelling middle-aged and older adults. A conceptual framework addressing the causal pathway of HL and health outcomes based on this review is concluded in figure 3. Though HL has been widely accepted to affect the health outcomes (Baker et al., 2007; Schillinger et al., 2002; Wolf et al., 2005), proper psychological and cognitive mechanism explaining the mediation of the relationship between HL and health outcome was still untracked. The empirical findings in this review suggested that future exploratory, interventional and cohort studies could focus more on the psycho-cognitive conceptual framework between HL and health outcomes of middle-aged and older adults.

This review also confirmed the importance of HL related studies on middle-aged and older population due to the existence of cognitive functional degradation led by the biological ageing.

2.5.1 Possible Mechanism Explaining the Causal Pathways between Health Literacy and Health Outcome

Based on the findings of this review, four possible theories and mechanisms explaining the relationship among HL, psycho-cognitive determinants and health outcomes were suggested. All these could be served as a ground for future clinical and community health practices and interventions.

In contradistinction to Bandura's Self Efficacy Theory (Bandura, 1989), the studies of this review suggested that middle-aged and older adults with limited HL did not merely require the perceived confidence or self-efficacy, but also the motivation, intention, perceptions and the actual ability (cognitive functions) for better health outcomes.

Self-Determination Theory (R. M. Ryan & Deci, 2000) addressed the autonomy (motivation, perceived barrier and benefits towards health behavioural changes), competence (self-efficacy and the actual competence towards health behavioural changes) and relatedness (attitude towards the health care system, health care professionals and health information) coincided with the psychological and cognitive determinants of this review. The psychological determinants of HL (self-efficacy; attitude or belief; perceptions; motivation and intention) coincided with all three domains of Self-Determination Theory while the capability of cognitive function and the capability to obtain, process and understand the basic information needed to make appropriate decisions regarding one's health (HL) empower the middle-aged and older adults with competence to take health actions.

Folk Concept of Intentionality (Malle & Knobe, 1997) embraced desire, belief, intention, awareness and skills was consistent with the findings of this review that intentionality, perceived abilities (self-efficacy) and abilities (cognitive function and HL) were important to make judgments of self-care behaviour on middle-aged and older adults with limited HL. For Folk Concept of Intentionality, the pro-attitude towards the health decision (desire, belief, intention, and awareness) would be changed to action only if the competence and the motivation were adequate (Wellman & Woolley, 1990). Moreover, intention, rather than desire and awareness, was regarded as the key step away from the action (Malle & Knobe, 1997). A study reviewed echoed this advocacy that patients of stage I and II primary breast cancer with higher HL (ability to make appropriate decisions regarding one's health) made preferences and intention for active participation in decision making (Lillie et al., 2007). Another article investigating the community clinical patients showed the agreement to the concept of intentionality on the causal pathway of HL and self-care behaviours. This study indicated that the patients with higher HL (competence to make appropriate decisions; explained 9% of behavioural intention) had a higher self-efficacy (perceived competence; explained 66% of behavioural intention) and hence increase the behavioural intention concerning the hormone therapy (Torres & Marks, 2009).

The HBM, one of the most frequently used health conceptual frameworks explains and predicts the individual belief and psychological readiness to take health actions. The constructs include perceived benefits and barriers, cues to action, perceived susceptibility, perceived severity, and self-efficacy (Champion & Skinner, 2008). Though three empirical studies have tried to look at the relationship between HL and HBM on patients with T2DM and colorectal cancer (Dolan et al., 2004; Peterson et al., 2007; Powell et al., 2007), systematic theoretical reviews on how and whether the psychological constructs of HBM mediated the relationship of HL and health outcomes have been missed. Self-efficacy, the most frequently measured construct, has been found to associate with HL in six out of eight studies reviewing the relationship between HL and self-efficacy on middle-aged and older adults with different chronic diseases (DeWalt et al., 2007; Ishikawa, Takeuchi, & Yano, 2008; S. H. Kim & Yu, 2010; Osborn et al., 2010; Peterson et al., 2007; Sarkar et al., 2006; Torres, & Marks, 2009; von Wagner et al., 2009). The negative relationship between perceived barriers of patients with colorectal cancer and HL was supported by two studies identified (T. C. Davis et al., 1996; Dolan et al., 2004). Low HL women perceived being painful, troublesome and harmful were the barriers of conducting mammogram (T. C. Davis, Arnold, et al., 1996) while low HL male veterans who received general medical care perceived that the FOBT is messy and inconvenient (Dolan et al., 2004). Perceived benefits, perceived susceptibility and perceived severity were under-investigated in studies with HL and health outcomes. Up to now, no validated and structural questions has been composed in different chronic diseases (Dolan et al., 2004; Gerber et al., 2005; Peterson et al., 2007; Plotnikoff et al., 2010). For cues to action, though patients with lower HL would ignore the recommendations from doctors and physician in general medical care setting (cues to action) (Dolan et al., 2004), technically, it is hard to generate dozens of cues which will lead to different behaviours. Proper questionnaires of cues to action on different chronic diseases and health behaviours should be drawn in the future HL studies.

The Conceptual Model of Health Literacy on self-care area was recommended in the present HL-oriented program due to two reasons a) the current study is a HL oriented study and hence the specific model designed for HL-oriented programs should be adopted and; b) this chapter optimize the casual pathways between limited HL and health outcomes in Conceptual Model of Heath Literacy (Paasche-Orlow & Wolf's model, 2007; Osborn et al., 2011) by systematically reaffirming the psychological determinants as mediating factors on the casual pathway between limited HL and health outcomes with a review on empirical studies. It is suggested that the future HL-oriented programs should include the knowledge and problem solving skills together with psychological determinants such as motivation and the components of health beliefs (Champion & Skinner, 2008; R. M. Ryan & Deci, 2000) explained and consolidated with the previous three theoretical psychological mechanisms into the Conceptual Model of Health Literacy to affect the health outcomes including the health behaviours (Osborn et al., 2011) of patients with chronic diseases.

To facilitate the future HL-oriented programs on health outcomes, the future studies should also review the mechanism social determinants (cues to action of HBM) explaining the relationship of HL and health outcomes (Fallowfield & Jenkins, 1999; Greene, Adelman, Charon, & Friedmann, 1989; Stewart, 1995), as the importance of patient-nurse communication, patient-families communication are important to affect the behavioural change of patients (Paasche-Orlow & Wolf, 2007).

2.5.2   Limitations

There were several limitations in this review. The review was limited to the 'full text' English articles published in six academic databases; hence, we were not sure that the other related articles did not exist. Knowing the inadequacy of studies, to maximize the amount of articles related to middle-aged and older adults, participants regardless the types of diseases, stage of illness and whether they had diseases or not were all included. Two possible shortcomings were, firstly, different natures of diseases would serve as a confounding factor affecting the relationship between the psycho-cognitive variables and HL; secondly, as the amount of articles on specific diseases were limited, it was hard to drew conclusion on particular diseases. Though all studies were extracted from academic database, some studies were conducted in local site with non-representative samples, so it limits the power to generalize the relationship between the psycho-cognitive variables and HL. Though most of the instruments measuring different outcomes were validated, reliable and comparable, different studies and different countries used different instruments, for example, when measuring HL, REALM, Korean Test of Functional Health Literacy in Adults (K-TOFHLA), S-TOFHLA, British Version of Functional Health Literacy in Adult (UK-TOFHLA) were used in different studies and countries; when measuring self-efficacy, general self-efficacy scale or self-efficacy scale for different diseases were used for different populations; when measuring attitude or belief, similar validated scales were used in different diseases, it was therefore not totally certain to what extent the results were affected by different instruments.

Despite of the methodological shortcomings, the message conveyed by the 19 empirical studies in this review was quite consistent. Future studies should center on the psychological mechanism, especially in the areas of perception, intention and motivation on the causal pathway of HL and health outcomes as only few studies in this review indirectly addressed these elements on middle-aged and older adults, however, the psychological elements are very crucial for self-care on the patient factor (Paasche-Orlow & Wolf, 2007).

MODIFIED HEALTH BELIEF MODEL OF EXERCISE FOR PATIENTS WITH TYPE 2 DIABETES MELLITUS-CHINESE VERSION: Translation and Cultural Adaptation, with assessment of Reliability and Validity

3.1 INTRODUCTION

To further understand the low adherence to PA behaviour among diabetic patients, the HBM has been widely applied in research. This health conceptual framework can predict health actions, such as PA behaviour, from individual health beliefs (Champion & Skinner, 2008). Brownlee-Duffeck et al. (1987) studied the impact of health beliefs on self-care behaviours of persons with T2DM, reporting that health beliefs accounted for 41% of self-care behaviours. This finding has recently been supported in a longitudinal study of the Health Aging program, which found that health beliefs were a leading predictor affecting PA behaviour in adults age 65 or above (Browning et al., 2009).

However, the proper HBM has not been applied to review the health beliefs of PA behaviour in Chinese T2DM patients. Several diabetic HBM-questionnaires covering self-care behaviours have been established and validated on middle-aged and older adults since the 1980s, but PA behaviour has yet to receive adequate attention (Given, Given, Gallin, & Condon, 1983; Hurley, 1990). PA behaviour has been examined as one subscale of self-care behaviours in previous diabetic HBMs. In order to measure different self-care behaviours thoroughly, Aljasem et al. (2001) suggested a focus on the relationship between health beliefs and specific self-care behaviours among T2DM patients. Researchers began advocating for the use of HBM to study PA behaviour in middle-aged and older adults with T2DM only after release of the Healthy People 2010 goals for diabetes patients in the US (Campbell et al., 2010). To our knowledge, no validated and reliable instrument has been constructed to measure the relationship between the Chinese HBM and PA behaviour among T2DM patients.

A simple, valid instrument to diagnose physical inactivity in patients is crucial to meet the heavy demand for healthcare in the HKSAR and to effectively identify the underlying causes. The HKSAR population of older adults has increased dramatically (from 0.6 million in 1995 to a projected 1.1

million in 2015), while Hong Kong is also experiencing a growing rate of nurse turnover (turnover 400 nurses in 2005 and 1007 nurses in 2010) and a declining number of nursing graduates (2,400 nurse graduates in 2013, only 2123 anticipated for 2015) (E. Chan, 2012). Diabetes centres in HKSAR provide only care from nutritionists and nurses with professional knowledge of medication, foot care and eye care, but these people are not necessarily experts in PA behaviour. The Modified Health Belief Model of Exercise for Patients with Type 2 Diabetes Mellitus-Chinese Version (HBME-T2DM-C) will therefore be useful for diabetes nurses with limited knowledge of PA behaviours.

The validated behaviour-, disease- and age-specific HBME-T2DM-C will be adjusted in Chapter 5 if the association between the HBME-T2DM-C and PA behaviour is further confirmed in this chapter (objective 2 of the thesis).

3.2 AIM OF THIS CHAPTER

This chapter is aimed to modify, translate and validate the HBME-T2DM-C.

The main purposes of the chapter are (a) to develop a reliable and validated behaviour-, disease- and age-specific instrument for Chinese T2DM patients and; (b) to examine the association between the HBME-T2DM-C and PA behavioural outcomes of this thesis as assessed by the International Physical Activity Questionnaire for Older Adults—Chinese version (IPAQ-C) and the subsequent mean daily number of steps per week taken by Chinese middle-aged and older adults with T2DM.

3.3 FRAMEWORK FOR THE DEVELOPMENT OF THE HEALTH BELIEF MODEL OF EXERCISE FOR PATIENTS WITH TYPE 2 DIABETES MELLITUS—CHINESE VERSION

The HBM prospectively relates individual health beliefs to health actions (Champion & Skinner, 2008). It was developed by social psychologists in the 1950s (Glanz, Rimer, & Viswanath, 2008). Although the scale of the Health Belief Model of Exercise (HBME) for general chronic diseases has been established in the US (Hayslip et al., 1996), specific HBME results for T2DM patients have not been investigated. Having been identified as an important component of HBM, perceived self-efficacy (Champion & Skinner, 2008), which has not been incorporated in HBME (Hayslip et al., 1996), has recently been found to be the strongest predictor of aerobic PA and resistant training of PA compared to demographic, medical and environmental predictors in patients with T2DM (Bandura, 1997; Plotnikoff et al., 2011; Rosenstock, Strecher, & Becker, 1988).

In order to incorporate all the components of health beliefs about PA behaviour among T2DM patients, so that clinicians can plan for interventions involving physical movement, this chapter seeks to develop the HBME-T2DM-C with reference to the updated constructs of the HBM, including perceived barriers, perceived benefits, perceived susceptibility to diabetes, cues to action and self-efficacy, in the health behaviour and health education areas (Champion & Skinner, 2008).

3.4 METHODS

3.4.1 Instrument Development

The HBME-T2DM-C was constructed to measure the health beliefs about PA behaviour of patients with T2DM. The initial design contained 52 items in five constructs: perceived exercise benefits (13 items), perceived exercise barriers (17 items), perceived susceptibility to diabetes (1 item), cues to action (12 items) and perceived exercise self-efficacy (9 items).

The development of the HBME-T2DM-C progressed over three phases: (a) construct and item identification and modification; (b) translation, back translation, expert review and pilot testing; and (c) series of validity and reliability tests.

3.4.2 Construct and Item Identification and Modification

There constructs of HBME-T2DM-C were extracted and one construct were modified from HBME in adulthood (Hayslip et al., 1996). The last construct, the perceived exercise self-efficacy was extracted from Perceived Exercise Self-efficacy—Chinese Version for older adults were included (L. L. Lee et al., 2009).

Three constructs (namely perceived exercise benefits, perceived exercise barriers and cues to action) were extracted from the HBME in adulthood (Hayslip et al., 1996). Participants responded via the five-point Likert scale (1 = strongly disagree; 5 = strongly agree). The internal consistency coefficient alphas were satisfactory (perceived exercise benefits: 89; perceived exercise barriers; .80; cues to action: .87) (Hayslip et al., 1996).

For the construct of perceived susceptibility, the HBME (Hayslip et al., 1996) was designed for 16 chronic diseases with one question for diabetes patients. To suit the diabetes context, we modified the one item of perceived susceptibility to diabetes component in HBME with the item from Diabetic Health Belief Model (Hurley, 1990). The question is 'I believe that my PA will prevent complications (e.g. heart problems, kidney problems, stroke, hypertension and amputations) related to diabetes'. Participants responded using the five-point Likert scale.

Furthermore, the perceived exercise self-efficacy construct in the HBM developed by Champion and Skinner (2008) was included in the HBME-T2DM-C. The nine items in the recently validated Perceived Exercise Self-efficacy—Chinese Version for older adults were included (L. L. Lee et al., 2009), with Cronbach's alpha at .75. The 10-point appraisal scale was adopted in both the English and Chinese versions, with higher scores representing greater PA self-efficacy. Moreover, construct validity was established between the high self-efficacy and perceived health status.

3.4.3 Translation, Back Translation, Expert Review and Pre-testing

In order to develop a cultural-, language- and country-specific scale for older patients with T2DM, four constructs of the HBME-T2DM-C (namely, perceived exercise benefits, perceived exercise barriers, perceived susceptibility to diabetes and cues to action), were translated according to the Guidelines for the Process of Cross-Cultural Adaptation of Self-Report Measures (Beaton, Bombardier, Guillemin, & Ferraz, 2000). The translation followed five processes: translation, synthesis of translations, back translation, expert committee review and pre-testing. First, two people who were proficient in both languages (one informed and one uninformed) translated the scales from English into Chinese. Next, the synthesis of translations settled the compromises made by translators. Then another pair of proficient speakers performed back translation (Chinese to English). Fourth, an expert committee of 4 members reviewed the translation to avoid information bias. The committee included: two sports psychologist from academia, one certified exercise educator, and one Advanced Practice Nurse in diabetes. These four experts rated and commented on the representativeness, clarity and comprehensiveness of the four constructs. Appendix 2 presents the experts' review form of HBME-T2DM-C. The experts' ratings on the content validity index (CVI) of the HBME-T2DM-C were calculated based on the most frequently cited article on content validity in nursing and health areas (Grant & L. L. Davis, 1997). Lastly, all items were tested with 20 older adults with T2DM. Feedback was collected on clarity and coherence of wordings and items. Three sentence structures were refined to suit patients' level of understanding after the pre-testing.

3.4.4　Samples

The quota sampling design was adopted in this study. The inclusion criteria of this study consisted of: (a) Chinese; (b) diagnosed with T2DM (patients had an HbA1c over 6.5%); (c) age 45 or above; (d) cognitively intact (scoring 8 or above on the Short Portable Mental Screening Questionnaire; SPMSQ); (e) able to communicate in Cantonese; and (f) able to do PA (walking without assistance).

The sample size of 324 participants was based on the effect size described in the study by Tudor-Locke et al. (2002), which measured the number of steps per day taken by diabetic patients with fair perceived health status. The average number of steps was 6,169 (SD = 2,713). This study assumed that the number of steps per day taken by the subjects in the intervention group would increase by an average of 1,000 steps (or slightly more than 15%) to 7,169 while the subjects in the control group would have no change in their number of steps. This expectation was reasonable as it would attain the minimal amount of moderate PA for adults (7,000 to 8,000 steps per day; see Tudor-Locke et al., 2011). The proposed number of steps complemented the recommended level of activity ageing (> 6,500 to 10,000 steps) for adults in a campaign initiated by the Department of Health of the HKSAR of China (Department of Health, HKSAR, 2010). Therefore, the 15% increase represented an appropriate target for the T2DM patients. This study assumed a small to medium effect size at 0.35. According to G-power 3.1.6, 260 participants are required to give the study appropriate statistical power to determine the effect of the intervention (5% alpha and 80% power). Assuming 20% attrition, 324 participants are to be recruited.

An extra 50 participants were recruited to check for test-retest reliability on the sub-scales of the HBME-T2DM-C. The sample size was based on a recommendation made by Johanson and Brooks

(2010) in their investigation of sample size for a pilot study on initial scale development. Using 50 participants in a scale-validation study was a conservative approach when compared to Johanson and Brooks's recommendation.

For the exploratory factor analysis (EFA), to find out the sample size that would be adequate for the effects, the recommended and validated subject-to-variable ratio for factor analysis was 5:1 or greater (Nunally & Bernstein, 1994). A minimum of 260 (52 variables x 5) participants were required to give the chapter appropriate power to test the construct validity. The analysis of this chapter was based on the baseline measurements of this study. The steps count collected at the first week after baseline measurement was based on the control group only. Therefore, 374 participants were adequate to meet the subject-to-variable ratio.

Figure 4 illustrates the recruitment process, beginning with the 438 T2DM patients who were approached. After the screening tests, 374 out of 421 eligible patients agreed to participate in the study, and 50 of the 374 participants were invited to complete the survey again after 14 days.

In all, 374 participants were recruited from two public hospitals HKSAR from January to August 2012. Of these, 324 participants were recruited for the validation of the HBME-T2DM-C (as described in this chapter) and also for the subsequent RCT (see Chapter 5). 324 participants were requested to wear a pedometer, a steps counting monitor, for seven days, however, for the predictive validity, this chapter only analysed the step counting for 162 participants from control group. The reason for choosing control group was that no training was provided for that specific group in the RCT (see Chapter 5), so that the number of steps per day would not be affected by any trainings provided. Another 50 participants were recruited to provide test-retest reliability on the sub-scales of the HBME-T2DM-C.

3.4.5   Other Measures

Two measures were included in this study: the seven-item IPAQ-C (Deng et al., 2008) and the number of steps counted by a pedometer (Yamax Digiwalker SW-200).

The 7-item Short Form International Physical Activity Questionnaire (IPAQ) is a validated and reliable instrument, which can be self-administered or conducted by telephone interview, used to measure health-related PA behaviour (IPAQ Committee, 2005) in different countries (Bassett, Schneider, & Huntington, 2004; Craig et al., 2003; Deng et al., 2008; Ekelund et al., 2006) and among different age groups (Arvidsson, Slinde, & Hulthen, 2005; Bassett et al., 2004; Deng et al., 2008; Kolbe-Alexander, Lambert, Harkins, & Ekelund, 2006; Macfarlane, Lee, Ho, Chan, & Chan, 2007). It records the frequency, intensity and duration of activities and the time spent in sedentary life during the last seven days (IPAQ Committee, 2005). The scale of IPAQ could then generate the METs value per minutes per week according to the compendium of PA behaviour (IPAQ Committee, 2005).

The IPAQ-C Hong Kong version was validated with a population of persons under age 60 (Deng et al., 2008; Macfarlane et al., 2007). The IPAQ-C for older adults (age 51-82) was validated by the Guangzhou Biobank Cohort Study (Deng et al., 2008) in Guangzhou, where residents speak the same language as Hong Kong. The level of test-retest reliability was adequate, with the intra-class correlation coefficient (ICC) ranging from 0.81 to 0.89. Significant moderate correlation was found between the IPAQ-C and the pedometer-measured number of steps taken by subjects (partial r = 0.33 adjusted for sex, age, and education; p < 0.001). Moreover, strong association was found between the walking domain of the IPAQ-C and the number of steps taken (partial r = 0.58, p < 0.001).

The number of steps was measured with a pedometer, an objective step-counting device that has been validated with a T2DM population. Pedometers (Yamax Digiwalker SW-200) were employed to monitor the number of steps taken by participants for 7 days. Concurrent validity was established between the pedometer and self-reported measures of activity with a moderate relationship (r = .15 to .36) and the reliability coefficient (Cronbach's alpha) was .87 (Strycker, Duncan, Chaumeton, Duncan, & Toobert, 2007). Pedometers have been widely utilized in different countries with middle-aged and older adults with T2DM (Bravata et al., 2007; Dasgupta et al., 2007, 2010; Kolt et al., 2009). Moreover, this objective instrument was employed as a standard tool to validate the IPAQ and the tools related to frequency of exercise (Craig et al., 2003; Deng et al., 2008). In order to standardize the calculation, the average daily number of steps for a week has been used in many studies (De Greef, Deforche, Tudor-Locke, & De Bourdeaudhuij, 2010; Manjoo, Joseph, Pilote, & Dasgupta, 2010; Tudor-Locke, 2009), and most studies have calculated the average based on three to five valid days per week for analysis (De Greef et al., 2010; Newton, Wiltshire, & Elley, 2009; Yates, Davies, Gorely, Bull, & Khunti, 2009). In this chapter, four or more valid days per week will be used to obtain the average daily number of steps per week.

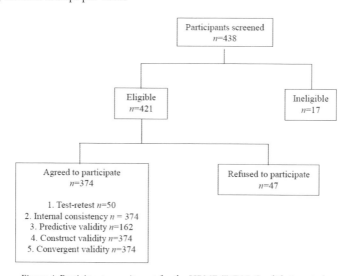

Figure 4. Participant recruitment for the HBME-T2DM-C validation study.

3.4.6   Data Analysis

Psychometric properties of the HBME-T2DM-C were assessed with content validity, construct validity, convergent validity, predictive validity and test-retest reliability. Cronbach's alpha was calculated to assess the internal consistency of each subscale of the HBME-T2DM-C. Content validity was calculated using CVI, with a four-point ordinal rating scale on different items of each subscale (1 = not relevant at all; 4 = very relevant). The proportion of responses of 3 or 4 was calculated and the item CVI was accepted only if the proportion was greater than 0.70 (L. L. Davis, 1992). The EFA with Geomin rotation was used to explore the factors of the 52 items of the HBME-T2DM-C. Any factors with eigenvalue greater than 1 will be considered. Any item with factor loading greater than 0.5 within each latent factor was considered for inclusion in the correspondent subscale of the HBME-T2DM-C. Pearson correlations were calculated to test the relationships between the HBME-T2DM-C, IPAQ-C (convergent validity) and subsequent mean daily number of steps per week (predictive validity). The ICC was calculated to determine test-retest reliability of scale over a two-week period. Analysis was performed using the Statistical Package for the Social Sciences (SPSS) version 20.0 (SPSS Inc. Chicago, IL). All significance tests were two-sided, with a p-value smaller than 5%. 95% confidence intervals (CI) were constructed. The EFA was analyzed with MPlus Version 7.

3.4.7   Procedure

Registered nurses specialized in diabetic care who worked in hospital diabetes centres assisted the recruitment process by identifying potential participants, who were then invited to meet the project's research assistant at the diabetes centre. The research assistant explained the processes and procedure and screened participants for eligibility. All participants who met the inclusion criteria were invited to sign the written informed consent and to complete the self-reported questionnaires (Appendix 3). In addition, one pedometer was given to 162 out of 374 participant (control group of this study), who were instructed to wear it at all times while awake for 7 days and record the daily number of steps taken by themselves. The reason for choosing the control group was because these participants were not affected by the intervention. The self-reported questionnaires were administered twice (test-retest) for the recruited 50 out of 374 T2DM patients, 14 days apart, so that the effects of learning, training or memory that could confound the reliability of the HBME-T2DM-C might be reduced. Patients could withdraw from the study at any time without giving a reason and with no prejudice to the health services that they received.

3.4.8   Ethnical Approval

Approval was obtained from the Institutional Review Board of the University of Hong Kong and Hospital Authority Hong Kong West Cluster and Kowloon East Cluster of the HKSAR (Appendix 4).

3.5 RESULTS

Of the 374 participating adults in the study, 49.7 % (n = 186) of them were male and 50.3% (n = 188) were female. The mean age of participants was 63.73 (SD = 10.47). The average number of chronic diseases was 2.37 (SD = 1.17). With regard to education, 13.6% of participants were illiterate, 17.6% had completed primary school level 1-3, 25.7% had completed primary school, 35% had attained secondary school level 1-5, 2.1% had completed secondary school and 5.9% had pursued further education. As for employment status, 32.4% were full-time or part-time employees at the time of the interview.

3.5.1   Content Validity

All 52 items of the five potential constructs of HBME-T2DM-C were accepted, with the CVI over 0.70.

3.5.2   Construct Validity

Table 5 shows the factor loadings in the four subscales of HBME-T2DM-C. The EFA resulted in four factors accounting for 65.96% of the total variance. All items in the subscales of perceived exercise benefits, perceived exercise barriers and perceived exercise self-efficacy were kept as the original subscale. Two out of 52 items were excluded because of the low factor loading. One item, i.e. doctor's recommendations (factor loading = 0.36), was dropped from the cues to action subscale because of the low factor loading (lower than 0.5). This item did not load adequately on any other subscales and hence was excluded from the overall questionnaire. Also, the item "I believe that my PA will prevent complications" in the subscale of perceived susceptibility to diabetes was excluded from the questionnaire due to inadequate loading on any subscale of the HBME-T2DM-C. As this was the only item on the subscale of perceived susceptibility to diabetes, the subscale itself was therefore excluded from HBME-T2DM-C. Consequently, the newly generated HBME-T2DM-C contained four subscales and a total of 50 items (13 in the perceived exercise benefits subscale, 17 in the perceived exercise barriers subscale, 11 in the cues to action subscale and 9 in the perceived exercise self-efficacy subscale).

Table 5

Factor Loading of HBME-T2DM-C Subscales			
F1	F2	F3	F4
Perceived exercise benefits	Perceived exercise barriers	Cues to action	Perceived exercise self-efficacy
Factor Loadings			
F1-1 (0.79)	F2-1 (0.62)	F3-1 (0.61)	F4-1 (0.84)
F1-2 (0.73)	F2-2 (0.68)	F3-2 (0.73)	F4-2 (0.94)
F1-3 (0.67)	F2-3 (0.74)	F3-3 (0.77)	F4-3 (0.91)

F1-4 (0.71)	F2-4 (0.67)	F3-4 (0.76)	F4-4 (0.78)
F1-5 (0.82)	F2-5 (0.73)	F3-5 (0.72)	F4-5 (0.94)
F1-6 (0.70)	F2-6 (0.65)	F3-6 (0.90)	F4-6 (0.93)
F1-7 (0.75)	F2-7 (0.71)	F3-7 (0.89)	F4-7 (0.94)
F1-8 (0.74)	F2-8 (0.67)	F3-8 (0.73)	F4-8 (0.84)
F1-9 (0.70)	F2-9 (0.75)	F3-9 (0.83)	F4-9 (0.83)
F1-10 (0.76)	F2-10 (0.78)	F3-10 (0.75)	
F1-11 (0.65)	F2-11 (0.65)	F3-11 (0.69)	
F1-12 (0.57)	F2-12 (0.65)		
F1-13 (0.73)	F2-13 (0.69)		
	F2-14 (0.60)		
	F2-15 (0.74)		
	F2-16 (0.70)		
	F2-17 (0.66)		
Eigenvalue			
23.26	4.50	3.53	3.01
Percentage of variance explained by the factors			
Individual 44.73%	8.66%	6.78%	5.79%
Cumulative 44.73%	53.39%	60.17%	65.96%

Note: n=374; HBME-T2DM-C = Modified Health Belief Model of Exercise for Patients with Type 2 Diabetes Mellitus-Chinese Version

3.5.3   Convergent Validity

Table 6 shows the correlation between each subscale of the HBME-T2DM-C and the IPAQ-C. We assessed the convergent validity of HBME-T2DM-C by reviewing its relationship with the IPAQ-C. Each of the four subscales of the HBME-T2DM-C showed low but significant correlations with the IPAQ-C (perceived exercise benefits: $r = 0.388$, $p < 0.001$; perceived exercise barriers: $r = -0.352$, $p < 0.001$; cues to action: $r = 0.249$, $p < 0.001$; perceived exercise self-efficacy: $r = 0.376$, $p < 0.001$).

Table 6

Pearson Correlation of HBME-T2DM-C Sub-scales with IPAQ-C	
Subscale	IPAQ-C (n=374)
Perceived exercise benefits	0.388***
Perceived exercise barriers	-0.352***
Cues to action	0.249***
Perceived exercise self-efficacy	0.376***

Note: ***p < 0.01

HBME-T2DM-C = Modified Health Belief Model of Exercise for Patients with Type 2 Diabetes Mellitus-Chinese Version; IPAQ-C = International Physical Activity Questionnaire for Older Adults—Chinese version.

3.5.4　Predictive Validity

We assessed the predictive validity of HBME-T2DM-C by reviewing its relationship with the subsequent mean daily number of steps per week. Each of the four subscales of the HBME-T2DM-C showed significant correlations with the subsequent mean daily number of steps per week (perceived exercise benefits: r = 0.307, p < 0.001; perceived exercise barriers: r = -0.274, p = 0.001; cues to action: r = 0.278, p =0.001; perceived exercise self-efficacy: r = 0.355, p < 0.001). Table 7 shows the correlation between each subscale of the HBME-T2DM-C and the subsequent mean daily number of steps per week.

3.5.5　Internal Consistency

Table 8 shows the internal consistency of subscales of the HBME-T2DM-C. All subscales of the HBME-T2DM-C showed high internal consistency (alpha > 0.9): perceived exercise benefits = 0.944, perceived exercise barriers = 0.948, cues to action = 0.955 and perceived exercise self-efficacy = 0.977. Hence, no item was eliminated from each subscale of HBME-T2DM-C due to insufficient consistency.

3.5.6　Test-Retest Reliability

All subscales of the HBME-T2DM-C showed high ICC: perceived exercise benefits = 0.911; perceived exercise barriers = 0.992; cues to action = 0.987 and perceived exercise self-efficacy = 0.996. Table 8 shows the test-retest reliability of each subscale of the HBME-T2DM-C.

Table 7

Pearson Correlation of HBME-T2DM-C Sub-scales with Subsequent Mean Daily Number of Steps per Week	
Subscale	Subsequent Mean Daily Number of Steps Per Week (n=162)
Perceived exercise benefits	0.307***
Perceived exercise barriers	-0.274***
Cues to action	0.278***
Perceived exercise self-efficacy	0.355***

Note: ***p < 0.01

HBME-T2DM-C = Modified Health Belief Model of Exercise for Patients with Type 2 Diabetes Mellitus-Chinese Version

Table 8

Internal Consistency and Test-retest Reliability of Subscales of the HBME-T2DM-C		
Subscale	Internal consistency (Cronbach's alpha) (n=374)	Test-retest reliability (ICC) (n=50)
Perceived exercise benefits	0.944	0.911
Perceived exercise barriers	0.948	0.992
Cues to action	0.955	0.987
Perceived exercise self-efficacy	0.977	0.996

Note: n = 374 for Internal consistency; n = 50 for Test-retest reliability

HBME-T2DM-C = Health Belief Model of Exercise for Patients with Type 2 Diabetes Mellitus—Chinese version; ICC = intra-class correlation coefficient

3.6 DISCUSSION AND CONCLUSION

3.6.1 Discussion and Conclusion

The results supported acceptable levels of content validity and construct validity with satisfactory internal and test-retest reliability on the HBME-T2DM-C. The four subscales all showed good convergent validity with the IPAQ-C, moreover, all four subscales of HBME-T2DM-C were found to have a significant predictive power with weak correlation on the mean daily number of steps per week.

The HBME-T2DM-C is the first health belief model to have addressed Chinese language, T2DM and middle aged and older adults in the assessment of health beliefs related to PA behaviour. The satisfactory psychometric properties of this new instrument verified the value of a specific tool for measuring health beliefs. Investigations of other self-care behaviours, especially other diabetic complication prevention self-care behaviours such as foot care and diet control, could similarly follow the recommendation by Aljasem et al. (2001) to pinpoint the specific health beliefs linked to the desired behaviour.

The findings of this chapter provided evidence to support the validity and reliability of the HBME-T2DM-C. Content validity was established by the expert panel, whose members agreed that the HBME-T2DM-C was a valid tool to assess the health beliefs related to PA behaviour among T2DM patients as the CVIs were all within the recommended acceptable range (L. L. Davis, 1992). Appropriate clarity, comprehensiveness and coherence of the instrument were noted by all recruited T2DM patients.

The EFA presented in this chapter showed perceived self-efficacy subscale is part of important components when we assess health beliefs in exercise (Champion & Skinner, 2008). Perceived exercise self-efficacy, the strongest predictor of aerobic PA and resistant PA training for T2DM patients (Plotnikoff, et al., 2011), was therefore incorporated in the HBME-T2DM-C. The HBME from Hayslip et al. (1996) included the cues to action subscale but not the perceived exercise self-efficacy subscale. However, results shown in this study revealed that perceived exercise self-efficacy accounted for 5.79% of total variance, or almost as much as the cues to action subscale (which accounted for 6.78%). This study sets another example to continue the inclusion of perceived exercise self-efficacy as a valid construct in the HBME, since the introduction of such inclusion suggested by Champion and Skinner's HBM (2008).

One item in the cues to action subscale suggested in the HBME of Hayslip et al. (1996), regarding doctors' recommendations, was deleted from the current model due to its low factor loading. This omission of a weaker item can increase the validity of the subscale and hence the overall model.

Though the previous studies reported correlation between individual subscales of the HBM and PA behaviour (Aljasem et al., 2001; Koch, 2002) on patients with chronic diseases in western countries, the convergent validity and the predictive validity of the HBME-T2DM-C add new and comprehensive, structural, disease-specific and language-sensitive information to the understanding of the associations between health beliefs and the IPAQ-C and the subsequent mean daily number of steps per week. All four subscales in the HBME-T2DM-C showed significant correlations with the IPAQ-C and the subsequent mean daily number of steps per week, indicating overall convergent validity and predictive validity of the whole instrument, which has not been found in the HBME for diabetes previously. This result gives new strength to the claim that the HBME-T2DM-C can serve as a more objective measurement of PA behaviour. Further studies could apply the HBME-T2DM-C to different PA behaviours among Chinese T2DM patients in varying regions of China.

This chapter successfully established the relationship between a newly developed HBM for Chinese T2DM patients, the HBME-T2DM-C, and the sophisticated IPAQ-C, which measures the frequency, intensity, time and type of PA so as to calculate the metabolic equivalence value per minutes per week according to the compendium of the PA behaviour completed by patients in the previous seven days. This convergent relationship further confirmed that more favourable health beliefs related to PA behaviour among patients with T2DM are linked to higher metabolic output achieved in the previous seven days. Results revealed the predictive value of the HBME-T2DM-C with regard to actual PA behaviour in middle-aged and older Chinese adults with T2DM. Better health beliefs regarding PA behaviour as measured by the HBME-T2DM-C predicted a higher number of steps over the subsequent week. These results reaffirmed the undeniable value of HBME-T2DM-C on diabetic PA behaviour by showing the association between HBME-T2DM-C and metabolic output achieved in the previous week and subsequent week of actual PA behaviour of Chinese older patients with T2DM.

In addition, the HBME-T2DM-C can be helpful in facilitating health belief intervention regarding exercise with T2DM patients. The four subscales can be used to diagnose the diabetic health beliefs that are specific to individual patients, so that appropriate follow-up actions can address each patient's personal health belief profile. For example, patients who scores poorly on the perceived exercise barrier subscale may still perceive the health benefits of PA. Health care professionals can inquire further about the barriers and take them into consideration when planning PA behavioural modification programs. Therefore, the HBME-T2DM-C helps to identify weaker aspects of health belief and enable effective PA behavioural modifications.

As the HBME-T2DM-C and IPAQ-C and the subsequent mean daily number of steps per week appear to be highly correlated, the HBME-T2DM-C will be adjusted as a covariate in the process of data analysis in Chapter 5.

3.6.2 Limitations

There are some limitations in this validation study. First, the sample recruited in this chapter is limited to one developed city, the HKSAR of China, of which citizens speak Cantonese as their native language, so the psychometric properties of the HBME-T2DM-C may not be appropriate to Mandarin-speaking populations. Second, quota sampling was adopted in two hospital diabetes centres of the HKSAR of China.

3.7 RELEVANCE TO CLINICAL PRACTICE

The data presented in this chapter provided evidence for the validity and reliability of the HBME-T2DM-C as an easily administered practical tool in diabetes care and PA behavioural modification. Health practitioners could use this scale to assess patients' perceived benefit, barriers, cues to action and self-efficacy of PA behaviour. With a better understanding of patients' health beliefs regarding PA behaviour, health practitioners could design specific programs to facilitate PA behaviour among diabetic patients. Moreover, as the health beliefs model of exercise was highly associated with PA behaviour among middle-aged and older Chinese patients, future HL-oriented programs should consider HBME-T2DM-C as a covariate when performing data analysis.

The Effect of A Health Literacy Oriented Program on Physical Activity behavioUr among Chinese Middle-aged and Older Adults with Type 2 Diabetes Mellitus

4.1 INTRODUCTION

It has been generally accepted that physical inactivity is one of the leading causes of T2DM (Mokdad et al., 2004), especially for people age 45 or older in both developed and developing countries (Hays & Clark, 1999; World Health Organization, 2004; Zhao, Ford, Li, & Balluz, 2011). PA behaviour has been shown to affect endogenous glucose production and lower HbA1c levels (Sigal et al., 2004). The causal relationship between PA behaviour and diabetes control has been further supported by a recent systematic review of 20 longitudinal cohort studies on T2DM patients age 24-84 in the US, the United Kingdom, Finland, Germany, China, and Japan, which found that PA behaviour is associated with a 20% to 30% reduction in diabetes risk (Gill, & Cooper, 2008). Knowing that PA behaviour is positively associated with glycemic control of T2DM, many organizations in various countries and cities have initiated different behavioural modification programs to maintain the PA persistency of patients with T2DM (Cardona-Morrell et al., 2010; Conn et al., 2003; Hudon et al., 2008; Jeon et al., 2007; Shigaki et al., 2010); however, PA behaviour has remained the least adhered-to diabetes self-care behaviour among middle-aged and older patients with T2DM (S. Kim et al., 2004; Rosal et al., 2005; Shigaki et al., 2010). An earlier study in the HKSAR confirmed that among five diabetic self-care behaviours, PA behaviour receives the lowest compliance rate from middle-aged and older adults with T2DM (Y. M. Chan & Molassiotis, 1999).

The flood of health information available in the 21st century (Cline & Haynes, 2001; McLeod, 1998; Sihota & Lennard, 2004) affects the effectiveness of self-care education and behavioural modification programs, as decision making regarding self-care behaviours is affected by the quality of daily health information at one's discretion (Coulter et al., 1999). Moreover, many older adults with T2DM face difficulties in understanding, processing and distinguishing health information due to degraded cognitive functioning and cognitive neuropsychological performance, the consequences of aging and peripheral neuropathy (Boulton, 2005; Nguyen et al., 2013; Yeung, Fischer, & Dixon, 2009; see Chapter 3) and the poor cognitive determinants (verbal working memory, processing speed and cognitive function) have been found closely associated with low HL (Baker, Gazmararian,

Sudano, & Patterson, 2000; Baker et al., 2002; Baker et al., 2007; Federman, Sano, Wolf, Siu, & Halm, 2009; Levinthal, & Wu, 2008; Morrow et al., 2006; see Chapter 3). In Hong Kong, the extent of cognitive functional degradation among older adults was noted in a study that applied age- and sex-stratified sampling to a population of 2,011 (Woo, Ho, Lau, Lau, & Yuen, 1994). HL, the ability to navigate, understand, interpret, analyze and communicate appropriate health information, which has been considered by the Office of Surgeon General of the USDHHS as one of the top four public health priorities (Galson, 2008), should be taken into consideration in the planning of any self-care behavioural modification programs (Edwards et al., 2009). HL has been found to be closely associated with diabetes self-care behaviours and diabetes outcomes among older adults (Schillinger et al., 2002; Yamashita & Kart, 2011). Moreover, some researchers contend that limited HL is the first concern on the causal pathways, with mediators including knowledge and psychological determinants (Osborn et al., 2011; Paasche-Orlow & Wolf, 2007), and that these factors eventually affect self-care behaviours such as diet, nutritional behaviour, medication diseases self-examinations (Coleman et al., 2003; Howard-Pitney, Winkleby, Albright, Bruce, & Fortmann, 1997; Hussey, 1994; Murphy et al., 1996).

HL-oriented programs has been found to affect PA behaviour of middle-aged and older adults with hypertension, coronary heart disease and patients using complementary and alternative medicine (Joyner-Grantham et al., 2009; M. H. S. Lam, et al., 2011; Long, 2009; Ussher et al., 2010; Wolf et al., 2007; see Chapter 3 above). The effectiveness of HL-oriented programs to modify the PA behaviour of T2DM patients was not properly reviewed (see Chapter 2).

Despite the importance of PA behaviour in controlling blood glucose, only six interventional studies were extracted from a systematic review of HL-oriented programs on PA behaviour from studies on self-care behaviours of middle-aged and older T2DM patients identified in eight electronic databases between January 1990 and April 2013 (see Chapter 2). However, the systematic reviewed showed that the effects of HL-oriented programs on PA behaviour of T2DM patients are uncertain (see Chapter 2).

Several conclusions and research implications for future HL-oriented programs on PA behaviour can be derived from the review of existing research. First, as all six extracted studies (four RCT studies and two one-group pre-test/post-test interventional studies) were conducted in the US, the generalizability to different countries, cultures (IOM, 2003; Zarcadoolas et al., 2005) and languages (IOM, 2004; Zanchetta & Poureslami, 2006) is questionable (see Chapter 2); hence, more research in countries other than the US is needed. Second, while PA behaviour was measured as one of the components of the self-care behaviours (also including diet, self-monitoring of blood glucose, foot care, and medication adherence) with subjective and non-validated self-reported measurements (Colberg et al., 2010), few studies have investigated the effect of PA behaviour on glycemic control independently from other self-care behaviours (Sigal et al., 2004); hence, studies of behavioural modification programs on PA behaviour alone are needed. Third, based on the findings of Chapter 2, future HL-oriented programs on PA behaviour should be planned with stronger theoretical support and larger sample size.

4.2 AIMS OF THIS CHAPTER

This chapter aims to investigate the effects of a HL-oriented program on PA behaviour of Chinese middle-aged and older adults with T2DM (objective 3 of the thesis).

4.3 METHODS

4.3.1 Samples

This chapter is a two armed (intervention and control group) RCT. Patients from two of the seven hospital clusters in the HKSAR. One diabetes centre that offers regular diabetic follow-up was selected in each hospital cluster.

The inclusion and exclusion criteria of the proposed study were set up with reference to the criteria used in a previous diabetic study (M. F. Chan, Yee, Leung, & Day, 2006) in the HKSAR. The inclusion criteria for eligible patients were: (a) diagnosed with T2DM (had an HbA1c over 6.5%); (b) age 45 or older; (c) able to communicate in Cantonese; and (d) able to do PA (walking without assistance). Patients with the following conditions were excluded: (a) diagnosed with Type 1 diabetes; (b) dementia or cognitive impairment; (c) concurrent illness such as acute infections, uncontrolled hypertension (blood pressure greater than 180/110), unstable angina, myocardial infarction in the past three months, late-stage diabetic complications, severe immunodeficiency, cirrhosis, or end-stage cancer; (d) inability to do PA (as advised by doctors); (e) inability to complete (with written and verbal assistance where needed) a questionnaire; and (f) SPMSQ < 8 (Appendix 5). A nurse-in-charge at each diabetes centre assisted with the recruitment of patients by identifying eligible T2DM patients from the medical screening list based on the inclusion and exclusion criteria and allowing the researcher to approach screened patients at the centre. The upper limit of HbA1c was not regarded as an exclusion criterion if the participants met all the inclusion criteria for this study, since a recent meta-analysis on the metabolic effects of PA behavioural modification programs for patients with T2DM showed that patients with high BMI and HbA1c could also benefit from programs to modify self-care behaviours (Conn et al., 2007).

Based on the previous study measuring the number of steps taken per day by diabetic patients with fair perceived health status (Tudor-Locke et al., 2002), the sample size was calculated as 324. The average number of steps of each patient was 6,169 (SD = 2,713). This study assumed a small to medium effect size at 0.35. According to G-power 3.1.6, 260 patients were required to give the study appropriate power to determine a small to medium effect of intervention. The level of significance for all tests were set at p<.05 with 80% statistical power. Assuming 20% attrition, 324 patients should be recruited.

4.3.2 Randomized Group Assignment

Trained research assistants screened for the eligibility of the patients at the diabetes centres of the two hospitals. Eligible subjects were invited to participate in the study. After taking the baseline assessment, patients were randomly assigned into two groups of 162 persons each (intervention group, n=162 and control group, n=162), based on a list of computer-generated random numbers.

4.3.3   Intervention

The intervention group received a six-week HL-oriented program (including one 60-minute lesson each week at a diabetes centre) and one telephone follow-up at week 10. The lessons, which included group meetings of a trained nurse, a certified exercise trainer and patients, were held at the diabetes centres of two hospitals. In the six lessons, the nurse and the certified exercise trainer encouraged patients to read at least one story from an HL-oriented program book on PA behaviour and listen to audiotapes from an HL-oriented program on PA behaviour at home at least once a week. The HL-oriented program book and audiotapes by Leung (2009) in consultation with family doctor, diabetes nurse, physical fitness expert, diabetes patients, with the assistance of social marketing experts (see Appendix 9). The book sought to empower older adults to obtain, understand and process health information (knowledge) so as to raise the patients' HL levels and to fortify their health beliefs and motivation (psychological determinants) to make their own health decisions about exercise. In the lessons, the nurse and certified exercise trainer (a) instructed the subjects, using information and examples, on the benefits, motivations and cues to action related to controlling blood glucose; (b) offered recommendations on how to reduce barriers to exercise; (c) tried to boost patients' perceived competence through references to the HL-oriented program books and audiotapes and reminded them to continue their pedometer reading, filling in PA behavioural self-record sheets and measuring their weight throughout the study period; and (d) asked the participants to look for significant others who could assist them to do exercise. Participants were encouraged to read the materials from the HL-oriented program on PA behaviour as frequently as possible and to set their own goals for walking every week. At week 10, the nurse conducted a telephone interview with each patient to assess progress in doing PA, clarify misconceptions, fortify health beliefs, reinforce the importance of walking exercise in diabetes care and remind patients to record the number of steps (as shown on the pedometer) on the record sheets. At 3-month and 6-month, the researcher met patients at the diabetes centres, conducted face-to-face interviews (on the second and third assessments) and collected all the pedometer record sheets. To ensure fidelity of the intervention, the nurse and the certified exercise trainer who provided the intervention underwent three days of training on the HL-oriented program on PA behaviour. Patients in the intervention group still received the standard care from diabetes centres, which included verbal health education by nurses in diabetes centres and distribution of leaflets. For the patients in the control group, they only received standard care from diabetes centres.

4.3.4   Primary Outcome Measures

The Short Form of International Physical Activity Questionnaire for Older Adults- Chinese Version

The seven-item Short Form IPAQ is a relatively validated and reliable instrument, which can be self-administered or completed by telephone interview, to measure health-related PA behaviour of people of different ages (Arvidsson et al., 2005; Bassett et al., 2004; Deng et al., 2008; Kolbe-Alexander et al., 2006; Macfarlane et al., 2007) and in different countries (Bassett et al., 2004; Craig et al., 2003; Deng et al., 2008; Ekelund et al., 2006). It records the frequency, intensity and duration of activities and the time spent in sedentary life in the last seven days. This scale could then be transformed to METs value per minutes per week according to the compendium of the PA behaviour (IPAQ Committee, 2005). The IPAQ-Hong Kong version has been validated with a population of persons under 60 years old (Deng et al., 2008; Macfarlane et al., 2007). The IPAQ-C for older adults was validated in Cantonese speaking ageing population (those age 51-82 with mean age 65.2) (Deng et al., 2008). The ICC for each domain of IPAQ-C (vigorous, moderate and light activities) ranged from 0.81 to 0.89. Significant moderate correlation is found between the IPAQ-C and number of steps taken (partial r = 0.33 adjusted for sex, age and education; p < 0.001). Moreover, a strong association is found between the walking domain of IPAQ-C and the number of steps taken (partial r = 0.58, p < 0.001) (Deng et al., 2008).

Number of steps

The number of steps was measured with a pedometer, an objective step-counting device that has been validated with a T2DM population. Concurrent validity was established between the pedometer and self-reported measures of activity with a moderate relationship (r = .15 to .36) and the reliability coefficient was .87 (Strycker, Duncan, Chaumeton, Duncan, & Toobert, 2007). Pedometers have been widely utilized in different countries to study middle-aged and older adults with T2DM (Bravata et al., 2007; Dasgupta et al., 2007; Dasgupta et al., 2010; Kolt et al., 2009). The Yamax Digiwalker SW-200, a simple model with relatively few features that has been recognized as the best step-counting device for research studies, was used (Crouter, Schneider, Karabulut, & Bassett, 2003). In order to standardize the calculation, the average daily number of steps for a week has been used in many studies (De Greef, Deforche, Tudor-Locke, & De Bourdeaudhuij, 2010; Manjoo, Joseph, Pilote, & Dasgupta, 2010; Tudor-Locke, 2009), and most studies have calculated the average based on three to five valid days per week for analysis (De Greef et al., 2010; Newton, Wiltshire, & Elley, 2009; Yates, Davies, Gorely, Bull, & Khunti, 2009). In this chapter, at least four valid days per week was used to obtain the average daily number of steps per week.

4.3.5　Other Measures

Chinese Health Literacy Scale for Diabetes (CHLSD)

CHLSD is a measurement of HL. The 34 items of CHLSD has been validated in older Chinese population (age 65 or above) (A. Y. M. Leung, Lou, Cheung, Chan, & Chi, 2013). It was developed on the basis of the revised Bloom's taxonomy model (Anderson & Krathwohl 2001). The CHLSD was found to correlate with the Diabetic Knowledge Scale (r = 0.398), the Diabetic Management Self-Efficacy Scale (r = 0.257) and the Chinese Value of Learning Scale (r = 0.303). The cronbach's alpha for CHLSD was 0.884 (A. Y. M. Leung et al., 2013).

Modified Health Belief Model of Exercise for Patients with Type 2 Diabetes Mellitus – Chinese Version

The 50-item HBME-T2DM-C has been translated and validated in Chapter 4. The CVI was at an acceptable level (> 0.8) (Grant & L.L. Davis, 1997). The EFA resulted in four subscales (perceived exercise benefits, perceived exercise barriers, cues to action, and perceived exercise self-efficacy) containing 50 items explained 65.96% of total variation. The subscale of perceived susceptibility to diabetes was deleted as there were no items loaded on this scale. The findings supported the association of the HBME-T2DM-C with the IPAQ-C ($p < .05$) (see Chapter 4). The HBME-T2DM-C's ability to predict the number of steps taken by middle-aged and older adults with T2DM was supported ($p < .05$) (see Chapter 4). HBME-T2DM-C demonstrated good internal consistency reliability (Cronbach's alpha of four subscales > .9) and test-retest reliability (ICC of four subscales > .9) (see Chapter 4).

Appendix 6 provides the Chinese questionnaire used at baseline, 3-month and 6-month; Appendix 7 shows the record sheet for pedometer readings.

4.3.6 Data Collection

Data were collected at baseline and at 3-month and 6-month follow-ups at two hospitals' diabetes centres by research assistants who were not involved in the intervention. Quota sampling method was adopted and for the categories, quota for gender was set at 1:1 ratio. A face-to-face interview was conducted at each of the three occasions to assess (a) demographic data such as age, gender, education, employment status and living arrangement; (b) clinical data such as diastolic and systolic blood pressure, HbA1c, number of chronic illnesses and BMI (in kg/m2); (c) HL for diabetes (using the CHLSD); (d) health beliefs of exercise for patients with T2DM (using HBME-T2DM-C); (e) PA behaviour using the IPAQ-C. Another objective PA behaviour, the average daily number of steps for a week taken in at least four valid days per week, was measured with the pedometer. Patients received an information sheet explaining the aims, process and arrangement of the study. Written informed consent (Appendix 8) was obtained from patients, who could withdraw from the study at any time without giving a reason. The protocol of this study was approved by the Institutional Review Board of the University of Hong Kong and the Hospital Authority Hong Kong West Cluster and Kowloon East Cluster (Appendix 4).

4.3.7 Data Analysis

Statistical analysis was conducted using SPSS version 20.0 (SPSS Inc. Chicago, IL). All significance tests were two-sided, with a p-value smaller than 5%. 95% confidence intervals (CI) were constructed. Baseline differences between the intervention and the control groups were measured using the two-independent-samples t-test and $\chi2$–tests for continuous and categorical characteristics, respectively.

The intention-to-treat approach would be adopted in this chapter. The attendance rate of the participants in this chapter would be calculated and all participants would be included in the analysis according to the protocol. The issue of drop-out would be tackled by the mixed model.

The between-group differences of two groups (the intervention and the control groups) on the mean estimated IPAQ-C and average daily number of steps for a week were examined by a generalized linear mixed model (GLMM), which accounted for the within patients' correlation due to repeated measures (baseline, 3-month and 6-month). The GLMM with an identity link was employed to analyse the normally distributed continuous variables of the IPAQ-C, while a negative binomial link was employed to analyse the counted data of steps with large variance. The advantage of GLMM is to deal with the missing value or dropouts. The factors to be examined were three times of measurement, the two groups and the interaction between the three times of measurements and the groups. Baseline characteristics including age group, sex, living arrangements, marital status, education level, employment, the number of existing illness, CHLSD, HBME-T2DM-C (perceived exercise benefits, perceived exercise barriers, cues to action, perceived exercise self-efficacy), diastolic and systolic blood pressure, HbA1c and BMI were controlled in the model according to previous meta-analyses and studies investigating interventions to modify PA behaviour among middle-aged and older patients with T2DM (Browning et al., 2009; Brownlee-Duffeck et al., 1987; Paasche-Orlow et al., 2005; Plotnikoff et al., 2013; Schillinger et al., 2002; see Chapters 1).

4.4 RESULTS

A total of 438 patients were approached at the two diabetes centres. 17 were excluded because they did not meet the inclusion criteria and 97 declined participation. The response rate was 73.97% (see Figure 5). The remaining 324 patients equally and randomly divided into intervention and control groups and they were requested to fill in the baseline assessment questionnaires. The attendance rate of the participants on the lessons provided was 82% and all of them were included in the analysis.The retention percentages were 91.4% (148 of 162) in the intervention group and 93.2% (151 of 162) in the control group at 3-month and 87% (141 of 162) in the intervention group and 88.9% (144 of 162) in the control group at 6-month. Except the readings of blood pressure, there were no differences on age, gender, marital status, height, weight, education standard, living situation, HL, health beliefs, number of comorbidities and HbA1c, between the participants remained and dropped out.

Among the 324 patients recruited, 50% were male and 50% were female. The mean age was 64 (24.7% age 45-54, 26.2% age 55-64, 30.6% age 65-74, 18.5% age 75-86). With regard to marital status, 21.3% of patients were not married while 78.7% were married. As for living status, 12% lived alone, 17.3% with children or people other than spouse, and 70.7% with spouse or with spouse and children. In terms of education, 14.8% reported no education, 17.6% had attained primary school level 1 to 3, 25.6% had attained primary school level 4 to 6, 17% had attained secondary school level 1 to 3, and 25% were at secondary school level 4 or above; 68.2% of the sample was not employed. The mean number of illness among the patients were 2.36 (SD = 1.16) and the mean years since diagnosis was 13.69 (SD = 9.03). HbA1c measurements averaged 8.49 (SD = 1.78) and the mean BMI was 25.72 (SD = 4.31). The demographic characteristics of the patients according to study group are shown in Table 9. At baseline, the intervention and control groups were comparable on nine characteristics including sex, living arrangements, marital status, education level, the number of existing illness, diastolic and systolic blood pressure, HbA1c and BMI; however, patients in the control group were younger and higher in HL and in HBME-T2DM-C when compared with the intervention group

Before fitting the models, multicollinearity was tested. None of the variance inflation factors of the independent variables were found to be close to or greater than 5; thus, multicollinearity of these variables was low and all variables were included in the regression analysis.

For the IPAQ-C, a significant interaction effect between time and group was found (F = 51.03, p < .001). The means of IPAQ-C of intervention group (with standard error in parentheses) at baseline, 3-month and 6-month were 891.65 (149.33), 2,197.13 (180.54) and 1,985.44 (167.88) respectively, while those of the control group were 1,073.29 (128.78), 969.78 (163.44) and 946.52 (149.63) respectively. As for within-group effects on the IPAQ-C, significant increases were found in the intervention group between the baseline assessment and 3-month assessment (increase of 1,305.48, 95% CI = 1,095.24 to 1,515.72, p < .001) and between the baseline assessment and 6-month assessment (increase of 1,093.79, 95% CI = 894.01 to 1,293.56, p < .001). For the control group, there were no significant changes on the IPAQ-C among the three time points (see Table 10). As for between-group effects on the IPAQ-C, there were no significant differences between intervention and control group at baseline, however, significant differences were found at 3-month (difference of 1,227.36, 95% CI = 797.07 to 1,657.64, p < .001) and 6-month (difference of 1,038.91, 95% CI = 649.79 to 1,428.04, p < .001). The IPAQ-C of the intervention group was significantly higher than control group at 3-month assessment and 6-month assessment (see Table 11).

Regarding average daily number of steps for a week, a significant interaction effect between time and group was found (F = 17.14, p < .001). The average (standard error) daily number of steps for a week at the three observation points (baseline, 3-month and 6-month) for the intervention group were 4,462.79 (342.66), 7,458.69 (520.36) and 7,581.57 (537.35), respectively, while those of the control group were 4,807.64 (327.65), 4,433.15 (270.88) and 4,351.39 (272.31), respectively. As for within-group effects on average daily number of steps for a week, significant increases were found in the intervention group between baseline assessment and 3-month assessment (increase of 2,995.9, 95% CI = 2,514.75 to 3,477.05, p < .001) and between baseline assessment and 6-month assessment (difference of 3,118.78, 95% CI = 2,469.45 to 3,768.11, p < .001). For the control groups, there was a significant decrease in average daily number of steps for a week between the baseline assessment and 3-month assessment (increase of -374.5, 95% CI = -706.08 to -42.91, p < .05) and between the baseline assessment and 6-month assessment (increase of -456.26, 95% CI = -860.04 to -52.48, p < .05) (see Table 10). As for between-group effects on the average daily number of steps for a week, there were no significant differences between intervention and control group at baseline, however, significant differences were found at 3-month (difference of 3,025.54, 95% CI = 2,002.56 to 4,048.53, p < .001) and 6-month (difference of 3,230.18, 95% CI = 2,173.42 to 4,286.95, p < .001). The average daily number of steps for a week for the intervention group was significantly higher than control group at 3-month assessment and 6-month assessment (see Table 11).

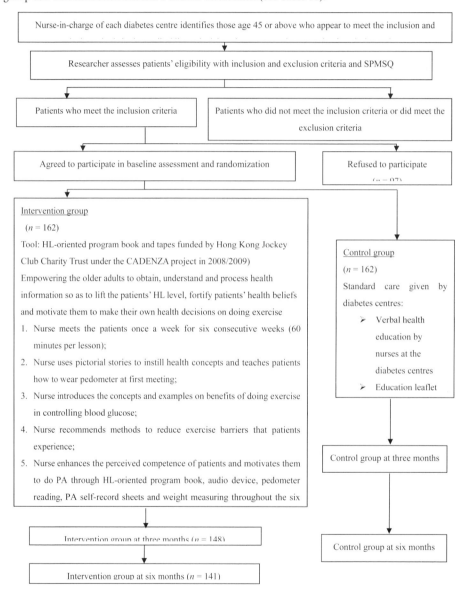

Figure 5. Flowchart of patients in the randomized controlled trial of the impact of a Health Literacy Oriented Program on Physical Activity Behaviour

Table 9

Demographic Characteristics of Patients in the Study of the Impact of a Health Literacy Oriented Program on Physical Activity Behaviour				
Characteristics	Total (%) (n = 324)	Intervention (%) (n = 162)	Control (%) (n = 162)	p
Age group (mean age=64)				
Age 45 to 54	80 (24.7%)	22 (27.5%)	58 (72.5%)	<.001***
Age 55 to 64	81 (26.2%)	40 (49.4%)	41 (50.6)	
Age 65 to 74	99 (30.6%)	60 (60.6%)	39 (39.4)	
Age 75 to 86	60 (18.5%)	36 (60%)	24 (40%)	
Sex				
Male	162 (50%)	80 (49.4%)	82 (50.6%)	.91
Female	162 (50%)	82 (50.6%)	80 (49.4%)	
Living arrangement				
Live alone	39 (12%)	17 (43.59%)	22 (56.41%)	.11
Live with children or people other than spouse	56 (17.3%)	22 (39.29%)	34 (60.71%)	
Live with spouse or spouse and children	229 (70.7%)	123 (53.71%)	106 (46.29%)	
Marital status				
Not married	69 (21.3%)	28 (40.58%)	41 (59.42%)	.10
Married	255 (78.7%)	134 (52.5%)	121 (47.5%0	
Education level				
No education	48 (14.8%)	21 (43.7%)	27 (56.3%)	.17
Primary school level 1 to 3	57 (17.6%)	34 (59.65%)	23 (40.35%)	
Primary school level 4 to 6	83 (25.6%)	47 (56.6%)	36 (43.4%)	
Secondary school level 1 to 3	55 (17%)	23 (41.8%)	32 (58.2%)	
Secondary school level 4 or above	81 (25%)	37 (45.7%)	44 (54.3%)	
Employment				
Working	103 (31.8%)	31 (30.1%)	72 (69.9%)	<.001***
Retired or not working	221 (68.2%)	131 (59.3%)	90 (40.7%)	
Characteristics	Mean (SD) (n = 324)	Intervention (SD) (n = 162)	Control (SD) (n = 162)	p
The number of existing illness	2.36 (1.16)	2.24 (1.02)	2.47 (1.26)	.077
Health literacy	45.73 (18.98)	41.35 (17.86)	50.1 (19.11)	<.001***

HBME-T2DM-C				
Perceived exercise benefits	3.38 (0.71)	3.23 (.64)	3.53 (.75)	<.001***
Perceived exercise barriers	3.34 (0.74)	3.52 (.68)	3.15 (.76)	<.001***
Cues to action	2.87 (0.83)	2.73 (.8)	3.01 (.84)	.002*
Perceived exercise self-efficacy	4.99 (1.89)	4.57 (1.75)	5.42 (1.93)	<.001***
Systolic blood pressure	143.18 (18.17)	144.67 (17.26)	141.68 (18.98)	.14
Diastolic blood pressure	76.37 (10.71)	76.27 (10.95)	76.47 (10.49)	.87
HbA1c	8.49 (1.78)	8.45 (1.73)	8.52 (1.83)	.73
BMI	26.72 (4.31)	25.38 (4.33)	26.06 (4.27)	.16

Notes: Data are presented as mean and standard deviation for continuous variables, and as total number for categorical variables; *p < .05; ***p < .001; HBME-T2DM-C = Modified Health Belief Model of Exercise for Patients with Type 2 Diabetes Mellitus-Chinese Version; HbA1c = glycated hemoglobin; BMI = body mass index.

Table 10

Adjusted Within-Group Differences on the IPAQ-C and Average Daily Number of Steps for a Week during a Health Literacy Oriented Program on Physical Activity Behaviour

Variable	Intervention			Adjusted change in 3-month vs baseline		Adjusted change in 6-month vs baseline		Control			Adjusted change in 3-month vs baseline		Adjusted change in 6-month vs baseline	
	Baseline	3-month	6-month	Estimate (95% CI)	p	Estimate (95% CI)	p	Baseline	3-month	6-month	Estimate (95% CI)	p	Estimate (95% CI)	p
	Mean (Standard Error)			Mean (Standard Error)				Mean (Standard Error)			Mean (Standard Error)			
IPAQ-C (METs/min/week)	891.65 (149.33)	2,197.13 (180.54)	1,985.44 (167.88)	1,305.48 (1,095.24 – 1,515.72)	<.001***	1,093.79 (894.01 – 1,293.56)	<.001***	1,073.29 (128.78)	969.78 (163.44)	946.52 (149.63)	-103.51 (-308.31 – 101.29)	.32	-126.76 (-320.18 – 66.66)	.28
Average daily number of steps for a week	4,462.79 (342.66)	7,458.69 (520.36)	7,581.57 (537.35)	2,995.9 (2,514.75 – 3,477.05)	<.001***	3,118.78 (2,469.45 – 3,768.11)	<.001***	4,807.64 (327.65)	4,433.15 (270.88)	4,351.39 (272.31)	-374.5 (-706.08 – -42.91)	.02*	-456.26 (-860.04 – -52.48)	.02*

Notes: *p <.05; ***p <.001; †Adjusted baseline characteristics included age group, sex, living arrangement, marital status, education level, length of illness, employment, HL, HBME-T2DM-C, BMI, diastolic and systolic blood pressure, HbA1c and time

Abbreviations: IPAQ-C = The International Physical Activity Questionnaire for Older Adults—Chinese version; HBME-T2DM-C = Modified Health Belief Model of Exercise for Patients with Type 2 Diabetes Mellitus-Chinese Version; HbA1c = glycated hemoglobin; BMI = body mass index; CI = Confidence Interval.

Table 11

Adjusted Between-Group Differences on the IPAQ-C and Average Daily Number of Steps for a Week during a Health Literacy Oriented Program on Physical Activity Behaviour

Variable	Baseline				3-month				6-month			
	Intervention	Control	Difference (95% CI)	p	Intervention	Control	Difference (95% CI)	p	Intervention	Control	Difference (95% CI)	p
	Mean (Standard Error)		Mean (Standard Error)				Mean (Standard Error)				Mean (Standard Error)	
IPAQ-C (METs/min/week)	891.65 (149.33)	1,073.29 (128.78)	-181.63 (-507.29 – 144.02)	.27	2,197.13 (180.54)	969.78 (163.44)	1,227.36 (797.07 – 1,657.64)	<.001***	1,985.44 (167.88)	946.52 (149.63)	1,038.91 (649.79 – 1,428.04)	<.001***
Average daily number of steps for a week	4,462.79 (342.66)	4,807.64 (327.65)	-344.85 (-1,172.3 – 482.59)	.41	7,458.69 (520.36)	4,433.15 (270.88)	3,025.54 (2,002.56 – 4,048.53)	<.001***	7,581.57 (537.35)	4,351.39 (272.31)	3,230.18 (2,173.42 – 4,286.95)	<.001***

Notes: *p <.05; ***p <.001; †Adjusted baseline characteristics included age group, sex, living arrangement marital status, education level, length of illness, employment, HL, HBME-T2DM-C, BMI, diastolic and systolic blood pressure, HbA1c and time

Abbreviations: IPAQ-C = The International Physical Activity Questionnaire for Older Adults—Chinese version; HBME-T2DM-C = Modified Health Belief Model of Exercise for Patients with Type 2 Diabetes Mellitus-Chinese Version; HbA1c = glycated hemoglobin; BMI = body mass index; CI = Confidence Interval.

4.5 DISCUSSION

4.5.1   Main Findings

This RCT demonstrated that an HL-oriented program on PA behaviour can be effective in modifying the least adhered-to self-care behaviour, PA behaviour, among middle-aged and older Chinese adults with T2DM. On average, the patients who received the HL-oriented program on PA behaviour improved significantly from under 1,000 averages METs-minutes/week to around 2,000 averages METs-minutes/week, and participants successfully maintained their PA behaviour three months after the intervention. Though the average METs-minutes/week after intervention was still lower than that of normal Hong Kong citizens age 40 and over after the intervention (P. H. Lee et al., 2011), the METs-minutes/week at 3-month post-intervention and 6-month follow-up assessments met (a) the international recommendations of moderate level of activity suggested by the guidelines accompanying the IPAQ (Arvidsson et al., 2005); (b) the standard of citizens in European countries with higher METs-minutes/week (Rutten & Abu-Omar, 2004), and (c) the definition of moderate activity level (1,860 to 3,780 METs-minutes/week) of middle-aged and older Chinese T2DM patients in another study (Qin et al., 2010). The results obtained through an objective and validated step-counting device, the pedometer, confirmed the positive effect of HL-oriented program on PA behaviour, as patients in the intervention group successfully lifted their mean daily number of walking steps per week from the sedentary level to meet international and local recommended standards (such as Department of Health, Hong Kong Special Administrative Region, 2010; Tudor-Locke et al., 2002; Tudor-Locke & Myers, 2001) by the 3-month assessment and continued to increase slightly at the 6-month follow-up assessment. This HL-oriented program on PA behaviour was able to motivate participants to regular walking, motivating from contemplation to action (Bulley, Donaghy, Payne, & Mutrie, 2007). The acceleration of PA behaviour throughout this HL-oriented program coincided with a previous study in which changes in PA behaviour kept accelerating throughout an HL-oriented program in the US (Rosal et al., 2005).

4.5.2   Contributions and Implications of This Chapter

This HL-oriented program on PA behaviour has been endowed with some new and unique features. According to the (IOM) of the US, training concerning HL must consider the context of language and culture. The IOM (2003; 2004) supplemented with an acknowledgement that training with HL has yet to be fully investigated in the late development of HL programmes since the 1990s. With reference to the systematic review of HL-oriented programs on PA behaviour in Chapter 2, unique features taking care of cultural specificity and characteristics of participants were incorporated in the HL-oriented program: (a) this is the first HL-oriented program that incorporated comics and Chinese language tailor made for middle-aged and older Chinese adults with T2DM; (b) pair trainer system (a nurse and a certified exercise trainer) was adopted in this program, with a view to take care of participants with different HL level for different enquiries; (c) special features (larger font size of the wordings in the comic; clear and purposively repeated message in the audiotapes) that take care of older adults' visual or auditory weakness, as these were the factors affecting HL (Osborn et al., 2011; Paasche-Orlow & Wolf, 2007; see Chapter 3).

This chapter addressed the Conceptual Model of Health Literacy (see Figure 1; M. H. S. Lam et al. 2011 and Chapter 3) of which health knowledge and psychological factors such as health beliefs, self-efficacy and perceived barriers were incorporated in the concerns of our HL intervention so as to modify the PA behaviour middle-aged and older patients with T2DM (Paasche-Orlow & Wolf, 2007; Osborn et al., 2011; see Chapter 3). Though limited HL is the first step on a causal pathway that affects the health behaviours and health outcomes (Paasche-Orlow & Wolf, 2007; Osborn et al., 2011) and it was known that PA behaviour of middle-aged and older patients with T2DM could be modified only through theoretically and systematically planned programs. Most of the previous interventional studies in the systematic review didn't put too much emphasis on theoretical designs of the programs (Plotnikoff et al., 2011; Tudor-Locke, 2004; Tudor-Locke et al., 2001; see Chapter 2). This chapter clearly showed that (a) theory based (with Conceptual Model of Health Literacy) interventions, especially theory related to HL, is appropriate for PA behavioural modification and (b) both knowledge and psychological factors such as health beliefs, self-efficacy and perceived barriers shall be addressed in HL-oriented program on PA behaviour.

This study was also the first to measure PA behaviour using two valid, objective instruments: the internationally validated IPAQ-C and pedometers, after the HL-oriented program. With these two measures, the frequency, intensity, time and type of PA were measured systematically, following guidelines presented in a joint position statement of the ACSM and the ADA (Colberg et al., 2010). Such method to assess PA is recommended for future studies and practices for three reasons: (a) PA was found to be one of the most important self-care behaviours to control HbA1c (Speer et al., 2008) and to reduce diabetes risk (Gill & Copper, 2008), however, it was the behaviour with the least compliance (S. Kim et al., 2004; Shigaki et al., 2010; Y. H. Tang et al., 2008). It is crucial to investigate the effect of HL-oriented program on PA changes; (b) Pedometers (model: Yamax Digiwalker SW-200) are popular, validated and objective measurements which are easy to be operated by older adults with limited health literacy and low education level. (c) IPAQ-C which records the frequency, intensity and duration of activities and the time spent in sedentary life during the last seven days, is a tool that can generate the METs value per minutes per week according to the compendium of PA behaviour (IPAQ Committee, 2005). Calculation is standardized and user-friendly.

4.5.3 Limitation

This intervention is conducted to Cantonese-speaking Chinese only, it may not be appropriate to Mandarin-speaking Chinese.

REFERENCES

Afridi, M., & Khan, M. N. (2003). Role of health education in the management of diabetes mellitus. Journal of the College of Physicians and Surgeons, 13(10), 558.

Albright, A. L, Franz, M., Hornsby, G., Kriska, A., Marrero, D., Ullrich, I., & Verity, L. S. (2000). American College of Sports Medicine position stand. Exercise and type 2 diabetes. Medicine and Science in Sports and Exercise, 32(7), 1345-1360.

Aljasem, L. I., Peyrot, M., Wissow, L., & Rubin, R. R. (2001). The impact of barriers and self-efficacy on self-care behaviors in type 2 diabetes. Diabetes Educator, 27(3), 393-404.

American Diabetes Association. (2010). Executive Summary: Standards of medical care in diabetes 2010. Diabetes Care, 33(Suppl.1), 7.

Anderson, L. W., & Krathwohl, D. R. (2001) A Taxonomy for Learning, Teaching, and Assessing: A Revision of Bloom's Taxonomy of Educational Objectives. Longman, New York, NY.

Andrulis, D. P., & Brach, C. (2007). Integrating literacy, culture, and language to improve health care quality for diverse populations. American Journal of Health Behavior, 31(Supplement 1), 122-133.

Arden-Close, E., Gidron, Y., & Moss-Morris, R. (2008). Psychological distress and its correlates in ovarian cancer: a systematic review. Psychooncology, 17(11), 1061-1072. doi: 10.1002/pon.1363

Arvidsson, D., Slinde, F., & Hulthen, L. (2005). Physical activity questionnaire for adolescents validated against doubly labelled water. European Journal of Clinical Nutrition, 59(3), 376-383. doi: 10.1038/sj.ejcn.1602084

Baker, D. W. (1999). Reading between the lines. Journal of General Internal Medicine, 14(5), 315-317.

Baker, D. W., Gazmararian, J. A., Sudano, J., & Patterson, M. (2000). The association between age and health literacy among elderly persons. Journals of Gerontology Series B: Psychological Sciences and Social Sciences, 55(6), 368-374.

Baker, D. W., Gazmararian, J. A., Sudano, J., Patterson, M., Parker, R. M., & Williams, M. V. (2002). Health literacy and performance on the Mini-Mental State Examination. Aging & Mental Health, 6(1), 22-29.

Baker, D. W., Wolf, M. S., Feinglass, J., Thompson, J. A., Gazmararian, J. A., & Huang, J. (2007). Health literacy and mortality among elderly persons. Archives of Internal Medicine, 167(14), 1503. doi: 10.1001/archinte.167.14.1503

Bandura, A. (1989). Human agency in social cognitive theory. American Psychologist, 44(9), 1175.

Bandura, A. (1997). Self-efficacy: The exercise of control. New York: Freeman.

Bassett, D. R., Schneider, P. L., & Huntington, G. E. (2004). Physical activity in an old order Amish community. Medicine and Science in Sports and Exercise, 36(1), 79-85. doi: 10.1249/01.mss.0000106184.71258.32

Beaton, D. E., Bombardier, C., Guillemin, F., & Ferraz, M. B. (2000). Guidelines for the process of cross-cultural adaptation of self-report measures. Spine, 25(24), 3186-3191.

「一帶一路」倡議與中國職業教育輸出

1. 林絢琛教授，博導，院士
2. 買琳燕女士

高等職業教育探索

HIGHER VOCATIONAL EDUCATION EXPLORATION

摘要

　　「一帶一路」倡議的積極回應呼喚中國職業教育輸出，對照職業教育輸出的 9 個基礎要求，即職業培訓的國際地位、職業教育的發展及普及程度、國家政策及配套支持力度、院校數量與人才支持力度、經濟體滿足需求的實力、職業培訓輸出的合作經驗以及輸出資源和當地的匹配度、學分學制課程開發能力、學歷與行業認證並重的制度保障程度、語言政策支援力度等，中國職業教育已具有輸出的優勢。針對「一帶一路」沿線國家對職業教育的需求，我國職業教育要提高與當地勞動力市場的相關性，增強職業教育標準在當地的通用性，提高技術技能人才對經濟技術變革及社會發展的適應性。

　　關鍵字：「一帶一路」；職業教育；教育輸出

Abstract

The positive response of Belt and Road's 9 initiative calls for the export of vocational education in China. In contrast to the nine basic requirements of the export of vocational education: the international status of vocational training, the development and popularity of vocational education, national policies and supporting efforts, the number of institutions and the strength of talent support, the strength of the economy to meet the demand, the cooperation experience of the vocational training export and the matching degree between the export resources and the local, the ability to develop the credit system curriculum, the degree of institutional guarantee of equal emphasis on academic qualifications and industry certification, the degree of language policy support, etc, and China's vocational education has the advantage of output.

In view of the demand for vocational education in the countries along the route, our vocational education should improve its relevance to the local labor market, enhance the universality of vocational education standards in the local areas and improve the adaptability of technical talents to economic and technological change and social development.

Key words："the Belt and Road"; vocational education; education output

為積極回應習近平主席於 2013 年提出的「一帶一路」倡議，中國教育部在 2015 年出臺的《高等職業教育創新發展行動計畫（2015-2018 年）》中，明確提出了支持優質產能走出去，擴大與「一帶一路」沿線國家職業教育合作的任務。在此背景下，越來越多的職業院校已開始結合自身情況，主動承擔起「一帶一路」倡議中的角色重任，推動著中國職業教育輸出。職業教育輸出需要發揮自身優勢和具備一些基礎性條件，更需要明晰「一帶一路」沿線國家對職業教育的具體需求而有所為。本研究就職業教育輸出的基本條件、中國職業教育具備的優勢以及「一帶一路」沿線國家的需求狀況進行分析，以期為職業院校的實踐發展提供參考和借鑒。

一、「一帶一路」倡議與職業教育

形式主義者或一些對職業教育尚未有深入認識的評論者曾提出將理念先進和發展充分的英、美、澳、德、瑞等國的職業教育體系整套引入中國，這越來越被證明是一種錯誤的思想。學習他國先進理念和經驗固然重要，但不加分析、不結合具體情況而貿然將其整套系統引入會走向另一個極端。每個國家的發展情況不同，發展週期不同，人口組成不同，自身教育系統以及體系間的銜接也不一樣，各自的優勢更不相同。中國地大物博，產業鏈比較健全，並不是所有國家的職業教育體系均可以與之匹配。「十一五」之後，中國職業教育體系發展迅速，甚至通過發展職業教育處理解決了不少社會問題，如貧窮問題。而截至目前，尚未有任何其他國家結合教育與民生問題出臺「培訓一人，就業一人，脫貧一戶」的政策。因此，中國在學習借鑒發達國家職業教育經驗的過程中，要充分認識自身的發展階段和發展情況，不能將自身已經具備的優勢抹掉，重新接受一套並未經過實踐檢驗是否適合自己的職業教育體系，而是要立足於自身的實際，既要參考別人的路，更走出自己的風格。當前，「一帶一路」建設正為中國職業教育提供了一個突出特色、高水準展示自我的重要發展機遇。「一帶一路」建設是時代要求和國家戰略下的產物，是職業教育發展面臨的重要外部發展環境。作為一項以中國為主導的全球範圍性的系統工程，「一帶一路」不僅大大拓展了我國職業教育的外延，也對其提出了全新的人才培養要求。[1] 隨著我國對外開放戰略步伐的不斷加快，尤其是我國產能和製造企業的「走出去」以及「一帶一路」

沿線國家重大專案建設的不斷推進，職業教育輸出問題已顯得越來越為急迫。一方面，「一帶一路」沿線國家亟需高品質的職業教育。這些沿線國家如印度等東南亞國家的職業教育普遍比較薄弱，產教融合和校企合作程度有限，當地民眾對高品質的職業教育有著較旺盛的現實需求。我國與東南亞、南亞和中亞等周邊國家有著天然的地緣優勢，這些國家不僅是「一帶一路」建設的優先推動區域，也是我國職業教育合作的重點方向。實施「走出去」辦學戰略，就是要充分發揮我國職業教育的資源優勢，輸出優質職業教育服務，打造中國職業教育品牌。[2] 另一方面，中國企業在推動「一帶一路」建設專案的過程中亟需人才支撐。由於「一帶一路」沿線國家大部分為發展中國家，職業教育水平偏低，職業教育資源嚴重匱乏，既沒有成熟的產業，也並未建立起較為完善的職業教育體系，中國企業亟需的一線技術技能人才缺口嚴重，導致許多項目建成後遲遲不能生產，生產後不能穩產，穩產後又不能有效降低成本創造效益等一系列問題。因此極需中國職業教育輸出，在當地有針對性地培養能滿足企業需求的技術技能人才，從而提高勞動生產率，降低人力資源成本。

二、職業教育輸出的條件及中國職業教育的優勢

根據系統性的收集與研究發現，職業教育輸出一般需要滿足 9 個基礎要求：職業培訓的國際地位、職業教育的發展及普及程度、國家政策及配套支持力度、院校數量與人才支持力度、經濟體滿足需求的實力（如是否有足夠的企業、設備支援及品牌支撐等）、職業培訓輸出的合作經驗以及輸出資源和當地的匹配度、學分學制課程開發能力、學歷與行業認證並重的制度保障程度、語言政策支援力度等。

（一）中國職業培訓的國際地位 2012 年 5 月，中國政府就與聯合國教科文組織在上海合作舉辦了第三屆國際職業技術教育大會（約 10 年舉行一屆），並通過《上海共識》提出了改進職業教育的技能培養水準、提升職業教育體系的投入產出效率、消除社會不公和排斥等多方面的措施，大大促進了國際職業教育的發展；2017 年 7 月，國際職業技術教育大會又在唐山通過了《國際職業技術教育大會唐山聲明》，發佈了「新版技能議程」，奠定了中國職業教育國際發展的新基礎。同時，中國還主辦了多場國際職業交流會議，包括世界職教院校聯盟（World Federation of Colleges and Polytechnics, WFCP）2014 年大會等。可見，中國職業教育在國際上有其影響力。

（二）職業教育的發展與普及程度 職業教育的發展及普及程度是一個國家職業教育發展成熟度的重要衡量因素，也直接決定著該國職業教育的輸出能力，如德國職業教育與培訓體系因其多元化的教育維度、對標的職業培訓和職業資格評價，並提供廣泛的職業前景和保證高水準的就業能力[3]，而在國際市場上具備了獨特的職業教育輸出資本。2008 年，

我國各地廣泛推廣各種形式的職業培訓,每年培訓勞動者 1.7 億人以上,這一數值比「一帶一路」地區總和還要大。2012 年,高等職業教育畢業生就業率為 90.4%,與學士學位課程畢業生的 90.9% 相當。透過經濟合作與發展組織(OECD)的網路資料庫,從職業教育與普通高中教育人數的比例以及 2014、2015 年普通高中教育與職業教育的學生人數比較看,在數量比例上,中國職業教育所佔比例雖然比 OECD 提供的平均值略低,但從人口數量較大、學生人數相對較多的國家來看,中國又與發達國家如英國、德國、荷蘭的比例大體 相當 [4]。因此,聯合國教科文組織第 38 屆大會主席在大會開幕致辭中高度讚揚中國職教發展的偉大成就,並列舉了中國職教一系列驕人的資料:如全國職業教育機構超過 1.2 萬個,提供的專業數量超過 1000 種,目前在校生超過 2600 萬人等。而這種普及程度正可以在「一帶一路」沿線國家中借鑒實施,也代表我國有足夠的人力、財力等輸出技術及人才。

(三)國家政策及配套支持力度圍繞共建「一帶一路」大局,我國政府加強頂層設計,勇於改革,設計中國標準,制定中國方案,積極主動地引領「一帶一路」倡議有效實施。2015 年 3 月,國家發改委、外交部、商務部經國務院授權,發佈了《推動共建絲綢之路經濟帶和 21 世紀海上絲綢之路的願景與行動》;國家發改委制定了《關於加快裝備走出去的指導意見》;工信部制定了《製造業「走出去」戰略規劃》;交通運輸部制定了《落實「一帶一路」戰略規劃實施方案》;國土資源部制定了《「一帶一路」能源和其他重要礦產資源圖集》;推進「一帶一路」建設工作領導小組辦公室發佈了《標準聯通「一帶一路」行動計畫(2015-2017)》;國家海關總署推出了服務「一帶一路」的 16 條措施;國家稅務總局制定了服務「一帶一路」建設的 10 條稅務措施等。各地區也按照中央的統一部署,創造條件,積極有效地對接國家的規劃和工作方案。同時,政府對企業「走出去」協助培訓,對院校外出合作搭建橋樑,對人才留學簽證等給予綠色通道。預計 2020 年,中國每年面向「一帶一路」沿線國家公派留學生將達 2500 人;2022 年,將建成 10 個海外科教基地,每年資助 1 萬名「一帶一路」沿線國家新生來我國學習或研修。由此可見,中國政府對「一帶一路」建設中的「走出去」已做好了充分的政策性準備。

(四)院校數量與人才支援力度從數量上看,中國職業院校數量、培養的學生數量均居世界第一,人才就業率也居於世界前列;同時職業院校覆蓋的行業性、地域性廣,具有很強的普適性。這為「一帶一路」建設的「中國創造」和「中國智力製造」打下了基礎。從培養品質上看,為了提高工業技術水準和學生的實戰能力,很多職業院校已局部或本土化汲取借鑒了德國製造、瑞士製造的雙軌制校企合作課程。隨著這種教學課程和方法的普及和深入,學生與企業、與就業崗位的對接能力將會進一步提升,將有更多的技術技能型人才支援「一帶一路」產業。

（五）經濟體滿足需求的實力是否有足夠的企業、設備支援以及品牌作支撐，也是職業教育輸出的重要條件。2011 年，中國生產了全球 90% 的 PC、80% 的空調、74% 的太陽能電池、70% 的手機，並在微支付、快遞、手機、家電、汽車、高鐵、路橋建造、石材、鋼材、水力發電等行業體現了中國企業在世界上的影響力。與此同時，中國職業院校與企業、行業協會間的「無縫合作」趨勢，也將為滿足「一帶一路」沿線國家需求提供更多的實戰經驗，從而由「中國製造」向「中國創造」轉變。同時，《推進共建「一帶一路」教育行動》也正式提出：「鼓勵中國優質職業教育配合高鐵、電信運營等行業企業走出去，探索開展多種形式的境外合作辦學，合作設立職業院校、培訓中心，合作開發教學資源和專案，開展多層次職業教育和培訓，培養當地急需的各類『一帶一路』建設者。」

（六）職業培訓輸出的合作經驗以及輸出資源和當地的匹配度中國職業教育與「一帶一路」沿線國家的合作在十多年前已經開始，且隨著服務「一帶一路」倡議實踐的逐漸深入，有著諸多實踐基礎與成功經驗。合作數量不斷增加，目前約有 50 餘所院校已和國外企業或院校共同培訓職業人才，培育教師數萬人，開展課程合作的國家也超過數十個，其中包括印尼、斯里蘭卡、坦桑尼亞、尚比亞、肯亞、埃及、新加坡、老撾、印度、泰國、柬埔寨、越南、韓國等；合作類型日漸豐富，如寧波職業技術學院承辦中國政府人力資源援外培訓專案，為印尼、坦桑尼亞、尚比亞、泰國等「一帶一路」沿線國家培訓政府和商業官員以及教師等，海南經貿職業技術學院利用區位優勢，與新加坡、阿聯酋等國的多個公司建立就業合作通道，中國有色金屬工業協會依託全國有色金屬職業教育教學指導委員會、中國有色礦業集團和南京工業職業技術學院等 8 所國內高職院校，在尚比亞先行實施，共同推動職業教育「走出去」等 [7]。經過多年的交流、合作與探索，無論在培訓標準上，當地文化及社會需求的差異標準上都經歷了很好的磨合。《推動共建絲綢之路經濟帶和 21 世紀海上絲綢之路的願景與行動》中更強調推進職業技能開發，調整專業設置，對接「一帶一路」國家與地區的國際化職教專業標準等。可見在合作經驗上，我國針對標準差異已作出了調整和應對。

（七）學分學制課程開發能力若向外輸出我國職業教育的學制及技術，就必須找到相同或可類比、可量度的單位，建立彈性的學分制度，如對中國職業教育學制的學科要求是否可探索建立可組合型的，為「一帶一路」沿線國家學生的學制在不同組合階段的進出創造條件等。此外，學分計算及互換也是一項值得關注和研究的領域。學分制在中國探索已有多年，在 2009 年的第十一屆全國人大常務委員會第八次會議上，時任教育部部長周濟就曾提出過學分計算及互換的論點，而在實踐中也有不少省市不斷在探索和優化。近年來，許多高職院校也開始了學分制的探索和實施並取得了一定的成績。由此說明，我國職業教育已在此方面為「一帶一路」教育輸出工作做了準備。

（八）學歷與行業認證並重的制度保障程度「一帶一路」沿線國家正處於發展或待開發時期，學歷輸出固然重要，但是行業職業資格認證對其來說更加實際。我國職業教育一直在探索學歷與資格的分合問題，如推行職業院校「雙證書」制度，創造條件使職業院校畢業生在取得學歷證書的同時獲得相應的職業資格證書，以期培養能迅速適應職業崗位需求的技術技能人才服務產業發展。近些年來，政府取消了許多國家層面的職業資格認證，使行業掌管其標準，讓行業發揮更大的自由度，更有效地建設行業標準。同時這也正是職業院校的發展契機，與各龍頭企業合作，共同設立行業協會或與現有的行業協會建構或強化屬於中國製造業品牌的標準，充分體現優質政策，把以前的中國製造品牌化，讓「一帶一路」沿線國家共同參與一個屬於「一帶一路」政策的「東方標準」。

（九）語言政策支援力度一方面，隨著我國國力的增強、國際影響力的不斷提升以及對外開放戰略的不斷深化，中文的橋樑作用越來越受到重視。越來越多的國外人員包括「一帶一路」沿線國家人員開始來華學習漢語，很多國家也已將漢語列為必修或選修課程。另一方面，現行政策也提供了相關支持。如教育部《推進共建「一帶一路」教育行動》中，明確要求發揮外國語院校人才培養優勢，推進多語種師資隊伍建設和外語教育教學工作。可見，伴隨著「一帶一路」建設的推進，語言在「一帶一路」教育發展上不會構成障礙。當然，中國職業教育輸出在具備優勢的同時，也存在著不容忽視的投資和辦學風險以及種種的政策不完善等問題，如職業院校的國際化能力尚待提高、政府不同部門間的協作機制尚不健全、他國「市場准入」資本的層層限制等，均需要在協作中逐步攻堅和解決。

三、「一帶一路」沿線國家對職業教育的需求

職業教育輸出為「一帶一路」沿線國家發展提供技術技能人才支持，勢必會受到沿線國家各種現實條件的影響。瞭解「一帶一路」沿線國家職業教育的現狀，研究探討「一帶一路」沿線國家對職業教育的需求，對職業教育應對「一帶一路」挑戰，推進「一帶一路」沿線國家職業教育的發展具有重要意義。地理版圖上的「一帶一路」沿線國家，有的是發達國家，有的是發展中國家，還有的是新興經濟體國家；有的地處內陸，有的地處海洋；有的是大國，有的是小國。從東到西、雙向對應的一條陸路和一條海路將沿線的各個國家鏈接了起來。由點到面、由線到片，再不斷向前延伸和向外輻射，形成了一個南北呼應、東西互濟、影響世界的整體概念。

依據地理區域特徵，「一帶一路」沿線 68 個國家（中國不包含在內）可分為 6 個區域，分別是東南亞（11 國），南亞 (7 國)，中亞及西亞（11 國），中東及非洲 (17 國)，中歐

及東歐（21 國）和亞洲東北部（1 國）。職業教育輸出活動不能盲目進行，需在對合作區域及目標院校的需求分析基礎上，結合院校地理位置及自身專業優勢有針對性開展。

下面以東南亞為例，梳理了「一帶一路」沿線國家對 GDP 貢獻大的 3 個主要產業情況，以期為職業院校在尋求哪一個國家的企業或院校合作的現實困惑提供些許參考。從東南亞 11 個國家的產業貢獻率來看，有 7 個國家的服務業貢獻率最高，且對國家 GDP 貢獻率均高於 40%，新加坡甚至達到 70%，其次是工業、石油及天然氣、電力；居於第二的是工業，覆蓋國家達到 6 個，貢獻率平均在 30% 左右，其次是服務業、製造業、林業、制衣、旅遊等；農業在所有 11 個國家中，貢獻率均排在第三位，平均值在 10% 左右。由此可見，東南亞國家的金融服務、商業服務、運輸服務、通訊服務、貿易服務、旅遊服務、製造業服務、零售業服務等領域，石油業、建築業、運輸設備業等行業均是這些國家經濟增長和民眾生活水準提高的關鍵因素，服務業、工業、農業等都需要大量受過良好職業教育的技術技能型人才。除了新加坡等職業教育和培訓發達的國家，絕大多數的東南亞國家如越南、緬甸、菲律賓等市場經濟不完善，與市場經濟發展密切的電子、建築、機械等行業的技術技能型人才缺口嚴重，資本以及技術技能人才的結構性短缺和市場需求使其國民經濟發展緩慢。因此在「一帶一路」倡議深化的進程中，極需加速職業教育的人力資本投入。

為此，中國在落實服務「一帶一路」倡議與行動精神，推動職業教育輸出資源的過程中，政府要在宏觀層面發揮統籌管理職責，整合優質教育資源，積極與合作國政府聯繫溝通，為職業院校提供資料資訊平臺，鼓勵教育機構圍繞「一帶一路」開展基礎性或前瞻性課題研究等；行業企業應在中觀層面發揮指導作用，推 動校企合作，聯合培養培訓人才；而廣大職業院校在以各種模式和途徑服務「一帶一路」建設的過程中，要在微觀層面圍繞「一帶一路」沿線國家產業發展的實際需求，重點採取如下措施：

一是要提高職業教育與當地勞動力市場的相關性。由於職業院校和普通高校畢業生之間勞動力市場回報的差異主要是由選擇驅動的 [8]，故職業院校在課程設置、師資選派、人才培養環節均需要關注和研究當地的實際經濟需求點以及工作領域的新發展和新要求，將學生的職業選擇與勞動力市場需求結合起來，將雇主對員工的實際技術技能需求連接起來，根據需求 特點和發展實際來制定人才培養計畫，創新培養模式；同時，需加強沿線國家建設所需人才的調查研究，提高職業教育與當地經濟發展需求的相關度和適應性。

二是要增強職業教育標準在當地的通用性。我國應儘快建立國家職業教育資格標準、國家職業技能教育品質保障機制等，增強職業教育培訓的針對性，並在職教標準、課程標準、實訓室一體化標準等方面加以研究、牽頭和完善，爭取獲得「一帶一路」沿線國家的認可並使用；同時應積極探索將職業資格證書標準和課程標准對接我國企業和「一帶一路」沿線國家相應企業的技術標準體系等。

三是要提高技術技能人才對經濟技術變革及社會發展的適應性。「一帶一路」沿線國家既有單一經濟體如沙烏地阿拉伯，欠發達經濟體如蒙古、緬甸、越南及非洲大部分國家，更有發達經濟體如土耳其、以色列等，以及新興經濟體如俄羅斯、印度等，除了傳統行業、低端手工技能人才需求的存在，新興經濟、高科技產業、高精尖技術技能人才需求業已成為，或將成為沿線國家的經濟發展推動力和需求點，這也是職業院校服務「一帶一路」建設的應關注的方面。

參考文獻

[1] 徐玉蓉，祝良榮．「一帶一路」倡議下職業教育發展的機遇、挑戰與改革方向 [J]. 教育與職業，2017, (16):7-13.

[2] 李建忠．「一帶一路」給職業教育帶來怎樣的發展機遇 [N]. 中國教育報，2015-05-28(13).

[3] BMBF. Report on vocational education and training 2016 [DB /OL]. 2016-04-15. http://www.bmbf.de/.

[4] The Dutch Ministry of Education, Culture and Science. Key figures 2009-2013 [DB/OL] [2014-07-02]. http://www. rijksoverheid.nl/.

[5][6] OECD(2017), Education at a Glance Database[EB/OL] [2017-12-28]. http://stats.oecd.org/.

[7] 劉紅．「一帶一路」戰略背景下我國職業教育發展機遇、挑戰與路徑一．「一帶一路」產教協同峰會會議綜述 [J]. 中國職業技術教育，2017(4): 20-23.

[8] Malamud 0, Pop – Eleches C. General education versus vocational training: edivence from an economy in transition[J]. Review of Economics & Statistics, 2010, (1): 43-6 0.

加拿大魁北克大學博士後研究班

Université du Québec à Montréal POSTDOCTORAL FELLOW

加拿大魁北克大學（蒙特利爾）簡介

加拿大魁北克大學創建於 1969 年，由分佈在魁北克的 10 餘所院校組成，目前註冊學生超過 86,000 名，按學生人數統計是加拿大最大的大學。魁北克大學蒙特利爾分校（Université du Québec à Montréal，簡稱 UQAM）是魁北克大學最大的分校，位於蒙特利爾市中心，作為加拿大綜合性公立重點大學之一，UQAM 由魁北克政府撥款，是本地發展最快的大學。

魁北克大學蒙特利爾分校開設管理、經濟、工程、語言、戲劇、歷史等 200 多個本科、碩士和博士專業課程，其管理學院為北美最大的管理學院之一，以優質的生源和卓越的教學品質著稱，並通過了國際高等教育權威 EQUIS 認證。UQAM 是加拿大第一個提供環境學博士學位的學校，世界衛生組織也將這裡的研究中心作為環境和健康的研究基地之一。

蒙特利爾分校與全球五大洲、40 多個國家的近 250 所大學院校建立了廣泛而深遠的合作，涉及 100 多種專業方向共 350 多個交流項目。學校有專門的國際關係服務機構，説明諮詢解答學生的所有問題。近年來，魁北克大學與中國各大院校在學術交流方面合作緊密，例如，與中山大學和廣州大學相關專業的學分互認、與北京語言大學的師生互換交流、與清華大學藝術專業的合作交流、與中國礦業大學合作舉辦工商管理碩士學位教育項目等。

博士後項目介紹

　　加拿大魁北克大學招收中國以及亞洲地區高級管理人員參加博士後專案，專案針對系統接受過專業教育的人士而開辦，為研究與實踐相結合的高端教育專案；具備研究和創新能力是高級管理人員必備的素質，參加博士後專案將取得研究經驗來提升個人的綜合能力。

　　博士後項目建立在傳統的博士學位教育之上，是更具有現實指導意義的高層次教育項目，該項目注重交流性的學習模式促使學生不斷地思考和創新，並建立終身學習提升的國際性平台。

項目特色

1、　亞洲唯一的歐美名校博士後研究專案；

2、　突破傳統的課堂教學模式，先進的導師指導管理理念；

3、　學員親赴加拿大，近距離感受魁北克大學等歐美名校學術氛圍；

4、　免語言成績，中英雙語培養模式。

入學條件

1、博士後申請人為獲得博士學位 5 年以內人員。

　　（申請人為獲得博士學位 5 年以上人員申請人可申請訪問學者）

2、未獲得博士學位的特別優秀人員需要特別申請。

專案責任教授

Prof. Dr. Prosper Bernard

Bernard 教授是魁北克大學蒙特利爾分校的前副校長、管理學院教學院長，MBA 和 EMBA 專案主管，研究生教育委員會會長及四所蒙特利爾大學聯辦的商業 PhD 專案主管。Bernard 教授為魁北克大學（蒙特利爾）管理學院經營策略系博士，曾獲加拿大魁北克大學優秀教師獎和管理學院最佳教師獎。

Prof. Dr. Michel Plaisent

工商管理博士，教授。蒙特利爾魁北克大學的資訊管理系主任。博士後專案負責人。

博士後指導教授

Ir Prof. Postdoc. Dr. PHILIP CHAN Postdoc
Supervisor

（陳勤業教授，博士後導師）博士後專案大中華地區負責人。

國家行政學院政府經濟研究中心博士後課題專家委員會副主任委員

加州州立大學 / 魁北克大學 博士後導師

國家行政學院博士後

斯坦福大學 / 伯克利加州大學 / 加州州立大學博士後

魁北克大學博士後

中國政法大學法學博士後研究員

天津大學管理科學與工程博士後

北京大學、天津大學、北京師範大學客座教授

廣州仲裁委員會首席仲裁員

香港國際仲裁中心 / （北京 / 上海 / 深圳 / 泉州）仲裁委員會 / 南京港澳仲裁院 / 珠海國際仲裁院仲裁員

加拿大魁北克大學博士後專案大中華地區負責人

Ir Prof. Dr. PHILIP CHAN Postdoc（陳勤業博士後導師，教授）

電郵： UQAMPostdocAdvisor@gmail.com /
philip@pcconsulting88.com

Université du Québec à Montréal

315 rue Sainte-Catherine East

Montreal, Quebec H2X 3X2

www.uqam.ca

加拿大魁北克大學博士後

（以下博士後香港導陳勤業教授工程師）

Université du Québec à Montréal POSTDOCTORAL FELLOW
(Ir Prof PHILIP KI CHAN FEng was below Postdoctoral Fellowship's Supervisor)

陳伯強博士後

世界財富500強企業新美亞集團 (Sanmina
Corporation, U.S.A.) 榮休主席及首席執行官
顧問及大中華區總裁兼董事

國際博士後協會美國分會主席

Prof Dr CHAN PAK KEUNG Postdoc

Former SANMINA Chairman & CEO's
Consultant and Greater China President

International Postdoctor Association (www.
IPostdocA.org) US Chairman

陳勤業博士後

魁北克大學博士後導師

國際博士後協會會長

Ir Prof CHAN KAN IP PHILIP FEng Postdoc

University of Quebec at Montreal (UQAM)
Postdoc Supervisor

International Postdoctor Association
(www.IPostdocA.org) President

邱柏民博士後

太古匯廣州發展有限公司 前副工程總監

國際博士後協會中國南部主席

Prof Dr Yau Pak Man Postdoc

Taikoo Hui (Guangzhou) Development Co
Ltd (Swire Group) Former Assisstant Director
— Projects (Construction)

International Postdoctor Association
(www.IPostdocA.org) South China Chapter
Chairman

黃淑儀博士後

黎治國博士後

現代集團高管

Eur Ir Dr LAI Che Kwok PhD Postdoc

Modern (Int'l) P&M Holdings Ltd
Management

廖寶城博士後

魁北克大學博士後導師

國際博士後協會會長

香港特區政府公務員

國際博士後協會香港分會主席

Dr Lu Po Shing Postdoc

HKSAR Civil Servant

International Postdoctor Association (www.
IPostdocA.org) Hong Kong Chairman

葉卓雄博士後

柯尼卡美能達香港有限公司
董事長

Dr Ip Cheuk Hung Robert
Postdoc

Konica Minolta Business
Soultion HK Ltd Managing
Director

高少偉博士後

伊麗嘉珠寶有限公司董事長

國際博士後協會香港分會
副主席

Dr Ko Siu Wai Postdoc

ERICA Jewellery Co Ltd.
Managing Director

International Postdoctor
Association
(www.IPostdocA.org) Hong
Kong Vice Chairman

蘇汝培博士後

匯輝控股有限公司董事長

Ir Dr So Yu Piu Fred Postdoc

Wu Fai Holdings Ltd
Chairman

曹光明博士後

江蘇幸運寶貝安全裝置製造
有限公司董事長

Dr Larry Cao Postdoc

Lucky Baby Group Founder
& CEO

胡驍博士後

上海聖瑞赦律師事務所
創辦人兼律師

Dr Hu Xiao Mike Postdoc

Shanghai Singrights Law
Firm Founder and Lawyer

美國加州州立大學博士後項目簡章

CALIFORNIA STATE UNIVERSITY MONTEREY BAY (CSUMB)– POST DOCTORAL FELLOW PROGRAM

加州州立大學簡介

加利福尼亞州立大學（California State University），簡稱加州州大（CSU 或 Cal State），是美國加州的一個公立大學系統。它是組成加州公立高等教育體系的三個大學系統之一。加州州大共有 23 個校區，學生總數超過 41 萬人，教職員總數超過 4 萬 4 千人。加州州大是全美最大的高等教育系統（不含社區大學）。加州 60% 的教師人力及 40% 的工程人才，及加州每年一半的大學畢業生及三分之一的碩士畢業生，出自加州州大系統。

自 1857 年以來，有超過 3 百萬的校友，從加州州大系統獲得學士、碩士及博士學位。目前加州州大系統共提供在約 240 種專業領域中，超過 1,800 種學位課程。

加州州立大學與全球五大洲、100 多個國家的近 300 所大學院校建立了廣泛而深遠的合作，涉及 200 多種專業方向共 100 多個交流項目。學校有專門的國際關係服務機構，說明諮詢解答學生的所有問題。近年來與中國各大學院校在學術交流與課程培訓方面合作緊密。

博士後項目介紹

加州州立大學招收中國以及亞洲地區高級管理人員參加博士後專案，專案針對系統接受過博士專業教育的人士而開辦，為研究與實踐相結合的高端教育專案；具備研究和創新能力是高級管理人員必備的素質，參加博士後專案將取得研究經驗來提升個人的綜合能力。

博士後項目建立在傳統的碩博學位教育之上，是更具有現實指導意義的高層次教育項目，該項目注重交流性的學習模式促使學生不斷地思考和創新，並建立終身學習提升的國際性平臺。

項目特色

1、亞洲唯一的美國三大名校合作博士後研究專案。

2、突破傳統的課堂教學范式，先進的導師指導管理理念。

3、學員親赴美國，近距離感受加州名校學術氛圍。

　　或選擇留在國內接受名校聘任之國內重點大學教授指導。

4、免語言成績，中英雙語培養模式。

5、雙證書：（1）參加美國培訓 1 週可得一張博士後論壇（Postdoctoral Symposium）出席證書。

　　　　　　（2）博士後研究員在 24 個月內完成學術文章發表可得博士後證書。

進修方式

專案總時長 24 個月，在國內安排有階段性中外指導教授見面會，項目期間統一前往進行美國一段時間的課程學習以及課題研究在加州州立大學，可安排到斯坦福大學與伯克利加州大學將為學生舉辦學術研討會，對學生進行研究方法指導，並安排有責任教授主講課程，還可前往歐美著名大學進行交流訪問。學業以個人研究為主，由加州州立大學與斯坦福大學、伯克利加州大學和亞洲區特聘教授共同指導，學生需撰寫一篇學術論义並在英文期刊上發表（中國學生可用中文撰寫後翻譯），發表期刊在加州州立大學學報（California Journal of Management and Technology）；由加州州立大學推薦的專業學術刊物或認可國際會議發表。學生個人也可在加州州立大學進行長期研究學習。

進修方式

項目總時長 6 天，1 天在斯坦福大學上課，1 天在伯克利加州大學上課，2 天美國加州州立大學，1 天在矽谷企業參訪，項目期間統一前往進行美國一段時間的課程學習，整個培訓過程由加州州立大學安排及輔導。

入學條件

博士後申請人為獲得博士學位 10 年以內人員。

入學時間及人數

　　每年限 10 人，赴斯坦福大學、伯克利加州大學及企業參訪研究學習 1 周，根據個人情況結合統一安排選定赴美具體時間。

證書授予

1) 參加美國培訓 1 周可得一張博士後論壇（Postdoctoral Symposium）出席證書（最終證書可能與樣本有少許差別）

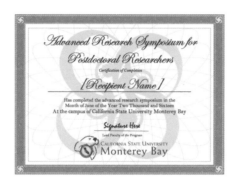

2) 完成項目要求後將獲美國加州州立大學（C3UMB）博士後證書和美國加州州立大學（CSUMB）。（最終證書可能與樣本有少許差別）

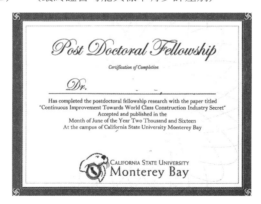

研究方向（包括但不限於）

Marketing 行銷

Finance 金融

Corporate Governance and Accounting 公司治理與會計

Logistics 物流

Tourism 旅遊

Urban Heritage 城市遺產管理

Human Resources 人力資源

Project Management 專案管理

Economics 經濟學

Business Administration 工商管理

Bio-industry and technology management 生物產業與技術管理

Economics of Technology 技術經濟

Information technology and Computer Science 資訊技術與電腦科學

專案責任教授

Ir Sr. Prof. PHILIP CHAN PhD Postdoc

陳勤業博士後導師，教授

博士後專案大中華地區負責人

蒙特利灣加州州立大學（CSUMB）特聘教授（Distinguished Professor）

和博士後導師（Postdoctoral Advisor）

蒙特利灣加州州立大學（CSUMB）博士後

法國北歐大學副校長（中國和香港）

Vice President (HK & China), SABI University, France

國家行政學院政府經濟研究中心經濟學博士後

Postdoctoral in Economics, Government Economics Research Center of National Academy of Governance, China

魁北克大學管理與科技博士後

Postdoctoral in Management and Technology of University of Quebec at Montreal (UQAM), Canada

天津大學管理科學與工程博士後

Postdoctoral in Management Sciences and Engineering of Tinanjin University, China

亞洲區特聘教授

　　根據博士後研究員專業方向及所在區域，在中國大陸、臺灣、香港、新加坡等亞洲地區選配最合適的指導教授，由亞洲專案辦公室協調。

美國蒙特利灣加州州立大學博士後

（以下部份博士後由香港陳勤業教授和美國陶翼青教授指導）

CALIFORNIA STATE UNIVERSITY MONTEREY BAY (CSUMB) POSTDOCTORAL FELLOWSHIP
(from 2015 – 2021)
(Below CSUMB Postdoctoral Fellowship Advised by Ir Sr Prof PHILIP KI CHAN PhD and Prof. Eric Tao PhD)

1) 陳天聰博士後（2016）

九龍倉：龍茂房地產開發（成都）有限公司 總經理

Prof. Chan, Tin Chung Argus PhD Postdoc
WHAFT: Long Mao Property Development (Cheungdu) Co. Ltd. General
Manager
International Postdoctoral Association (www.IPostdocA.org) Central China
Chairman

2) 陳勤業博士後（2016）

美國蒙特利灣加州州立大學 博士後導師

國際博士後協會會長

Ir Sr Prof Chan, Kan Ip Philip PhD Postdoc
Distinguished Professor and Postdoctoral Advisor of CALIFORNIA STATE
UNIVERSITY MONTEREY BAY (CSUMB)
President of International Postdoctoral Association (www.IPostdocA.org)

3) 廖寶城博士後（2016）

香港特區政府公務員

國際博士後協會香港分會 主席

Prof. Liu Po Shing PhD Postdoc
HKSAR Civil Servant
International Postdoctoral Association (www.IPostdocA.org) Hong Kong
Chairman

4) 潘偉駿博士後（2016）

威發國際集團有限公司 執行董事

國際博士後協會英國分會 主席

Dr Poon Wai Chun William PhD Postdoc

Perfectech International Holdings Limited Executive Director

International Postdoctoral Association (www.IPostdocA.org) UK Chairman

5) 仇玲玲博士後（2018）

中江亞威經濟作物種植有限公司

國際博士後協會中國中部 副主席

Dr. Qiu Ning Ning PhD Postdoc

Zhongjiang Ya Wei Economic Crops Co., Ltd.

International Postdoctoral Association (www.IPostdocA.org) Central China Deputy Chairlady

6) 黃國強博士後（2019）

香港某大型中央企業集團

Dr. WONG Kwok Keung keith Postdoc

China Central Government stated own enterprise in Hong Kong

7) 陳偉龍博士後（2019）

天龍人力資源管理有限公司 董事長

Dr. Raymond Chan Postdoc

Chairman, DRACO Human Resources Management Limited

8) 蒙嘉輝博士後 (2019)

中國港灣工程有限責任公司 高級項目經理

Dr. Alan Mong PhD Postdoc

Senior Project Manager, China Harbour Engineering Company Limited

9) 曾惠珍博士後 (2020)

迪德施管理顧問有限公司 創辦人及董事長

Prof. Dr. TSANG Wai Chun Marianna Postdoc

Founder and Chairman, TWC Management Limited

10) 趙芝勝博士後 (2020)

承創互連科技有限公司 創辦人及董事長

Dr. Chiu, Chi Shing Cook Postdoc

Founder and Chairman, CCSC Interconnect Technology Limited

11) 林煒彬博士後 (2021)

香港特區政府建築署 建築師

Prof. Dr. Lam, Wai Pan Wilson Postdoc

Architect, Architectural Services Department, HKSAR Government

12) 馬偉成博士後研究員 (target completed on 2022)

駿盈建築有限公司 創辦人兼董事

Founder & Director, Billion Smart Construction Co. Ltd.

回顧國際博士後協會參與亞洲質量網組織大會及展望

競爭激烈的世界中，品質依然是最重要的成功因素。香港的企業必須如書名「持續改善」以滿足市場不斷變化及需求。本人（香港品質學會會長）介紹「亞洲質量網」組織（ANQ），以及感謝陳勤業教授工程院士和國際博士後協會在 2012 年起一直支持「亞洲質量網」大會及博士後們在大會中發表文章交流。

「香港品質學會」（HKSQ）成立於 1986 年，屬非營利性組織，組建目標是進一步推動香港對品質創新演化的需求意識，通過對品質創新和顧客滿意的持續改進，達致卓越的產品和服務，同時向品質創新和可靠性專業人員提供持續的教育服務。我們是「亞洲質量網」組織（ANQ）的創始成員。在第十六屆亞洲質量研討會（AQS）指導委員會主席邀請各亞洲區的品質代表出席是次在東京舉辦的會議；香港的 HKSQ 與韓國的 KSM、台灣的 CSQ、中國的 CAQ、印度的 ISQ、泰國的 SQAT 和日本的 JSQC 在會議中決議成立亞洲質量聯盟的事宜並命名為「亞洲質量網」組織（ANQ）。其網頁為「www.ANforQ.org」。自 2002 年成立至今，已有來自中國，香港，日本，印度，韓國，新加坡，泰國等國家和地區的 17 家非盈利質量組織加入其中。它堅持以「提升亞洲人民的生活質量」為目標，倡導和推行各項質量活動。每年召開一次區域性大會，表彰亞洲地區為質量做出突出貢獻的組織和個人，分享和交流各國家和地區在質量領域取得的研究成果和最佳實踐。從而促進亞洲各國和地區的繁榮與團結，通過品質及有關的管理的研究及推廣活動促進不同行業發展，提升區內質量。這正好配合一帶一路的國策，與其他亞洲國家地區達致共同發展、雙贏局面。

博士後們在亞洲質量網組織大會發表的文章是經過香港品質學會論文推薦委員會推薦的。論文推薦委員會委員包括本人（香港品質學會會長），錢桂生副教授（香港城市大學）及潘杰輝教授（西印度大學，聖奧古斯丁分校（千里達））。評審的四個準則有質量相關性（Quality Relevancy）、技術優點（Technical Merits）、貢獻（Contributions）及適用性（Applicability）等。

本人及大會期望有更多專家參與及發表文章交流，推動國家的質量發展達到「雙循環」策略。本人提議「質量內循環」以粵港澳大灣區為首，再經亞洲質量網組織推動「質量外循環」在一帶一路的質量建設中發揮作用。

ANQ Paper Review Panel

- Local Society (e.g. HKSQ) nominate paper and poster to ANQ Congress Secretary
- ANQ Congress Organizer (Rotation based of Board members and General members) review and make final decision
- Best paper award is nominated by local society (e.g. HKSQ) and review by organizer and usually only <10% of papers awarded.

HKSQ Paper nomination panel

- Coordinator: **Dr. Lotto Lai** (Adjunct Professor, SEEM Dept., CityU of HK)
- Reviewer: **Dr. K.S. Chin** (Associate Head and Associate Professor, SEEM Dept., CityU of HK)
- Reviewer: **Prof. Kit Fai Pun** (Professor, Department of Industrial Engineering, University of the West Indies)

圖為香港品質學會論文推薦委員會委員

圖為第十屆亞洲質量網年會 2012 在香港舉辦，陳勤業教授工程院士支持及發表文章。

> Philip K.I. Chan, Prosper Bernard, Michel Plaisent and James Ming-Hsun Chiang (2012) "Evaluation of the Practice of Quality Assurance in Greater China Construction Industry", Proceedings of ANQ 2012, pp176-183.

　　圖為第十二屆亞洲質量網年會 2014 在新加坡舉辦，四名博士後支持及發表文章。Dr. Fred So（右五）、Dr. Eddie Liu（右三）、Dr. Robert Ip（右二）同 Dr. Kiev Ko（別一天發表文章），本人（左二）、錢桂生副教授（香港城市大學，左三）是 HKSQ 的論文推薦委員會委員。

> Frederick Y.P. So & Philip K.I. Chan (2014) "Environmental Considerations in the Construction Industries in Hong Kong and China" Proceedings of ANQ 2014. EN1-1.1.

> Mike Hu, Philip K.I. Chan & Prosper Bernard (2014) "Analysis of China Private-Equity and Venture Capital Market", Proceedings of ANQ 2014, OB1-1.3.

> Ko Siu Wai, Kiev & Philip K.I. Chan (2014) "Marketing Strategies of Jewelry Manufacture in Hong Kong: An Analysis" Proceedings of ANQ 2014, BP1-1.8.

> Po Shing Liu, Eddie & Philip K.I. Chan (2014) "Developing C2C Retail Entrepreneurship with Opportunities and Challenges: A Hong Kong Example", Proceedings of ANQ 2014, OB1-1.5.

> Robert Ip, Philip Chan & Prosper Bernard (2014) "Business Transformation: Case Study of Hardware Company", Proceedings of ANQ 2014, OB1-1.1.

圖為第十三屆亞洲質量網年會 2015 在中華台北舉辦，二名博士後支持及發表文章。

➢ Guang Ming Cao, Prosper Bernard, Michel Plaisent and & Philip K.I. Chan (2015) "A Research Report on the Necessity of Popularizing the Use of CRS (Children Restraint System) in China", Proceedings of ANQ 2015, HK-03.

➢ Zhang Jing, Chan Kan Ip Philip and James Ming-Hsun Chiang (2015) "How KFC allocates the 4 Ps Marketing Strategies in China", Proceedings of ANQ 2015, HK-04.

圖為第十四屆亞洲質量網年會 2016 在俄羅斯舉辦，四名博士後支持及發表文章。

➢ William W.T. Poon & Philip K.I. Chan (2016) "Poon's Formationof Seru-Flow Production on Manufacturing Medium-Large Quantity", Proceedings of ANQ 2016, HK-03.

➢ Argus T.C. Chan & Philip K.I. Chan (2016) "Strategic Issues and Direction of China's Innovative One Belt One Road Strategy" Proceedings of ANQ 2016, HK-04.

➢ Philip K.I. Chan (2016) "Continuous Improvement towards World Class Construction Industry Secret", Proceedings of ANQ 2016, HK-05.

➢ Po-Shing Liu & Philip K.I. Chan (2016) "A Study on Innovation and e-Service Quality for Developing e-Retailing Mass Entrepreneurship", HK-06.

　　圖為第十六屆亞洲質量網年會 2018 在哈薩克斯坦的阿拉木圖舉辦。段永剛（中國質量協會副會長，左一），狩野教授（Prof. Kano，左二）及黎劍虹博士工程師（左三）。

➢ Aaron Sum Chan & Philip K.I. Chan (2018) "Factors affect to Customer Loyalty in Logistics Industry" Proceedings of ANQ 2018, HK-02.

➢ Qiu Ning Ning & Philip K.I. Chan (2018) "Brief Discussion of Returnees' Employee Loyalty" Proceedings of ANQ 2018, HK-03.

圖為第十七屆亞洲質量網年會 2019 在泰國舉辦，四名博士後支持及發表文章。

➢ Lam Kong Ngan Kenny and Philip KI Chan (2019) "PRC (Greater Bay Area) Guangdong-Hong Kong-Macao Construction Culture Difference" Proceedings of ANQ 2019, ANQ 1-004.

➢ Wong Kwok Keung, Keith and Philip KI Chan (2019) "Sustainable Development Water Resource in Hong Kong Special Administrative Region, China" Proceedings of ANQ 2019, ANQ 2-006.

➢ Mong Ka Fai and Philip KI Chan (2019) "New Mechanically Automatic Inclinometer System" Proceedings of ANQ 2019, ANQ 2-048.

➢ Lam Kong Ngan Kenny and Philip KI Chan (2019) "Building a Team of Third-party Inspectors Integrity and Self-love Lead by Example Law Enforcement, Prevention and Education "Three-pronged"" Proceedings of ANQ 2019, ANQ 7-002.

圖為第十七屆亞洲質量網年會 2019 在韓國以線上方式舉辦，二名博士後支持及發表文章。

圖為香港代表及參與者

> Tsang Wai Chun Marianna and Philip KI Chan (2020) "Enhancement of Corporate Governance in Small to Medium Size Non-Governmental Organizations (NGO) in Hong Kong from the Board's Perspective" Proceedings of ANQ 2020, ANQ-024.

> Chiu Chi Sing, Cook and Philip KI Chan (2020) "Exploratory Study on Synergy of Sustainability Model between Time Bank and Innovative Elderly Healthcare Platform" Proceedings of ANQ 2020, ANQ-025.

作者：

黎劍虹博士工程師

香港品質學會主席

香港品質學會（HKSQ）會士

美國品質學會（ASQ）會士

國際質量研究院（IAQ）院士

城市大學生物化學系特約教授（2014-2015）

城市大學系統工程及工程管理學系特約教授（2017-2022）

FASQ, FHKSQ, MHKIE, ASQ CMQOE, IRCA QMS 首席稽核員

中國質量研究院簡介

　　中國質量研究院（CHINA QUALITY INSTITUTE）成立于 2015 年，是中國質量俱樂部下屬的質量研究機構。

　　在「中國製造 2025」實施製造強國戰略行動綱領的指導背景下，中國質量研究院聚焦行業質量管理特點，致力于行業質量管理範式、大數據時代質量管理、全生命周期質量管理及質量成熟度測評等方面的研究工作，幫助行業質量變革、提升企業整體競爭力。

　　中國質量研究院匯集了行業質量專家、高校專家教授、資深培訓講師、實戰職業經理人，經過多年的積澱，出版發行質量相關著作十餘部、擁有數十項知識產權和專利證書、聯合權威機構推出多項職業認證體系……

　　中國質量研究院的使命是，打造具有影響力的質量研究基地，爲産業降本提質增效；爲行業主管部門專項質量提升研究；爲國家質量決策提供理論方法及技術支撐，服務于質量強國建設，爲世界級的中國質量貢獻專業力量。

研究領域：

官方網站：www.cqi.org.cn
聯繫電話：021-52961817
微信公衆號：zgzlyjy

中国质量研究院
电话：021-52961817
邮箱：service@qualityclub.cn

新華社亞太總分社

原社長俱孟軍金句精華集

氣質比年齡重要
微笑比顏值重要
開心比感情重要
Temperament is beyond age
Smile is beyond outlook
Happiness is beyond relationship

一貫是堅持
三觀是思想
五官是顏值
Persistence depends on consistency
Thoughts depends on world view
Beauty depends on facial features

遇事別糾結
遇險別慌張
遇人別逞強
Be decisive with things
Stay calm with danger
Be humble with others

學會獨處
減少依賴
堅持學習
Learn to stay alone
Be less dependent
Insist on studying

勞動者最光榮
學習者最聰明
思想者最偉大
Labor is of great honor
Learner is of great intelligence
Thinker is of greatness

競爭必須正當
暴利必須遏制
貪婪必須打倒
Competition must be fair
Profiteer must be banned
Greed must be forbidden

天外有天
樓上有樓
人中有人
Broader sky always be there
Higher tower always be there
More intelligent people always be there

光芒萬丈是太陽
光陰似箭是時間
光怪陸離是社會
Radiating sun brighten
Fleeting time flies
Sophisticated world is grotesque

兔子不去水中游
烏龜不吃窩邊草
人生不用比賽跑
Rabbit hardly go to swim
Turtle hardly eager for grass beside
Life is not for race

縱觀舊時代
把握今朝事
應對大變局
Throughout the old days
Seize the present
Deal with great change

有孩子把孩子當孩子
沒孩子把自己當孩子
好孩子把爹媽當孩子
Parents regard kids as children
Someone without kids regard themselves as kids
Good kids regard parents as children

哲學離不開焦慮
數學離不開苦惱
文學離不開失落
Philosophy lives with anxiety
Mathematics lives with agony
Literature lives with desperation

吃飯讀書旅行
只要每天成長
何必追求成功
Dine read travel
Grow for each day
That is the success

大局需要洞察
微觀需要觸感
現實需要面對
Big picture needs insight
Micro needs sensation
Reality needs courage

求之不來是遙遠
揮之不去是眼前
買之不到是時間
Out of reach is remoteness
What lingering is the present
Money cannot buy you the time

團團夥夥是勢力
親親疏疏是關係
彎彎曲曲是道路
Power is among gangs
Relationship can be close and alien
Road can be winding and twisting

站在十字路口

看見燈紅酒綠

不問東西南北

Stand still at crossroad

With sophisticated world in sight

Without asking for direction

學最好的別人

許最好的願景

做最好的自己

Learn the best of others

Wish for the best

Be the best of yourself

人以德為貴

物以用為值

錢以花為好

Harmony of human lies in virtue

Value of things lies in usage

Value of money lies in spending

熱烈慶祝博士後論文集出版
分享企業管理心得和經驗

主賀

TWC MANAGEMENT LIMITED
迪德施管理顧問有限公司

信誠工程國際有限公司
SHUN SHING ENGINEERING INTERNATIONAL LIMITED

感謝黃煒權博士CSUMB訪問學者的大力支持！

聯賀

聯賀

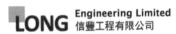

國際爭議解決及專業談判研究院
Academy of International Dispute Resolution &
Professional Negotiation

IDRMI International Dispute Resolution & Risk Management Institute
國 際 爭 議 解 決 及 風 險 管 理 協 會

中國貿促會/
中國國際商會調解中心

內地—香港聯合調解中心
Mainland - Hong Kong Joint Mediation Center

香港和解中心

专访国家行政学院政府经济研究中心博士后课题
专家委员会副主任、杰出教授、工程院院士

陈勤业
立足粤港澳发展趋势，
打造新硅谷中国梦

文/本刊记者 符国涛 李楚欣

陈勤业

陈勤业教授是一位土生土长的香港人，但是从小的家庭教育让他从未忘记自身体里流着的中国血脉。从未忘记自己是中国人。经过多年的努力奋斗，终于有契机于1997年香港回归祖国后被外资公司派驻上海让他能在祖国内地工作，最后他进入了国家行政学院政府经济研究中心的博士后课题专家委员会，成为该委员会的副主任。

2018年3月，中共中央印发了《深化党和国家机构改革方案》，将中央党校和国家行政学院的职责整合，组建新的中央党校（国家行政学院），实行一个机构两块牌子，作为党中央直属事业单位。

陈勤业教授向记者介绍了国家行政学院主要职责，分别是：组织培训与教学；紧密围绕政府工作和培训与教学的需要，重点开展对策性研究与咨询；参与制定有关培训的方针政策，与政通合作，负责培训教材和有关参考书的编写、编译；组织地方行政学院和有关培训机构开展广泛的业务交流，合作办学，科研协作及师资培训，根据需要委托地方行政学院或国务院有关部门的培训机构举办部分培训班；开展对外交流与国际合作；适量培养以行政学、管理学为主的硕士生、博士生；向国家机关和有关单位推荐人才。

当问及陈勤业教授如何看待最近的粤港澳大湾区发展时，他有着自己独特的见解。

陈勤业教授回忆说，他于2016年8月带领一批高端博士到美国斯坦福大学出席博士后论坛，斯坦福大学教授斩钉截铁地说硅谷是全球唯一的，不可以复制的，因为硅谷的独特性包括三方面：一，硅谷有全球大学排名第二的斯坦福大学，吸引着全球顶尖创新科技人才做研究；二，大量风险资本在硅谷寻找投资项目，特别垂青斯坦福大学生科研项目；三，硅谷的创业文化俗生大量最新兴公司，有一大部分新兴公司是斯坦福大学学生科研项目被风险资本看中而投入，后来更发展成为全球科技巨人，如惠普公司和雅虎，等等。当时的陈勤业教授听到了这一介绍之后，心里既可惜又无奈。

但是当2017年8月，陈勤业教授再次带领高端博士到斯坦福大学出席博士后论坛时，斯坦福大学教授兴高采烈透露他研究发现中国的粤港澳大湾区内新兴公司5年后（即2022年）会超越美国旧金山市大湾区成为全球最多新兴公司的一个区。奇怪的是教授只有提及量但没有提及质，陈勤业教授思考，教授的话意味着我国粤港澳大湾区的新兴公司质量不一定能超越旧金山市大湾区，无论如何，这也是香港能够成为"新硅谷"的一个很好的机遇，可以发挥香港所长，令中国梦提前实现。斯坦福大学教授有这理研究主要源自，粤港澳大湾区建设是习近平总书记亲自谋划、亲自部署、亲自推动的国家战略。

陈勤业带领美国博士后到国家行政学院政府经济研究中心出席博士后期师委员会交流

陈勤业获聘为国家行政学院政府经济研究中心博士后课题专家委员会副主任委员

陈勤业带团出访

陈勤业带团出访

陈勤业教授表示，2017年4月，国家发改委制订和印发了《2017年国家级新区体制机制创新工作要点》，其中包括了广州的工作重点为深化探索粤港澳合作，之后，由特区政府主要官员组成的考察团前往粤港澳大湾区多个城市进行实地考察，其间会见了广东省副书记、省长马兴瑞等人。他们表示非常重视粤港澳大湾区城市群发展规划，在2017年7月，国家发改与粤港澳三地政府签署了《深化粤港澳合作，推进大湾区建设框架协议》。香港特别行政区行政长官林郑月娥指出，香港将为粤港澳大湾区提供服务，而大湾区亦将有助多元发展，特别是在科技创新方面。陈勤业教授引述十九大报告强调，要支持港澳融入国家发展大局，以粤港澳大湾区建设、粤港澳合作、泛珠三角区域合作等为重点，全面推进内地同港澳互利合作，制订完善便利港澳居民在内地发展的政策措施。这种重大利好，使香港共得更大发展空间。

"打造粤港澳大湾区、建设世界级城市群，推动粤港澳三地经济社会发展迈上新台阶，引领中国实现从经济大国向经济强国的转变，既是粤港澳地区加快经济社会深度调整与转型、实现可持续发展的需要，也契合我国提高全球竞争力和影响力的客观要求，有利于拓展港澳地区的发展新空间、保持港澳长期繁荣稳定；有利于培育可以与纽约、伦敦、东京比肩的世界级城市群，建设高水平参与国际经济合作的新平台；有利于推动'一带一路'建设实施、通过区域双向开放，构筑经丝之路经济带和21世纪海上丝绸之路对接融汇的重要支撑区域；有利于探索建立高标准贸易规则，通过充分发挥港澳作用，探索建立与国际接轨的开放型经济新体制；有利于构建区域合作新机制、辐射和带动泛珠三角地区发展。过去大家一直讲粤港合作，但我觉得看得很，刚好有一个契机，就是港珠澳大桥、大桥一连通，我区的感觉还真是出来了。"陈勤业教授说。

随着湾区建设、城市的行政界会进一步模糊，形成一个密不可分且均衡发展的生态系统。例如，港珠澳大桥通车后，粤港澳大湾区内部各单位的地理距离大幅缩短。陈勤业教授于大桥通车后11月初带领儿位嘉宾到访珠海高新区考察，从香港至珠海高新区只需一小时，可推进粤港澳大湾区城市发展规划和推动广东和香港澳区城经济一体化发展，实现优势互补，创新驱动，都具有十分重要的意义。对腹地更强的辐射带动能力，是世界一流湾区必备素质，粤港澳大湾区要成为带动广东及我国沿海其他地区加快发展的强劲引擎，以此为平台推动珠江口两岸世界级城市群建设，并以强大的辐射力带动西至北部湾、东至油头湾的我国南沿海地区加快发展，打造中国湾区经济标杆。

除此之外，24名在港的中国科学院院士、中国工程院院士于2017年6月

写信给国家主席习近平，表达希望为国家发展创新科技。习近平主席高度重视并作出重要指示："香港拥有较雄厚的科技基础，拥有众多爱国爱港的高素质科技人才。""支持香港成为国际创新科技中心。"国家科技部及财政部已推出措施，准许国家科研项目经费资助香港相关科研项目，国家科技部副部长黄卫于2017年访问香港，与香港各大学校长，以及在港的中国科学院、中国工程院院士会面，商讨加强内地与香港进一步科技合作的具体措施。香港高级教育界争取国家科研"资金过河"已有一段时间，此次措施是一大突破，亦确定香港大湾区规划定位为国际创新科技中心的角色。

陈勤业教授提出，要有序扩大深化两地合作、科技部、财政部已先后召开专门会议，研究加强内地与香港科技合作的举措。目前在港两院院士主要信反映的国家科研项目经费过境香港问题、科研仪器设备入境关税优惠等问题。已基本解决，国家重点研计划已对香港十六个国家重点实验室香港伙伴实验室直接给予支持，并在试点基础上，对国家科技计划直接资助港澳科研活动作出总体制度安排，另外，香港在内地设立的科研机构，均已享受到支持科技创新的进口税收政策，在澳门的两个国家重点实验室与港澳伙伴实验室，也得到了国家科技计划直接支持，下一步将支持爱国爱港科研人员深入参与国家科技计划，有序扩大和深化内地与香港科技合作。

陈勤业教授认为，对于公司的发展而言，有梦想是重要，研发一样新产品，由无变有的过程，靠的就是思想力、有梦想，才可以由材大吹而不放有。否则，急功近利的人，必定会选择走捷径，不外乎互利财技，或为短期利益牺牲长远发展。太多人选择眼前利益，一直坚持理想的人实是寥寥，物以罕为贵，所以少数成功的追梦者才能走遠。

粤港澳大湾区成为"新硅谷"，陈勤业教授说这是指日可待的。现在粤港澳大湾区万事具备，只欠"斯坦福大学"，作为国家行政学院政府经济创新研究中心博士课题专家委员会副主任，陈勤业教授希望能在内地和香港之间搭起桥梁、优化资源，让内地和香港的大学能整合出一所中国的"斯坦福大学"，吸引全球顶尖人才进驻，投入创新科技研究项目，打造真正的"新硅谷"。 🔘

KanlP Philip Chan

Adviser

2017 Advanced Research Symposium @ Stanford, UC Berkeley, CSUMB

斯坦福大学、伯克利大学和蒙特利湾加州州立大学博士后导师2017年出席证

陈勤业著作

TWC MANAGEMENT LIMITED
迪 德 施 管 理 顧 問 有 限 公 司

迪德施管理顧問有限公司簡介
TWC Management Limited

專業企業財務服務 優質核心業務

迪德施管理顧問有限公司
創辦人及董事長
曾惠珍教授

迪德施管理顧問有限公司「迪德施」，於 2000 年創辦，成立已超過 20 年，以商德為本，提供專注的服務包括：

- ·會計、公司秘書、核數、稅務
- ·一般企業架構的顧問服務
- ·財務相關的服務
- ·私募基金集資顧問及管理服務
- ·成立香港及各類離岸公司及私募基金公司
- ·上市公司秘書職責
- ·其他企業相關顧問服務

迪德施的客戶，分別來自不同行業，如製造業、貿易、建造業、工程、運輸、資訊科技、投資、飲食、印刷、娛樂、物業管理及各專業人士，如律師、工程師、財務策劃師等。

迪德施多年來為不同中小企及大集團提供全面可靠而專業的企業管治服務。傳統服務範圍包括會計，財務報告，設立香港和離岸公司架構，公司秘書及相關的法規顧問服務等。另外，迪德施的核心私募基金業務提供基金管理及財務顧問服務，以香港本土的規模，在國際舞台上與國際金融集團分一杯羹。迪德施亦為機構提供首次公開發行股票（IPO）相關的服務，及協助客戶申請證監會的許可

我們的客戶分別來自世界各地，當中包括美國、加拿大、德國、英國、日本、台灣、印度、星加坡、印尼、菲律賓、越南、澳洲等地，而本地的則有香港、澳門及中國大陸。我們的客戶也包括海外及本港上市公司，上市公司的集團公司，上市公司的高層管理層等。

愛心回饋社會

迪德施亦會提供服務予一些慈善或非牟利機構，當中只會收取象徵性的行政費用或甚至完全免費，藉以回饋社會。

誠信為先的「三心」

由於香港位於亞太地區的心臟地帶，設有健全的法律監管制度。可信的銀行體系，完善的基礎設施，故吸引很多內地和外國投資者到此開展業務。雖然香港市場環境優越，但挑戰在於競爭激烈。迪德施正可利用他們的專業知識，靈活套用國際和本土的做法，具競爭力的費用，個人化的服務，幫助他們熟悉香港的營商環境。迪德施以客為本，誠信為先是他們的信念；亦正因如此，迪德施的客戶主要均由口耳相傳的推介而來。「公司向來提倡『三心』，即務求客戶對迪德施的專業能力放心，對我們的辦事效率及優質服務投以信心，對我們操守及可靠品質感到安心。」

oneAs1a
亞 洲 脈 絡

以「工匠精神」打造先進數據中心產業，引領行業前沿

亞洲脈絡（OneAsia）母公司於1992年成立，擁有30年數據中心產業經驗。作為一家植根香港於亞太地區首屆一指的中立數據中心、網絡及雲端解決方案運營商，OneAsia數據中心據點覆蓋中國、日本、韓國、泰國和新加坡等地，致力提供高效能超穩定數據中心、跨數據中心連接、智慧多雲端解決方案及IT外包服務。

OneAsia一直以創新科技迎合市場及客戶需求，精益求精，使客戶無論身在任何角落，都能獲得無間斷的優質服務。回望過去30年發展歷程，創辦人兼行政總裁李松德先生在OneAsia發展過程中深深領略到「工匠精神」營商理念的重要性。

古今中外「工匠精神」都是事業的基礎，中國《莊子》中的「庖丁解牛」、日本「職人精神」、瑞士鐘錶手藝都引證了工匠的堅持專注。總括而言，「工匠精神」是一種職業精神、執著及實踐。李先生如何以「工匠精神」六大要素「專業、標準、精準、完美、創新及人本」帶領OneAsia發展？

如今，IT技術、人工及機器智能、大數據都是新經濟發展的重要一環。作為這些資訊科技產業的基本要素，政府及各機構對數據中心產業制定嚴格標準及法規。OneAsia始終秉承專業精神，嚴格遵守行業標準，並加入不同業界組織，亦創立香港數據中心協會，與業界保持緊密交流。

數據中心、網絡及雲端解決方案講求各方面的精準計算，稍有失誤即會影響服務交付。因此，OneAsia的每一個項目，從建構、運營到管理，每一步都力臻完美，方可提供令客戶滿意的解決方案。

面對需求急增及市場競爭，OneAsia 不斷改革創新。由基礎設施設計建造承辦發展成數據中心運營商，全面發展數據中心、網絡連接及雲端計算。要配合業務演變，人才是重要基石之一。李先生著重人才培訓，掌握資訊科技行業技能和知識，幫助員工與公司同步成長，與時並進，現時公司超過一半的員工擁有超過 20 年行內經驗。

中國總理李克強先生曾於 2016 年全國「兩會」上的《政府工作報告》中提出「培育精益求精的工匠精神，加快培育大批具有專業技能與工匠精神的高素質勞動者及人才，推動經濟轉型和消費升級」。於這個數碼化新時代，OneAsia 將繼續秉持「工匠精神」，持續改善，引領數據中心行業前沿，推動行業發展。

卓越商界奇才
國際博士後協會

廖寶城博士後。

廖寶城博士後
與陳勳棠博士
後合照。

廖寶城博士後
及國際博士後
協會團到訪加
拿大魁北克大
學及會議留影。

廖寶城博士後

學無止境
提供進修機會

所謂「學海無涯」，學問好比汪洋大海，無邊無際，唯有抱着「活到老，學到老」的精神，才可以更了解世界、緊貼時代的步伐。深明此道理的國際博士後協會香港區會長廖寶城博士後，多年來一直致力進修，增進知識，並為年青人提供學習及建立人際網絡的平台，讓年青人一起「學而不已，闔棺乃止」。

撰文：何烱城　攝影：鄧志輝，部分相片由被訪者提供

PRIME AWARDS FOR OUTSTANDING LEADERS 2014

PD
POSTDOC

年青人是社會未來的棟樑，我們希望可以提供機會，讓他們持續進修，傳承香港及中國的經濟發展。

現時為國際博士後協會香港區會長的廖寶城博士後，相當明白學習的重要性。他由文憑讀起，曾就讀香港理工大學、香港城市大學、香港中文大學、香港大學，甚至是英國大學的文憑、學士、碩士、博士及後博士課程，當中大多與電子工程及工程管理有關。

現在，他正就讀一個調解訴裁員的課程，「現時香港與內地，以及東盟地區（包括：文萊、柬埔寨、印度尼西亞、老撾、馬來西亞、緬甸、菲律賓、新加坡、泰國，以及越南）的經貿往來漸趨頻繁，當中可能出現不同類型的糾紛，將會為法庭帶來很大負擔。」而廖寶城博士後則希望可以透過成為調解訴裁員，協助減輕法庭負擔和為訴訟雙方節省金錢及時間。

廖寶城博士後表示，國際博士後協會與不同機構合作，讓年青人持續進修。

廖寶城博士後到訪加拿大首都渥太華國會山莊留影。

團結大中華博士後　助博士後持續發展

除了個人會不斷進修，廖寶城博士後亦相當積極的為國際博士後協會服務。國際博士後協會由陳勳華博士後成立，是第一個將大中華地區的博士後，聚集在一起，推動大中國研究的協會。協會透過提供資源，進行資源整合，加強問責制；制定和實施改善博士後經歷的政策；促進收集，分析和傳播博士後學者的數據，以及建立一個自我維持，民主組織授權的博士後組織四個方面，關注博士後的事業發展和生活狀況，凝聚博士後人才的力量。協會現時分別於美洲、南中國及香港設有分會。

最近，協會更與中國企業家博士俱樂部合作，提供交流機會予博士後學者，讓他們可以與中國企業家博士俱樂部會員、企業董事長、總裁及企業高管等，互相了解及切磋，擴大自己的人際網絡，發展自己的專業。

深明「學海無邊」的道理，多年來一直持續進修。除了於香港多間大學獲得學士及碩士學位，亦於外國多間大學考獲碩士、博士及後博士學位。

提供課程　助年輕人裝備自己

其實協會不僅着力於為博士後發聲及凝聚他們的力量，同時亦會為年青人提供交流、分享及建立網絡的機會。「年青人是我們社會未來的棟樑，我們希望可以常領他們、持續進修，並為他們提供機會，讓他們好好裝備自己，傳承香港及中國的經濟發展。」

故此，協會與不同大學及機構合作，推出多種課程予青年人，讓他們可以持續進修。同時，協會亦與中國的高校保持緊密聯系，讓中國學生亦有機會加入協會。「內地的教育水平己越來越高，高校更會為學生提供不同類型的訓練課程，以提高他們的技能。這些改變令內地學生對我們這類型協會的需要更大。」

進軍東盟、華中、華北　提供專業人才

有見東盟國家發展迅速，其中由東盟國家及中國所組成的自由貿易區，「中國—東盟自由貿易區」（或稱東盟10+1），更成為世界上三大區域經濟合作區。協會計劃在未來於東盟地區開設分會，為當地提供更多專業及高學歷的人才，協助發展當地經濟。

在未來，國際博士後協會亦希望可以將分會拓展至華北及華中地區。廖寶城博士後表示，「中國擁有高知識水平的人才越來越多，我們希望可以透過成立更多分會，凝聚他們的力量，並為年青人帶來一個持續進修及建立自己人際網絡的平台。」M

信誠工程國際有限公司簡介

筆者年少時得到已故爺爺的忠告，叮囑「做人要將目光放遠一點，不要害怕吃虧」，並且筆者的已故叔叔亦送了一本筆記簿給筆者，上面寫著「努力學習，天天向上」，然而筆者中學會考只有 5 科合格成績，當時筆者認識了人生第一位啟蒙老師羅Sir，他是一位在中學任教工科的老師，在家裏設有工場，工餘時間在家創作一些與機械及電子有關的小發明，筆者當時幫他做學師，開始接觸工程界，當其時 90 年代初，社會繁榮，筆者在工業學院修畢工程文憑後，在一間空調工程當繪圖員，晚上繼續在理工學院進修機械工程高級證書課程，受到同事的渲染，認為

羅秉強博士、訪問學者

「今朝有酒今朝醉」，所以不思進取，完成高級證書課程後，並沒有繼續進修，及後不停轉工，最長一份工作是維持了 5 年，其餘都是一至兩年左右，甚至乎幾個月都有，思前想後，發覺學歷不夠是最大致命傷，於是在 2000 年於城市大學修讀消防工程學士夜間課程，並於 2004 年取得學士學位，及後轉往澳門參與興建賭場工作至 2009 年，然後回港到一間機電顧問公司工作，老闆是一位香港工程師學會註冊工程師，並且不斷考取有關工程的牌照，由於當時這間公司是新成立，所以工資低於市場價格，猶記得當時每天中午與老闆一起捱麵包的日子，當時老闆有見及此，將有關營商的思維及方法傳授給筆者，並提醒筆者一定要仿效他，考取工程牌照才是皇道，如水喉匠牌，電業工程人員牌，消防三級牌等，並要朝著專業工程師的方向進發，他是筆者人生第二位啟蒙老師，於是筆者馬上修讀機電工程碩士夜間課程，並於 2014 年獲得碩士學位，期間並考獲上述工程牌照，筆者當時雄心壯志，於 2014 年 5 月成立了第一間工程公司名為信卓工程有限公司，當時筆者想申請消防一及二級牌，於是筆者又繼續進修電機及電子學位夜間課程，並於 2016 年修畢課程，於 2017 年成功取得消防一及二級牌及電業工程承辦商資格，公司取得有關牌照後，生意續有轉機，並且得到父母，現任太太及朋友滙哥的背後支持和鼓勵，壓力得以舒緩，而當時有幸認識人生第三位啟蒙老師陳勤業博士後導師，繼續「持續改善」，隨即修讀土木工程博士，並於 2020 年修畢博士課程，並加入國

際博士後協會，現在正參與美國加州州立大學聯合斯坦福大學，伯克利加州大學三大名校聯合網上訪問學者項目，並朝著博士後進發，同時加入成為大舜基金智囊團成員，另外筆者現正申請通風專門承建商牌照及修讀土木工程課程，打算為公司考取通風承辦商資格及一般承建商資格，個人方面亦已申請註冊專業工程師資格，貫切本書「持續改善」的方針。

有幸得到公司彼鄰大師指點，於 2020 年 11 月，公司正式改名為信誠工程國際有限公司，希望用信心，信譽及熱誠，誠意，繼續在工程界為客戶服務。信誠工程國際有限公司為一間迅速發展及成長的綜合式屋宇設備專門承建商，為客戶提供全面的機電工程服務，其中包括不同類型樓宇的各類空調系統，電器裝置，消防系統，給排水，保安系統，低壓電安裝，弱電安裝，水泵工程及樓宇自動管理等系統的設計，供應，安裝及維修保養。客戶包括不同的政府部門，公共機構，學院，非牟利團體及私營機構。

信誠工程國際有限公司是香港少數擁有多項專業認可牌照及同時擁有自己工程隊伍的工程公司，憑著直接採購減低分判成本及運用先進系統管理，為客戶提供優質及價錢具競爭力的工程服務。

事實上，隨著我們擁有的專業團隊及不斷精益求精的敬業態度，令我們對香港的業務發展充滿信心。所有員工均按照品質管理程序的要求，於合約指定的要求質量及工期內準時完成工程。信誠工程對客戶服務極為重視，為確保水準達至理想程度，我們積極地為員工提供培訓計劃，令各員工潛能得以盡量發揮，務求以客為尊。展望未來，信誠工程國際有限公司期望成為香港主要具規模機電工程及維修保養服務公司。

可喜的是近來我公司的員工，身邊的朋友，同學，也受到筆者的渲染，仿效筆記考取個人的工程牌照，而筆者也樂意給他們提供相關的考牌資料。

栢昇規模化背後的營商成功關鍵

李家良博士

栢昇企業有限公司
創辦人兼董事
油尖旺愛心基金會 大會副主席
北區消防安全大使名譽會長會 副主席

栢昇企業有限公司

　　成立於 2008 年，栢昇提供一站式專業電機工程服務，同時亦是香港註冊消防裝置承辦商。栢昇業務涵蓋顧問、安裝、設計、保養及維修等，在工商業大廈、數據中心、住宅、醫院以及購物中心工程皆有豐富經驗。近年栢昇致力擴展業務，包括滅火筒環保回收及進行水壓測試，亦有為客戶提供水力工程服務，務求鞏固更長遠互信的客戶關係。

創業時期

　　作為栢昇的創辦人，我從事消防系統行業有逾二十年經驗。回想創業時期（2008 年）正值金融海嘯，眼見經濟不斷下滑，我決定背水一戰，把握理想時機出外闖一番事業。然而，在創業路上一方面要面對激烈行業競爭壓力，另一方面亦要應對不同難關，包括資金周轉問題、生意下滑、員工有工傷等情況。幸得太太全力支持及一班同事不懈努力，公司最終得以走出低谷。在數年間，由一人代理氣體滅火用具，到不斷進修消防工程知識，栢昇得以逐步建立客戶群，擴充規模，穩步發展。

尋求突破

由一人公司到現時過百名員工，栢昇經歷重大轉變，而我自己的角色亦要因應公司管理規模而調整。起初我需要凡事親力親為，處理公司大小事務，每日接見近七名客戶並要

聯繫不同廠商確保供貨穩定。隨著業務成型，公司開始聘請更多專才負責不同的工作範疇，太太負責日常營運管理，自己則轉為主力開拓業務。我認為這便是應用了「因時而變，隨事而制」的管理哲學。作為營商者需要根據時期而改變自己的策略並隨著事物不同的發展方向確立不同的管理方法。

營商成功關鍵

我一直信奉兩個重要營商價值，「以客為尊」以及「與時並進」。論及「以客為尊」，公司最大的目標是為客戶提供務實、可靠的方案，發展可持續關係。縱使公司業務不一定出現爆發性增長，但機電消防工程屬於必不可少的服務。因此，必須更用心、誠懇、踏實地一步步經營，一絲不苟，取得客戶信任，方能持續穩定發展。栢昇的一站式服務便體現了有關理念，當客戶需要為建築物設置全新系統時，栢昇為減低工程成本和便利客戶進行管理，能夠從設計、施工、安裝、保養及維修全方位跟進。有些時候，客戶會提出新需求，栢昇會本著「幫到就幫」的務實理念，開拓新服務板塊，轉化為公司吸納新客源的動力，同時滿足客戶需要。至於「與時並進」，我經常鼓勵同事運用新思維解決問題，建立不斷進步的企業文化。在營商方面，栢昇配合市場發展，開拓嶄新服務。例如，察覺到傳統建築方法效率低，栢昇於是積極發展預製技術，務求達到製造及建築一體化。李博士亦充分展現行動力，短時間內於元朗設置工場，大力發展球墨鑄鐵水管（D.I）預製件技術。

展望將來

市場和科技正不斷變化，每一位營商者都需要具備前瞻性和勇於創新的精神，應用創新概念，不斷更新成長。在未來的日子，我將繼續帶領栢昇擴展服務板塊、引入新元素，為客戶提供更優質一站式服務。

李明漢 CPA, CGMA

- 亞太區公用事業、新能源與綜合性大企業證券分析師
- 特許全球管理會計師
- 美國注册會計師
- 佛蒙特州注册會計師
- 香港証監會（四號牌照人士）
- 香港醫療輔助隊顧問
- Bridge Foundation 董事及創立成員
- 香港青年會（2014-2017）副主席

在再生能源領域工作已有 20 年的分析經驗，了解過無數能源環保議題和研究過無數能源公司如何發展可持續再生能源，各國機構都不斷在能源轉化和環保上精益求精，達到減低碳排放。氣候變化（Climate Change）對世界的影響已經日趨普及明顯，最大原因導致全球溫度上升就是「碳排放」，因此人類生活若然想達到永續性，絕對有賴於循環天然資源的應用。

早在 2020 年，中國已提出爭取在 2060 前達到「碳中和」，當中存在複雜的路徑及規劃。

挑戰包括儲能技術、氫氣應用、完善電網建設、「碳捕捉」（Carbon Catch）等，面對種種的困難都需要時間讓成熟的技術量產，才能達致完全取代化石燃料發電。中國是第一個國家提出「碳中和」目標的國家，無疑亦令其他國家加速減排的進程。而 2021 年 4 月的時候各國也就環保議題上達到重要的共識。這讓我意識到，既然世界各國對這個議題都十分重視，並且目標一致的，可見中國領先的減排領域與其他發展中的國家展開經濟技術合作的可能性和機遇，在未來的 40 年的經濟發展目標上是絕對不容忽視。

當然，利用再生能源並達到減少碳排放及減慢氣候變化是一個世界性及專門的技術議題，不是一般大眾可以左右的事情。因此，我一直思考著，在我工作上如何讓接觸我的投資者對這個再生能源領域有更大的了解及關注，建議適當的投資，是我首當其衝的任務。我相信經濟上的關注及對相關機構的投資，方可喚醒更多公眾對關注減碳及氣候轉變的重要。

　　既然「持續性」發展是我一直研究的議題，在工作以外的範圍，如何做到「持續改善」生活習慣而幫助減碳也是我無時無刻提醒自己的事。作為一般市民的我，對環保減碳可作出的承諾及呼籲：

（一）乘搭電力推動的交通工具

（二）關注氣候議題，傳播分享減碳的重要

（三）留意能源政策，作理性投票

（四）減少使用塑膠

（五）選擇良心畜牧企業

（六）推廣中小學的環保教育

　　「氣候轉變」帶來的影響和「碳中和」計劃，看似與生活息息相關又遙不可及的事情，這個議題，已經不是我們多用或少用一兩個塑膠產品可以逆轉的事情（當然我不是在這裏宣傳大量浪費塑膠），「氣候轉變」對地球的而且確有著非常深遠的影響和對我們的環境造成巨大的改變。在我研究領域中，了解到國內一些城市已經實實在在對「氣候改變」作出及進行重新規劃。相信這個減碳議題在經濟、民生及政治上，未來十年至廿年間將會佔非常重要的部份。

徐俊英 先生

管理學博士，研究生導師，英國擎天藝術產業基金主席，在推動「藝術＋金融發展」之外，他與北京大學、中央黨校、斯坦福大學、牛津大學等合作，主研藝術產業金融發展方向。

藝術，是一種反映現實、寄託情感的文化，它具有欣賞價值、收藏價值、文化價值、哲學價值、歷史價值等。

金融，是經濟生活中，銀行、證券、或保險從業者從市場主體募集資金，並借貸給其他市場主體的經濟活動，從廣義上說，政府、組織、個人等市場主體通過募集、配置和使用資金而產生的所有資本流動都可稱之為金融。

從定義及涉及面看，二者之間似乎並無關聯，然而突然間的某一天，它們碰撞、交融，有了同行的軌道，於是，一個新興的詞語——藝術金融開始出現在大眾視野。

經過十多年對藝術、金融的研究及沉澱後進入藝術金融領域，並創立了英國擎天藝術產業基金，基金目前主要分為金融及藝術品資產管理。

我們深知藝術市場的金融化探索不是一個孤立的資本運營模式，它關係著藝術市場的運行，關係著藝術家的創作，關係著藝術品藝術價值的實現。

而從目前中國藝術品市場的發展現狀來看，必須經歷也正在經歷藝術品的商品化、資產化和金融化過程。在此過程中，商品化是基礎，資產化是關鍵，金融化是戰略方向。

「然而中國的藝術＋金融的發展不夠快速，也不算完善。」這是中國藝術金融市場的現狀，也是需要借助民間的力量，一方面促進其快速成長起來，另一方面為實現中華民族偉大復興提供些許助力。這是本公司心中最大的願景。

也因此，我從二十年前鍾情於中華文化藝術，到五年前創立英國擎天藝術產業基金，一直不曾停歇的是致力於推動英國擎天藝術產業基金的成長、成熟與完善。

管治身思維

許華達博士

高級營養保健師

北歐大學公共行政學系 榮譽博士

健型堡 Wellness Treasure 創辦人

中正會計師事務所

中正顧問有限公司

許華達顧問有限公司 董事

香港成長教育協會 主席

　　能與一群頂級學者及企業管理翹楚聯賀「持續改善－頂級企業實戰秘笈」出版，甚覺榮幸！並獲在此分享一篇個人體驗，更深感殊榮！感謝陳勤業教授錯愛邀請。

　　A.I. 人工智能、5G 面世，世界正以另一個更 Cyber 面貌發展。「持續改善」作為不斷提醒自己、企業是要持續而不斷以「新思維」改善才能續領風騷。心興讀者們在此書內以可觀摩多位頂級大師的管理秘笈，增見廣益。

　　我看見，良好的企業，不論是超級巨企或大中小企業，其管理層必定構建出一套健全管理「法、規」以督導推進企業前行，縱使其中有何重大人事變換或面對危機，堅實的管理「法、規」都能成為企業保護盾牌。然而，我又看見，眾多管理者對自身管理都顯示得鬆懈，有些更覺得次要！終於，巨人倒下，他的健康都與他揮手道別！這一下子他一切心血建立的事業、財富、更甚的是生命也離他而去。企業幸存，唯是個人遠去！而最為人惋惜的莫過於蘋果教主 Steve Jobs 史提夫 · 喬布斯辭世！賭王何鴻燊博士因病而放下其事業王國！還有更多鮮為人知的因管理自身健康不善而喪失一切的抱憾者！

優良管治「法、規」會嚴謹要求企業遵守，但對自身的管理卻不以為然。他們對於事業發展，企業及科技創新有超百份百投入。但對自己健康往往高估自身能力，同時又低估慢性疾病對身體的破壞力和殺傷力！身為營養保健師，我有責任提醒大家：根據世界衛生組織 WTO 調查結果全球人口只有 5% 是健康、20% 左右處於疾病狀態、大部分若 75% 在亞健康圈內。同時香港大學在 2013 年初一項調查更顯示 97% 香港人曾出現最小一項亞健康症狀。身體管理殊不簡單，小小心臟每天為身體輸送 8000 升血液、全身血管全長超過 10 萬公里，可環繞地球 2.5 圈、腸道含菌量超過 100 兆。切莫等到癌症、心血管疾病、糖尿病、柏金遜症等慢性殺手找到上門時才抱憾！

各位優秀的企業領導者，在這超世代時刻，「管治身思維」才是對自己、對企業、對人才的第一身保護。

健型堡 Wellness Treasure——您的體健協調夥伴。提倡讓食物取代您的藥物，以營養建構您的健康。

對個人及企業團體提供營養、健體專業意見服務。

支持中小企業可持續發展，不僅可以提供就業機會，更可為社會創造更多經濟效益。

事實上，國內外許多著名的巨頭企業都是從中小企業不斷成長壯大起來的。一直以來中小企業對社會的貢獻確實功不可沒。

企業在追求可持續發展的過程中，企業和管理人員都必須要與時並進、持續學習、持續改善，方能持續發展。

過去 30 多年我一直在一家擁有 56 年歷史製造行業的內地港企任職副總裁，負責企業管治的工作，見証從一家小型工廠發展到今天一家多元化的大型企業，期間經歷過行業的盛衰，兩次金融危機，2008 金融海嘯，工廠轉型，2019 新冠狀病毒等等的問題，面對過不少困難和挑戰。但我們都能迎難而上，克服困難，渡過每一個難關，並達致可持續發展的目標，而且賺取了不少珍貴的管理經驗。雖然現在未能在這裏將我們每一個實戰經驗作出分享。但我非常樂意將我個人認為最重要的三點先行與大家分享：1. 現金流管理；2. 營造和諧職場；3. 避免打官司。

郭慧芳博士
香港知識產權顧問有限公司（IPAS）大灣區企業顧問、會計師（英國、澳大利亞）、保險師、仲裁員、調解員

首先，我會嚴謹監控全集團的現金流量預算及現金流量的實際情況，用以協助公司制訂投資或發展策略。（事實上有研究報告顯示有 80% 企業倒閉原因，就是現金流管理不善，導致資金周轉不靈所致）。因此，現金流量報告比損益表更為重要。其次，我把調解技巧應用於日常管理，發現對維持客戶、供應商、上司、下屬、工人等，對保持有效溝通和良好協作關係具有明顯改善。並利用調解理論和技巧營造和諧職場，例如：企業與員工彼此互相尊重，互相欣賞，關心認同，關注各方的利益和需要，訂立共同目標，同心協力追達各項生產指標。團結互助，戰勝每一個困難。建設企業和員工都有更好的經濟效益。員工做得開心，自然不會轉工。當員工樂意為你效勞，便會增加歸屬感和忠誠度，生產力自然會提升。企業能夠留住人才，減少員工流失是可以減低因招聘及培訓新人的成本，從而可提升企業利潤。（事實上有研究報告顯示一個員工的流失給企業帶來的財務成本相當於他一年的薪酬的 213%）。

最後，我建議不論企業或個人避免打官司。因為打官司會帶給各方當事人極大的精神壓力，因而影響日常工作和生活。打官司須將所有資料公開，故不論審判結果贏或輸，對雙方企業或個人都有負面的影響。萬一企業輸了官司，更可能因此而陷入財困。

我為了配合公司發展，一直持續學習，持續改善，並已發展到其他的專業。近幾年，我更有機會參與深圳前海法院港籍代理人陪審員，以及在內地法院參與跨境商事糾紛及勞動爭議的訴前調解工作，在處理案件過程中，發現若企業能夠加強管理，加強與員工溝通保持良好勞資關係，這是可以避免許多爭議發生的。有見及此，我近期應香港貿易發展局邀請，在他們網上平臺開講公益性的研討會，並以實務和案例形式與內地港企分享管理經驗（內容包括：港人在內地發展稅務安排；勞動法須知、勞動爭議預防及解決方法；如何善用中港兩地相關保險保障企業和員工等）。我希望籍此能夠幫助企業和個人風險預防及制定更有效的管理制度。此外，我於 2014 年以調解代理人身份之一參與深圳國際仲裁院「首宗以香港促進式調解，並以和解協議書作出仲裁裁決書的跨境商事糾紛的調解成功過案」，涉案金額 5000 萬人民幣。另於 2018 年應東莞第二法院邀請以線上方式調解達至和解「首宗涉港人跨境勞動爭議」成功個案。並於 2019 年應東莞市勞動仲裁院邀請與全市 300 多名勞動仲裁員授課，介紹香港勞工法例及分享粵港澳大灣區的調解成功個案。

我目的是希望透過經驗分享或培訓，除可幫助中小企業「可持續發展」外，更可以促進兩地的企業和個人溝通交流，互相瞭解，減少誤會分歧，減少爭議，減少訴訟，共同建造更美好的營商環境以及和諧工作生活環境。

LONG Engineering Limited 信豐工程有限公司

管理人的商道及五事

信豐工程有限公司建立於 2013 年 至今八載 承蒙各方友好行業前輩委託信賴 業務涉獵不同公營及私營土木工程項目 營收越億 聲譽日隆 建立起 200 人施工團隊 機械設備投資超越千萬 在近年波動的市場環境氣氛當中 能夠自強掙扎 發展至今實屬萬幸

筆者亦為信豐創辦人 一直學習及思考管理哲學 其中孫子兵法有云 兵者詭道也 在瞬息萬變及市場競爭激烈的情況下 管理人如何在詭道中 思想提升 創造價值 建立商道 實為重要

商道不只是為賺取利益 應是經商過程獲取人心 賺取信任 祈求合作共贏 令到客戶滿意而公司亦能夠成功營運 火中取栗 實屬為大原則

如何實踐大原則 應當關注執行細節 是否對於每一次的機會客觀分析 筆者認為可以參考孫子著之五事 道 天 地 將 法 作為綱領 量度每個機會虛實 審視每個處事方法

道就是指出天地萬物 每人每事有它的自然規率 跟宇宙自然配合而發生的事物就是道 違反了自然界及道德的事 就是違反了道 不得道者 成功的機會杳然 就算成功僥倖獲得 爭取成本也異常高昂 例如希望獲得一個商業機會 做違反道德犯規之事 最後只會得不償失

天就是指天時 四季轉換 天陰天晴 月全月缺 下雨還是颱風 都是大自然現象 如果將工作安排跟四季轉換配搭適宜 就會事半功倍 例如將室外工程安排在旱季時節舉行 進度定必會有幫助 否則亦然 在雨季風季大興土木 定必影響進度 效益減半

地是指地形地勢 筆者立足工程行業 對此感受更深 例如工程在偏遠地區 交通不便 水電不至的地方 成本一定高昂 由於相關配套極難安排 偶爾遇到不順利的情況 連救亡也非常困難 所以要對於自己團隊的狀態非常了解 能夠克服什麼地形地勢 選對戰場

將就是人才選用 就算是多麼完美的系統 人的參與及操作也是關鍵的一點 將相之才不單止要德才兼備 也要有強大的心理質素 對於突如其來的困難以及不幸 有所擔當 不半途而廢者為優 筆者認為 真正的人才將相 要有遠大的目光 廣闊的心胸 也要有靈巧的手腕 表現精準 亦要持之以恆

法就是戰術以及方法 做事的方法 處事的手腕十分重要 方向是對了 方法做錯了 效益自然減少 所以論述方法最好是集思廣益 根據以往的經驗 以及創新的思維 定立破格的方法 對於特定事件 更需要因時因地制宜 筆者曾經面對合約爭議 最後由於了解對方心中所冀 建議雙贏的商務合作方案 代替爭議內容 結果合作共贏 各取所需

筆者與信豐團隊過去八載 一直與各方持分者於行業友好協商 順勢而立 發展奮發有為的工程公司平台 當中體驗了孫子之著 對問題多角度檢視 用心推進 每事定當迎刃而解 創造自己獨特的商道 與客戶並肩向前 抵勵前行 創造未來

ETR
Law
Firm
廣信君達律師事務所

中國股權投資之「對賭協定」藝術

伴隨中國市場「大眾創業，萬眾創新」時代的深化與演變，股權投資中設置「對賭」與「回購」條款已成為平衡投融資雙方資訊不對稱和不誠信等風險的一種常見交易安排。

從某種角度而言，對賭協定實際上是投融資雙方對目標公司未來估值偏離預定軌道的一種預設調節機制，其獨特之處在於該機制一般須在投資談判階段便由雙方進行商定並形成具有法律約束力的既定規則。

對賭條款的核心目的是對沖股權投資市場化風險，調整目標公司將來某一個時間點的整體估值，當對賭條款被觸發時，投資方可以通過回購條款退出目標公司或通過股權補償規則進一步控制目標公司，保障自己的投資收益和控制風險。

羅國蔚 博士研究生
執業律師
中國註冊金融分析師
上市公司獨立董事資格

對賭協議中最常見的可量化指標通常包括經營性指標（指營業額、淨利潤、扣非淨利潤等）與非經營性指標（指企業上市、市場佔有率、研發成果、產量等）。在對賭期限內，投融資雙方將已審計的公司財務資料與之前預設的經營性指標或非經營性指標進行對比，當目標公司超額完成預設指標時，說明目標公司股權現估價值與實際價值相符甚至低於實際市場價值，投資方獲得超額回報；當目標公司財務數據未能如期達到預設指標時，說明目標公司股權現估價值高於股權實際市場價值，投資方預期收益目標存在減損風險，此時融資方應當以現金補償、股權回購或者股權補償等方式向投資方進行補償。

2019 年 11 月，中國最高人民法院第九次全國法院民商事審判工作會議形成會議紀要（簡稱《九民紀要》），《九民紀要》的出臺以及近年來司法實踐中形成的一系列指導性案例，明確了對賭協議效力、回購條款適用的裁判規則，為私募股權投資多元化、創新化發展提供重要法律保障。

駿盈建築有限公司 **BILLION SMART** CONSTRUCTION CO., LTD.

駿盈建築有限公司
一站式助工商企業太陽能發電，
共建綠色經濟

馬偉成博士

博士後研究員

英國皇家特許測量師

駿盈建築有限公司創辦人兼董事

Sir Dr Ma Wai Shing, Kenny KTS KCMA

Founder and Director, Billion Smart Construction Company Limited

- 特許建築測量師—皇家特許測量師學會

- 屋宇署註冊一般建築承建商之獲授權簽署人

- 綠建專才—香港綠色建築議會

- 資深會員—香港建築業仲裁中心

- 認可建築調解員—香港建築業仲裁中心

- 專業會員—香港策劃工程師學會

- 國際博士後協會會員（MIPostdocA）

- 百位傑出華人楷模

　　隨著近年全球暖化問題迫在眉睫，已有不少媒體、非政府組織已不斷提倡綠色生活，減少浪費等等。以香港而言，政府近年亦不斷推行不同的環保政策，例如膠袋稅、「藍天行動」計劃，年前更設立上網電價政策之上網電價計劃，向電力公司售賣所生產的可再生能源。

特區政府推動全民產電的「上網電價」計劃自 2019 年初起接受申請，節能顧問業務頓成市場機遇。駿盈建築有限公司（簡稱駿盈）董事馬偉成（Kenny），認為港人對可再生能源依然不太熟悉，為提倡更多人使用綠色能源，公司在數年前建構專業團隊，進軍節能市場，為客戶度身訂造全方位的安裝太陽能發電系統的專業服務，提供最具效益的解決方法，為保護環境盡一分力，締造社會共贏。

隨著上網電價計劃出爐，根據可靠數據，截至 2020 年，已收到超過二萬宗申請，據悉當中包括鄉郊的小型屋宇、私人獨立屋等已率先參與計劃，而不少企業、學校與機構市場相繼安裝太陽能板，市場逐步擴大。本公司於 2014 年成立時主力建築工程業務，為屋宇署註冊一般建築承建商，承接各類型大小工程，在行業內早已建立起一定的聲響；隨著政府推出上網電價計劃，公司決定組織為數逾 30 多人的專業團隊，決心進軍這個藍海市場。

自 2018 年底上網電價計劃推出，節能公司像雨後春筍般在市場出現；駿盈建築有限公司憑著專業專注態度立足市場。香港地小人多，樓高密集，相比其他歐美國家，在本港架設太陽板的難度可謂不低，如夏季受到颱風吹襲、商廈林立、天台的面積狹小，及日照角度等，為搭建工程帶來挑戰。本公司擁有自己的專業團隊，如結構工程師、電力工程師，而本人則建築測量師，若客戶在安裝方案上遇到任何問題，我們無須把工作外判，由報價、其後的勘察、設計、安裝及測試，本公司的專業團隊為客戶詳細分析，能為客戶真正度身訂造，為其選擇最適合、效能最高的解決方案，讓他們安枕無憂。因此，本公司在行業上累積了不俗的口碑。

駿盈建築有限公司成立至今，已累積逾數百的客戶。本公司先後獲得 ISO 9001 質量管理，ISO 14001 環境管理，ISO 45001 安全管理，ISO 50001 能源管理體系標準的認證，確保所進行的項目達致最高系統管理標準。隨著公司在市場上口碑載道，除一般小型屋宇市場外，公司已進軍低密度住宅、獨立屋市場，逐步開拓商業樓宇，大型工廠及工廈天台也是公司主要目標。相比如新界小型屋宇，工廈及大型樓宇涉及安裝支出及難度（如業權問題）等較高。舉例說，曾有客戶在設計解決方案時，發現天台防水層出現滲漏問題，本公司的專業團隊便迅速解決問題，為客戶免費贈送及更換防水塗層，提供了貼心的增值服務，無形中幫助了業主減輕不少壓力。

另一方面，本公司為客戶提供不同的融資方法及合作方案，解決創建系統龐大的資金問題。不少業主會考慮到太陽能發電系統的前置成本而卻步，成本會因系統發電容量及太陽能版技術等而有所不同，金額由數十萬至千多萬元不等。公司備有不同的融資方案，無論由業主以全部整付、電費共享分成、或以月供形式購入太陽能發電系統。業主可自由選擇。透過各式靈活及融資方案，既可為環保出一分力，又無需動用流動資金，輕鬆踏出低碳生活第一步，共建可持續發展的未來。

本公司在完成安裝工程後，項目團隊時刻與客戶保持緊密聯絡，確保太陽能供電系統維持穩定。而客戶更可透過手機應用程式，讓用家可隨時監察收入及節能報告，檢查發電效能及系統是否運作正常，客戶更可透過公司提供不同介面的客服平台，獲得全方位支援。

至於未來一年的大計，本公司仍以本地市場為重，除建立網上平台外，在線下物色工廈天台設置小型太陽發電實驗場，及在地舖開設實體店，讓市民親身體驗再生能源發電的效果。在市場開拓方面，除了一般商廈外，酒店、學校及機構都是其服務對象。

除積極爭取市場的同時，本公司會肩負起社會責任，向如學校及機構加強在環保教育、上網電價的資訊教育，與市民齊心共建綠色香港，為保護地球有限資源出一分力。

劉銘豪（Ritz Lau）管理心得及經驗分享精華

眨眼間已從事創科，機械人，自動化設備及產品設計已經三十多年，管理及技術的執行都離不開一個「彈」字。這個彈字有極其豐富，深度及闊度的意義。

彈斥（criticize，accuse，impeach）。還記得很清楚約十年前我從美國回流香港幫前任公司負責開發及量產最新一代電腦資料貯存磁頭，當時在現有舊廠房加建了一個無塵實驗室，經過九個月的艱苦鑽研開發及試驗，成功率維持不變停留在絕對零的水平，我的百人工程團隊已感乏力，所有人對我的技術及管理能力抱有懷疑，批評及建議聲天天不絕。最後顧客（Oracle）首席技術官 H 君（也是我的前任顧主）從美國飛到中國商討應對。當時心知不妙，依循美國模式，最可能發生的就是彈斥及換人（先彈議後彈黜）。開會前段就是解釋所有試驗及結果，後段只有 H 君的一個問題「What's next?」。我已預備的答案也只有一個，就是構建另外一個新無塵廠房，將所有設備搬往新場繼續試驗。這新無塵廠房會比舊無塵實驗室貴二十倍。當時 H 君立刻在我老闆前拍台指着我大罵，You want me to loss my job? 我只抱着執包服心態對他說 You have a better option? 我解釋晶片內有最少一層約 30nm 厚的濺射薄膜（sputtering film）在後加工制程中經原子力顯微鏡及掃描式電子顯微鏡觀察下好像消失了，懷疑是被空氣中的負離子侵蝕。新無塵廠房會裝置多層化學過濾，隔絕化學侵蝕。最後，我的分析對了，一個月後有零的突破成功率為一個巴仙，一年後達到七成。

創科路上被彈斥是少不免，態度上須要冷熱兼備來面對問題，那就是冷靜分析及熱愛科研。一點傲氣有時是需要的。在美國工作謙虛是不需的。

彈藥（ammunition）。一般是指資金。攪創科的第一考慮點，就是有無足夠彈藥苦戰兩年。

很多時候跟創科初哥朋友分享創科心路時我會先了解他是屬於大公司還是小男人（個體戶）的情況。大公司彈藥充足，創科程序有規有矩，不需深入討論，唯一要注意是將自己的創科意念或樣板要用書面文字或圖片記錄及公開傳閱，儘量避免口頭建議，減少日後創科或升職或權益爭奪。

如果是你是個體戶，有一份正常收入工作，除了有創意又有點技術外，又無財又無勢又無富爸媽，但是很渴望創科，這樣要首先尋找方法籌集彈藥。自己儲蓄，太太／外父儲蓄，眾籌，政府／大學基金，天使／創投基金，百樂武士，樓宇加按，工作加班，朋友／股東集資，專利交易……都可考慮。彈藥供應鏈未清楚前，決不可遲工尋夢。借錢或買樓攬創科絕不可取，如果採用朋友／股東集資，股權分配必須有一個大股東作決策。平均股權就等如創科失敗。

小男人攬創科因彈盡糧絕而終止創科項目非常普遍，所以彈藥儲備是整個創業夢成真的催化劑。彈票可以嚴重影響聲譽，切忌。

彈是「弓」在「單」（bow + oneself）。弓是由你單人控制的工具，別人幫不到你。弓加箭要適當配合才會實現預期效果。弓就是你個人科技能力，箭就是彈藥，是發出去的。在籌集彈藥時除了要不斷加強個人功／弓力外，還要另外人知道你的功力水平，及保護你的功力避免偷走。這樣就要靠知識產權保護了。

最後要提的就是「英雄有淚不輕彈」。個人創科的成功率不高是現實，全球皆是，不只香港。一兩次失敗是正常的，成功往往都是累積多次失敗的成果，創科人一樣有血有淚，因失敗而流淚甚至彈淚是健康的，值得支持。與家人及老友分享成敗得失可能會獲得意想不到得益。

想做事不錯很容易，就是不做。做多錯多是自然的。需然老闆不愛聽，但是你也不需要告知所有錯誤。現今世界講求效率成果。所以我的管理哲學是，要多做多試，可接受多錯，但是要錯得快，然後改得快。

版面所限，未能在此分享彈性（flexibility），反彈（reaction），流彈（stray bullet），彈奏（flicking）等等，有機再續。

智慧城市科技 (AI+eVTOL) 塑造 新時代新都市！

全球都市人口正趨集中化，一連串城市問題將愈加嚴重。新時代智慧城市解決方案將會有甚麼新興科技可以更有效地處理這些都市病呢？

陳德明
駿盈數碼有限公司主席

在智慧城市解決方案中，最亮眼的當然是 5G 網絡莫屬。智慧交通或智慧工廠等應用均需依靠 5G 網絡支撐運作，全因 5G 不但連線速度高，更擁有低至 1 毫秒（千分之一秒）的延遲時差，無人自駕車或智能機械人可在 1 毫秒間向雲端的 AI 人工智能回饋現場數據，然後 AI 又可在 1 毫秒內將運算結果發送回來，讓用戶端系統可在最短時間內作出應變反應。

可是，當雲端運算普及後，科創界隨即發現，建基於雲端的物聯網架構已追不上了，所以近年有人提出「邊緣運算（Edge Computing）」技術概念——在雲端與用戶端之間，加多一層更靠近數據來源的「Edge」端運算層，把運算工作分散到路由器、閘道器等設備去處理。

邊緣運算的具體實現，正是「Artificial Intelligence of Things」的概念。這其實是 AI（Artificial Intelligence）與 IoT（Internet of Things）的高度融合，除可協助企業發展出新的營運模式外，還可優化智慧城市各層面的運作，包括：智慧出行、智慧家居、智慧零售、智慧工廠、智慧醫療等。

有了 5G 與邊緣運算後，當中智慧出行、無人車的技術基礎便可變得更加鞏固。不過，無人車只是智慧出行發展的起步點而已，未來都市交通工具的最終表現應該是垂直電動起降的飛天車（eVTOL Flying Car）。韓國車廠現代汽車（Hyundai）與 Uber 已經聯合發表飛天的士「S-A1」，目標是在 2023 年投入服務。垂直電動起降的飛天車（eVTOL Flying Car）估計不久將來在粵港澳大灣區投入服務，為大灣區同城化帶來強大的支撐，不少城市亦正規劃可供 eVTOL 飛行器升降的設施。

5G 網絡、邊緣運算、AIoT 智聯網、無人車，以至飛天車等新興科技，將會為下一波智慧城市革命帶來轉折點，接下來，人才的培育成為最關鍵。新一代青少年需要突破傳統框架，以智慧城市應用為目標進行創新，來豐富今天的傳統教育模式，這作能為科創產業向前推進。

　　除了跟學校合作、輔導鼓勵學生們發明專利，更需要引進全球更多的人才，香港更需要發揮超級聯絡人的關鍵角色，讓國際產業人才、科技創新人才，落戶香港再引進內地，加快企業成長，為國家數碼產業引進國際資源要素，讓國家數字碼產業人才集聚。只有這樣，新時代智慧城市科技才能可持續地做福全人類。

聯合教育投資集團簡介

集團 顧景

聯合中外教育集團，合作共贏，創造價值。

集團 價值觀

聯合力量，利他（客戶，合作夥伴）共強，回歸教育本質，為客戶解決根本性教育的需求。

集團 業務範圍

聯合教育國際管理基金開發，發掘，投資，營運優質的教育項目，並對有短板的教育體系作診斷，對接聯合，並提供資金優化產業，繼而提升其效益。我們致力發展學前教育，中小學教育，職業教育，高等教育及海內外教育產業併購，務求以全產業鏈佈局優化內部教育資產。

集團 展望

聯合教育國際管理基金銳意收購，引進國外優質的教育品牌及資源並提升及整合國內教育資源，以併購，對接聯合或投資融資的形式協助中國內地教育機構提升至國際水準，達到產業升級的目標。

為了強化投資項目管理，被併購企業診斷及矯正，教育產業研發，師資，課程，教材素質管理，從而達到系統性增值，基金下面會設立聯合教育管理集團技術性統籌及整合產業。

我們展望未來會繼續投放資源，投資國內外優質的教育產業，更牢固的抓緊全球教育機遇，併購，整合，研發教育產業，提升教育產業佈局，為中國的教育產業走出中國奠定基礎。

U 聯 - 馬可孛羅會

United Platform Marco Polo Club

為聯合教育投資集團旗下福利平台與人力資源平台

1. 致力打造一個營銷聯盟，為聯盟會（包括集團股東，董事，員工，合作夥伴）提供福利、創造品牌，招募人才。

2. 致力為聯盟會員鞏固基礎，強化客戶的黏性。

3. 致力為聯盟會員提供政策性及前瞻性的一個閉環系統，保障其會員在其業界的權益。

4. 致力為聯盟會員強化個人品牌及營銷魅力。

5. 致力為聯合教育投資集團 2200 股東，董事與同事，U 聯馬可孛羅會 4,000 個保險及金融營銷團隊及 400 個健康產品團隊會員提供福利。

6. 致力為集團打造一個獨家和閉環的福利平台。

7. 致力為 U 聯馬可孛羅會的同事以及客戶提供爆炸性優惠。

8. 致力提供人力資源平台以閉環形式輔助各聯盟團隊招募人才。

9. 致力提供免費或以成本價製造一些具業界前瞻性、政策性以及方向性的技巧、資訊及課程，以助其制定目標。

10. 致力提升團隊及平台的名氣、活動及知名度以協助聯盟團隊招募人才及強化客戶忠誠度

企業網頁：http://united-edu.org

黎家良先生經驗分享

　　孩子的健康成長是家長們最關心、最大的事業。香港管理諮詢有限公司的核心理念是通過深度瞭解人的性格特質定義孩子的長處、才能和天賦，幫助父母親們為孩子建立一個理想的、能給孩子們健康成長的學習環境與條件。校園大五（The SchoolPlace Big Five Profile™）是一種基於人格五因子模型（FFM）的人格評估方法，多年來已成為世界各地心理學家、職場教練、教育工作者和專業輔導顧問的標準。校園大五專門採用校園專用術語為 8 至 22 歲的中、小學和高等教育孩子編寫的測評工具與分析報告。校園大五分析報告根據孩子的個性來定義自己的長處、才能和天賦。全球很多中小大學通過校園大五（SchoolPlace big five Profile™）向孩子與家長提供大五模式，以幫助孩子進行個人發展、學習方式和課程決策、為實習和職業選擇提供準備充分信息。企業更可以利用校園大五工具來參考雇傭直接從教育環境進入職場的優秀學生。這種增強的自我意識能播下希望的種子，建立信心，並使孩子能够從小學、中學到大學，建立更開放思維與想像，並開始規劃他們的未來職業目標。

　　能更瞭解自己性格的孩子更有能力回答以下問題：「我是誰」、「我能做什麼」、「我做得最好的是什麼」、「我如何判斷事物」，並能更積極參與課堂作業，在決策上變得更自信。校園性格評估適用於 8 至 22 歲的青少年，幫助他們提高學術水平，培養積極的自我意識，探索不同的學業選科與未來職業選擇考慮。在學的孩子們通常還沒有準備好有效地評估所有可供選擇的學科或專業。此外，父母親們也不確定該如何幫助孩子。大多數孩子都會告訴你，他們在為選擇最適合自己的學科專業而掙扎。孩子們通常面對的問題和挑戰是：學習興趣、學科選擇、認識自我、提高自信、與人溝通、職業規劃、如何成功、與壓力管理等。校園大五的優點特色是：1、讓家長更瞭解孩子心理、學業需求、職業傾向與價值觀；2、讓家長能更清晰瞭解孩子的性格特質與潛能；3、幫助孩子選擇理想學科、幫助孩子完成學業規劃；4、幫助家庭解決溝通問題；5、幫助孩子規劃職業動向；6、指導孩子培養獨立能力、情感智能與領導力發展；7、讓家長更能接受孩子在人生成長過程中有各自不同的目標方向。校園大五幫助孩子塑造更完美人格，創造未來。

鳴 謝

Experience Sharing 經驗分享

1) 李松德先生

亞洲脈絡有限公司創辦人兼行政總裁

Mr. Charles Lee

Founder and CEO, OneAsia Network Limited

2) 李家良博士

栢昇企業有限公司創辦人兼董事

Dr. LEE, Ka Ling Alex

Founder and Director, Pacific Sense Enterprises Limited

3) 李明漢博士

香港上海匯豐銀行亞太區公用事業、新能源與綜合性主管

Dr. Li Ming Hon, Evan

Director, Global Research, HSBC Global Banking and Markets

4) 徐俊英博士

英國擎天藝術產業基金主席

Dr. Choi, Chon-Ieng Benny

Chairman, Sandalphon Pillars Skyhigh Culture and Industries Fund

5) 許華達博士

高級營養保健師

健型堡 Wellness Treasure 創辦人

Dr. Hui, Wah Tat Anthony
Senior Nutritionist
Founder, Wellness Treasure

6) 郭慧芳博士

會計師（英國、澳大利亞）、仲裁員、保險師

香港知識產權顧問有限公司（IPAS）大灣區企業顧問

Dr. Kwok, Wai Fong Kitty、FAIA、FFA/FIPA、Arbitrator、C.I.P.
Enterprise Consultant of Greater Bay Area, Hong Kong Intellectual
Property Advisory Service Limited

7) 陳智敏博士、工程師

信豐工程有限公司創辦人兼董事

Ir Dr. Chan Chi Man
Founder and Director, LONG Engineering Limited

8) 羅國蔚律師、博士研究生

廣信君達律師事務所律師

Mr. Guo Wai Luo, PhD candidate
Attorney at Law, ETR Law Firm

9) 馬偉成博士
駿盈建築有限公司董事

Dr. Ma, Wai Shing Kenny

Director, Billion Smart Construction Co. Ltd.

10) 劉銘豪博士、工程師
立人智研科技有限公司董事總經理

Ir Dr. Lau Ming Ho, Ritz

Managing Director, Standbot Robotics Limited

11) 陳德明先生
駿盈數碼有限公司主席

Mr. Chan Tak Ming

Chairman, Cheer Champion Cyber Limited

12) 黎家良先生
香港管理諮詢有限公司

Mr. Eddie Lai

President, Hong Kong Management Consultancy Limited

13) 黃燁權博士、CSUMB 訪問學者
洹博綜合管理顧問有限公司項目副總裁

Dr. Patrick Wong, CSUMB Visiting Scholar

Vice President, Projects, Envpro Integration Management

14) 羅偉雄博士

國際博士後協會榮譽會長調解顧問

香港和解中心主席

Dr. LAW Wai Hung, Francis

Mediation Advisor of the International Postdoctoral Association.

President of Hong Kong Mediation Centre

Forward 前言

F1) 那士榮博士

國際博士後協會榮譽會長

加拿大魁北克大學教授及前副校長

Dr. Prosper Bernard

Honorary President of the International Postdoctoral Association.

Professor and Former Vice Vector at the University of Québec in Montréal.

F2) 陶翼青博士

蒙特利灣加州州立大學教授兼創新和經濟發展院院長

國際博士後協會榮譽顧問

Dr. Eric Y Tao

Professor and Director

Institute for Innovation and Economic Development at

California State University Monterey Bay

Honorary Advisor of the International Postdoctoral Association.

F3) 施俊輝先生

教育育局局長政治助理

香港特別行政區政府教育局

Mr Jeff SZE

Political Assistant to Secretary for Education

Education Bureau, 港 The Government of the HKSAR

Preface 序

P1) 盧偉國博士工程師 GBS, SBS, MH, JP

國際博士後協會榮譽顧問

中國人民政治協商會議全國委員會成員

香港特別行政區立法會議員（工程界），2012- 現在

香港經濟民生聯盟主席

Ir Dr the Honorable LO WAI KWOK, SBS, MH, JP

Honorary Advisor of International Postdoctoral Association (www.IPostdocA.org)

Member, CPPCC National Committee

Member of Legislative Council (Functional Constituency – Engineering)

Chairman, Business and Professionals Alliance for Hong Kong

P2) 何鐘泰博士工程師 SBS, MBE, SBStJ, JP

國際博士後協會榮譽顧問

香港特別行政區立法會議員（工程界），1996-2012

全國人大代表（第十屆和第十一屆）

Ir Dr RAYMOND HO CHUNG-TAI, SBS, MBE, SBStJ, JP

Honorary Advisor of International Postdoctoral Association

Member of Legislative Council, HKSAR (1996-2012)

Deputy to the National People's Congress, PRC (10th and 11th terms)

P3) 初志農教授

中央人民政府駐香港特別行政區聯絡辦公室

教育科技部前部長

Professor Chu Zhu Nong

Former Director General, Education, Science and Technology Department,

Liaison Office of the Central People's Government in the HKSAR

P4) 俱孟軍先生

新華社亞太總分社原社長

Mr JU Meng Jun

Former President of XINHUA NEWS AGENCY Asia Pacific Region Bureau
and HKSAR Branch